# Advanced in Dewatering and Drying Processes

# Advanced in Dewatering and Drying Processes

Guest Editor

**Jan Havlík**

MDPI

Basel • Beijing • Wuhan • Barcelona • Belgrade • Novi Sad • Cluj • Manchester

*Guest Editor*
Jan Havlík
Department of Energy
Engineering
Faculty of Mechanical
Engineering
Czech Technical University in
Prague
Prague
Czech Republic

*Editorial Office*
MDPI AG
Grosspeteranlage 5
4052 Basel, Switzerland

This is a reprint of the Special Issue, published open access by the journal *Processes* (ISSN 2227-9717), freely accessible at: https://www.mdpi.com/journal/processes/special_issues/Dewatering_Drying.

For citation purposes, cite each article independently as indicated on the article page online and as indicated below:

| Lastname, A.A.; Lastname, B.B. Article Title. *Journal Name* **Year**, *Volume Number*, Page Range. |
|---|

**ISBN 978-3-7258-4119-6 (Hbk)**
**ISBN 978-3-7258-4120-2 (PDF)**
**https://doi.org/10.3390/books978-3-7258-4120-2**

Cover image courtesy of Jan Havlík

# Contents

# About the Editor

**Jan Havlík**

Jan Havlík is an assistant professor at the Department of Energy Engineering, Faculty of Mechanical Engineering, Czech Technical University, in Prague, Czech Republic. His major research areas include heat exchanger design, heat transfer, advanced drying technologies for energy systems, condensation technologies, and thermal process optimization. His teaching activities involve subjects such as renewable energy sources, heat processes and heat exchangers, experimental issues, and fundamentals of energy conversions. He is currently a member of the research team of several research projects on the topics of drying energy fuels, flue gas condensation, and increasing the efficiency of biomass boilers. He received the Best Scientific Presentation Award at the 3rd Nordic Baltic Drying Conference 2019 for the following contribution: "Indirect Dryers for Biomass—Experimental Characteristics for Design and Scale Up".

processes

Editorial

# Special Issue on "Advanced in Dewatering and Drying Processes"

**Jan Havlík**

Department of Energy Engineering, Faculty of Mechanical Engineering, Czech Technical University in Prague, Technicka 4, 16600 Prague, Czech Republic; jan.havlik@fs.cvut.cz

Drying and dewatering processes have many industrial applications such as in the agricultural, energy, food, chemical, pharmaceutical, paper, and textile industries. In most drying applications, there is a need to control or improve the quality of the output product [1,2]. At the same time, since the drying process is an energy-intensive operation, any improvements to the existing dryer design are desirable in order to save energy [2,3]. This is a motivation for further research on process intensification and energy consumption reduction, while ensuring a high-quality final product [3]. Therefore, it is important to have a good understanding of the process and the underlying mechanisms.

In this Special Issue, current knowledge and new trends in drying and dewatering techniques have been described, both in terms of experimental and theoretical approaches to the design of new drying and dewatering systems. A total of 11 submissions with different focuses were published.

The focus of the papers can be divided according to the used drying method, the application, and the material dried (see Table 1), with most papers relating to food drying. Energy (fuel drying) and biological applications are also represented.

**Table 1.** Analysis of the contributions in the Special Issue.

| No. | Drying Type | Application | Material |
|---|---|---|---|
| 1 | freeze drying | food | porous materials |
| 2 | convection, fluidized bed and solar drying | food | curcuma longa |
| 3 | indirect drying | fuel | bark |
| 4 | droplets natural evaporation | biological | water droplet |
| 5 | gas–liquid separation | fuel | NG |
| 6 | microwaves and freeze drying | food | tomato |
| 7 | conventional air drying | food | beans |
| 8 | microwave/hot air drying + ultrasound treatment | food | hawthorn |
| 9 | conventional air drying—pretreatment | fuel | lignin |
| 10 | droplet natural evaporation | biological | blood droplet |
| 11 | convective drying + pulse electric field treatment | food | mushrooms |

Drying in the food industry is traditionally the most widespread application. Eight contributions have been published on this topic for drying different materials. Different drying methods were investigated, ranging from conventional hot-air drying to freeze

Received: 9 April 2025
Accepted: 14 April 2025
Published: 16 April 2025

**Citation:** Havlík, J. Special Issue on "Advanced in Dewatering and Drying Processes". *Processes* **2025**, *13*, 1201. https://doi.org/10.3390/pr13041201

drying or microwave drying. A crucial parameter of drying in the food industry is the preservation of the quality of the dried product. Ndisanze and Koca investigated the product quality and fruit characteristics of tree tomatoes using a combination of microwave and freeze drying. A study by Levin et al. presents a prediction of the heat and mass transfer behaviour during freeze drying of porous food particles with the aim of optimising the process, concluding that an increase in porosity is associated with a reduction in mass transfer resistance, but at the expense of reduced heat transfer through the dried particle layer. Llano et al. studied the effect of different drying methods (convection drying, fluid bed drying, and traditional sun drying) on the quality of Curcuma longa (turmeric) powder. The results showed that convection drying and fluidized bed drying, unlike solar drying, did not have a significant effect on the quality of turmeric. Dadan et al. analysed the use of a combination of convection drying with pulsed electric field for drying mushrooms. Kaveh et al. discussed the effect of ultrasonic pretreatment and microwave hot-air drying on drying time, energy requirement, and quality characteristics of hawthorn fruits.

Unlike food applications, where the aim is to maintain product quality, energy applications are all about improving the material properties for maximum energy gain in subsequent use. Three papers have been published on this topic. Havlík and Dlouhý deal with indirect drying of biomass in energy systems and analyse the influence of the degree of final drying of wet biomass on the required size of the dryer. They conclude that for drying wet bark, the optimum degree of drying in indirect dryers is between 31 wt% and 13 wt%. Ji et al. proposed a new design of combined gas–liquid separator for natural gas transportation and storage. Saadon and Osman investigate the effect of drying pretreatment on the hydrolysis of lignin from Napier grass when using conventional air drying.

Drying of biological materials is a rather peripheral part of the application of drying, on which two papers have been published. Here, it involves the dewatering of droplet solutions of various biological materials, using natural evaporation to subsequently biologically analyse the resulting product. Pal et al. investigate the drying kinetics of droplets containing globular protein, phosphate-buffered saline and thermotropic liquid crystals. Ancheyta-Palacios et al. present an experimental study that investigates pattern formation in dry blood droplets with different concentrations of ultrapure water.

In general, drying has a wide range of applications. The research in all published papers focused mainly on describing the kinetics of the drying process. The theoretical description of the process is relatively complex and is often limited by several boundary conditions; therefore, all research papers are solved experimentally. In particular, the current trends in research topics are the combination of different drying methods [2,4–7], the use of material pretreatment to modify the material properties before drying [1,3,8], and the reduction in the energy consumption of the process [4,9]. A major challenge for the future is the development of methods that, while maintaining the highest possible quality of the dried product, will at the same time achieve the lowest possible energy consumption, also taking into account the economic aspect of the process [1] and the topics on the use of renewable energy sources and their environmental impact [9], which are nowadays very intensively addressed.

**Conflicts of Interest:** The author declares no conflicts of interest.

**List of Contributions**

1. Levin, P.; Buchholz, M.; Meunier, V.; Kessler, U.; Palzer, S.; Heinrich, S. Comparison of Knudsen Diffusion and the Dusty Gas Approach for the Modeling of the Freeze-Drying Process of Bulk Food Products. *Processes* **2022**, *10*, 548.
2. Llano, S.M.; Gómez, A.M.; Duarte-Correa, Y. Effect of Drying Methods and Processing Conditions on the Quality of Curcuma longa Powder. *Processes* **2022**, *10*, 702.

3. Havlík, J.; Dlouhý, T.; Pitel', J. Drying Biomass with a High Water Content—The Influence of the Final Degree of Drying on the Sizing of Indirect Dryers. *Processes* **2022**, *10*, 739.
4. Pal, A.; Gope, A.; Iannacchione, G.S. Hierarchical Exploration of Drying Patterns Formed in Drops Containing Lysozyme, PBS, and Liquid Crystals. *Processes* **2022**, *10*, 955.
5. Ji, L.; Zhao, Q.; Deng, H.; Zhang, L.; Deng, W. Experimental Study on a New Combined Gas–Liquid Separator. *Processes* **2022**, *10*, 1416.
6. Ndisanze, M.A.; Koca, I. Dehydration and Rehydration Kinetics Modeling in the Phytochemical, Aroma, and Antioxidant Capacity of Tree Tomato Fruit Dried with Microwaves and Freeze Driers: A Comparative Study. *Processes* **2022**, *10*, 1437.
7. Amadeu, L.T.S.; Queiroz, A.J.d.M.; Figueirêdo, R.M.F.d.; Ferreira, J.P.d.L.; Silva, W.P.d.; Gomes, J.P.; Paiva, Y.F.; Costa, C.C.; Moura, H.V.; Santos, D.d.C.; et al. Controlled Germination of Faba Beans: Drying, Thermodynamic Properties and Physical-Chemical Composition. *Processes* **2022**, *10*, 1460.
8. Kaveh, M.; Nowacka, M.; Khalife, E.; Imanian, K.; Abbaspour-Gilandeh, Y.; Sabouri, M.; Zadhossein, S. Hawthorn Drying: An Exploration of Ultrasound Treatment and Microwave–Hot Air Drying. *Processes* **2023**, *11*, 978.
9. Saadon, S.Z.A.H.; Osman, N.B. Effect of Drying Pretreatment on Cellulolytic Enzymatic Hydrolysis of Lignin from Napier Grass. *Processes* **2023**, *11*, 1092.
10. Ancheyta-Palacios, M.; Velasco-Terán, I.G.; Carreón, Y.J.P.; González-Gutiérrez, J. Dried Droplets of Diluted Blood to Detect a High Concentration of Lipids. *Processes* **2023**, *11*, 2047.
11. Dadan, M.; Barańska, A.; Matys, A.; Rybak, K.; Witrowa-Rajchert, D.; Wiktor, A.; Nowacka, M. Impact of Pulsed Electric Field Treatment on the Process Kinetics and Selected Properties of Air and Dehumidified Air-Dried Mushrooms. *Processes* **2023**, *11*, 2101.

## References


1. Bhattacharjee, S.; Mohanty, P.; Sahu, J.K.; Sahu, J.N. A critical review on drying of food materials: Recent progress and key challenges. *Int. Commun. Heat Mass Transf.* **2024**, *158*, 107863. [CrossRef]
2. Fathi, F.; Ebrahimi, S.N.; Matos, L.C.; Oliveira, M.B.P.P.; Alves, R.C. Emerging drying techniques on food safety and quality: A review. *Compr. Rev. Food Sci. Food Saf.* **2022**, *21*, 1125–1160. [CrossRef] [PubMed]
3. Chojnacka, K.; Mikula, K.; Izydorczyk, G.; Skrzypczak, D.; Witek-Krowiak, A.; Moustakas, K.; Ludwig, W.; Kułażyński, M. Improvements in drying technologies—Efficient solutions for cleaner production with higher energy efficiency and reduced emission. *J. Clean. Prod.* **2021**, *320*, 128706. [CrossRef]
4. Havlík, J.; Dlouhý, T. Improving the energy effectivity of biomass drying for utilisation in energy systems by combining convective and contact drying. *Drying Technol.* **2024**, *42*, 622–635. [CrossRef]
5. Ndisanze, M.A.; Koca, I. Dehydration and Rehydration Kinetics Modeling in the Phytochemical, Aroma, and Antioxidant Capacity of Tree Tomato Fruit Dried with Microwaves and Freeze Driers: A Comparative Study. *Processes* **2022**, *10*, 1437. [CrossRef]
6. Kaveh, M.; Nowacka, M.; Khalife, E.; Imanian, K.; Abbaspour-Gilandeh, Y.; Sabouri, M.; Zadhossein, S. Hawthorn Drying: An Exploration of Ultrasound Treatment and Microwave–Hot Air Drying. *Processes* **2023**, *11*, 978. [CrossRef]
7. Dadan, M.; Barańska, A.; Matys, A.; Rybak, K.; Witrowa-Rajchert, D.; Wiktor, A.; Nowacka, M. Impact of Pulsed Electric Field Treatment on the Process Kinetics and Selected Properties of Air and Dehumidified Air-Dried Mushrooms. *Processes* **2023**, *11*, 2101. [CrossRef]
8. Saadon, S.Z.A.H.; Osman, N.B. Effect of Drying Pretreatment on Cellulolytic Enzymatic Hydrolysis of Lignin from Napier Grass. *Processes* **2023**, *11*, 1092. [CrossRef]
9. Kherrafi, M.A.; Benseddik, A.; Saim, R.; Bouregueba, A.; Badji, A.; Nettari, C.; Hasrane, I. Advancements in solar drying technologies: Design variations, hybrid systems, storage materials and numerical analysis: A review. *Solar Energy* **2024**, *270*, 112383. [CrossRef]

processes
MDPI

Article

# Comparison of Knudsen Diffusion and the Dusty Gas Approach for the Modeling of the Freeze-Drying Process of Bulk Food Products

Patrick Levin [1,*], Moritz Buchholz [1], Vincent Meunier [2], Ulrich Kessler [3], Stefan Palzer [4] and Stefan Heinrich [1]

[1] Institute of Solid Process Engineering and Particle Technology, Hamburg University of Technology, Denickestrasse 15, 21073 Hamburg, Germany; moritz.buchholz@tuhh.de (M.B.); stefan.heinrich@tuhh.de (S.H.)
[2] Nestlé Research, 1000 Lausanne, Switzerland; vincent.meunier@rdls.nestle.com
[3] Nestlé Product Technology Centre, 1350 Orbe, Switzerland; ulrich.kessler@gmx.ch
[4] Nestlé, 1800 Vevey, Switzerland; stefan.palzer@nestle.com
* Correspondence: patrick.levin@tuhh.de

**Abstract:** Freeze-drying is generally used to achieve high quality products and preserve thermal sensitive components; however, it is also considered as a high energy and costly process. Modeling of the process can help to optimize the process to reduce these drawbacks. In this work, a mathematical model is presented to predict the heat and mass transfer behavior for freeze-drying of porous frozen food particles during freeze-drying to optimize the process. For the mass transfer, a comparison between Knudsen diffusion and the more complex dusty-gas approach is performed. Simulation results of a single particle are validated by experiments of single-layer drying to extend the usage of this model from a single particle to a particle bed. For the moisture transfer, adaption parameters are introduced and evaluated. A comparison shows a good agreement of the model with experimental results. The results furthermore suggest a strong correlation of the drying kinetics with pore size and particle porosity. An increase in the pore diameter strongly improves the overall mass transfer rates and hence is a suitable parameter for an effective increase of the drying rates in freeze-drying.

**Keywords:** freeze-drying; drying of frozen particles; modeling; dusty gas model; improvement of mass transfer; internal porous structure

**Citation:** Levin, P.; Buchholz, M.; Meunier, V.; Kessler, U.; Palzer, S.; Heinrich, S. Comparison of Knudsen Diffusion and the Dusty Gas Approach for the Modeling of the Freeze-Drying Process of Bulk Food Products. *Processes* **2022**, *10*, 548. https://doi.org/10.3390/pr10030548

Academic Editor: Jan Havlík

Received: 16 February 2022
Accepted: 9 March 2022
Published: 11 March 2022



## 1. Introduction

Freeze-drying is used to dry sensitive products in the food and pharma industry [1] and is seen in comparison to other drying processes as a rather energy and cost intensive technology [2]. Especially in the food industry, it needs to be ensured that in addition to the quality, the drying times are held as short as possible to be competitive with other drying technologies. Freeze-drying consists of three steps: freezing, primary and secondary drying [3,4]. For liquid food materials, e.g., soluble coffee or tea, a part of the water crystallizes and the solution is concentrated during the freezing step. The remaining water is called non-freezable or bound water and remains in the glassy matrix. During primary drying, the crystalline water is sublimated, while during secondary drying the bound water needs to be desorbed from the amorphous food matrix. Both drying steps follow different drying mechanisms and for both, the dried layer forms a resistance for the heat and mass transfer within the drying particle. The understanding of these limiting effects allows for a prediction of the drying times and therefore the possibility to predict the drying kinetics by modeling, which is a powerful and cost-effective tool for process optimization.

Several model approaches were used in the past to describe the freeze-drying process. These mainly focused on drying in vials used predominantly in pharma industry or on tray freeze-drying used for the drying of bulk materials. The approaches range from single particle models to complex pore network models of particle beds on trays.

Many authors studied the freeze-drying process in packed beds: Liapis and Bruttini [5] considered also the freeze-drying of particles in packed beds, where they extended the application to spray-freeze drying which received broad interest [6,7]. The authors used the dusty gas model for the mass transfer, already applied by Song and Yeom [8] and Sheehan and Liapis [9]. Here, a mixture of inert gases and vapor was considered during the drying approach. In the work of Liapis and Bruttini [10], this model was further extended and validated by experimental data derived by Song et al. [7] and Her et al. [11]. The authors found out that the gas flux of inert gases is negligible compared to the vapor flux, during primary drying. The recent approaches of Warning et al. [12] considered the importance of a sublimation width instead of a single sublimation front, adding more precision, but also more complexity to the model. Gruber et al. [13] and Vorhauer-Huget et al. [14] confirmed experimentally that the sublimation front is not homogeneous, but affected by local changes in heat and mass transfer. The complexity applies also to the recent progresses of pore network models as applied by Vorhauer et al. [15]. The pore network model allows the coupling of pores with various sizes, allowing a more precise prediction of the vapor flux and its dependency on pore geometry and arrangement.

The influence of structure parameters on the heat and mass transfer rates were only insufficiently investigated and validated and therefore need further investigation. As discussed in previous publications [3,16], changes in the internal porous structure can have a significant impact on the drying time. Modeling approaches in this direction were already applied and vary strongly in complexity. In general, complex models are able to achieve higher precision, mostly because of a higher degree of freedom. These advantages usually come at the cost of increased computing power and simulation time. Furthermore, in each model different experimental results were used. A comparison between a simple Knudsen diffusion approach and an approach with higher complexity such as the dusty gas model can help to judge the experimental results according to existing models, which was done in this work. For the validation, the structural parameters such as porosity and pore size were varied to observe the change in drying time. Since the focus lies on the change of internal porous structure, the internal mass transfer limitation is of high interest, which is modeled and validated on a single-particle scale.

## 2. Methodology

### 2.1. Model Structure

A shrinking core approach is used to capture the main heat and mass transfer mechanisms of the single-particle freeze-drying process. Therefore, the model considers three distinct regions: a frozen core (index fc), a porous dried layer (index dl) and an infinitesimal particle surface region (index p). The shrinking core model is schematically presented in Figure 1. When the particle dries, the frozen core region shrinks and its boundary moves towards the center of the particle. The water vapor is transported through the pores of the growing dried layer region from the frozen core to the surface. The sublimation process is divided into two different stages: The primary stage considers a pore diffusion transport of the water vapor through the porous layer. For the second stage, a simplified approach of a first-order reaction kinetic is assumed to remove the remaining ice from the particle. The one-dimensional structure of the model allows only for a resolution of the heat and mass transfer along the radial direction. The radiative heat transfer from the bottom and top shelf as well as radiation from the environment to the particle surface are considered. Due to the low pressure environment, convective heat transfer at the particle surface is neglected. From the surface, heat is transferred by conduction through the porous dried layer to the sublimation interface of the frozen core region.

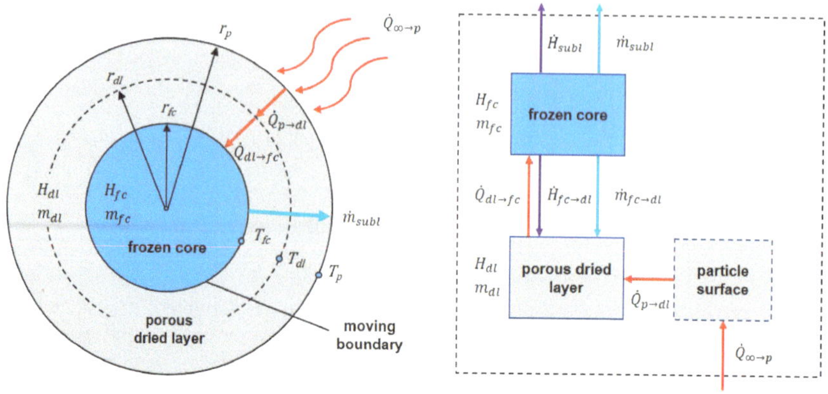


**Figure 1.** Model structure for the freeze-drying process of a single particle using a shrinking core approach. The different model regions, i. e. frozen core, porous dried layer and particle surface, are shown on the left hand side. The image on the right hand side shows the different transfer flows between the model regions.

**The main assumptions and simplifications of the model are listed below:**

- Only the vapor transport within the porous structure of the particle is investigated, while the vapor transport within the drying chamber is neglected.
- The material is homogeneously distributed within each respective model region.
- The pores all have the same pore diameter and are connected to the outside of the particle. Closed pores are neglected.
- Material properties are independent of the temperature.
- Particles have a spherical shape.
- The influence of the sample holder is neglected and only radiative heat transfer is considered.
- Only radiative heat transfer in the drying chamber is considered. The particle is modeled as a gray body.

The frozen core and the dried regions are assumed to be ideally mixed, neglecting possible gradients in composition or temperature. The transient behavior of the mass and enthalpy is modeled by a CSTR (continuously stirred tank reactor) approach and are shown below:

$$\frac{dm_{fc}}{dt} = -\dot{m}_{subl} - \dot{m}_{fc\to dl}, \tag{1}$$

$$\frac{dH_{fc}}{dt} = -\dot{H}_{subl} - \dot{H}_{fc\to dl} + \dot{Q}_{dl\to fc}, \tag{2}$$

$$\frac{dm_{dl}}{dt} = \dot{m}_{fc\to dl}, \tag{3}$$

$$\frac{dH_{dl}}{dt} = \dot{H}_{fc\to dl} - \dot{Q}_{dl\to fc} + \dot{Q}_{p\to dl}. \tag{4}$$

$m$ and $H$ are the mass and the enthalpy of the respective region, $\dot{m}$ the mass flow, $\dot{H}$ the enthalpy flow and $\dot{Q}$ the heat flows between the regions. The mass flow and the enthalpy flow of sublimation, $\dot{m}_{subl}$ and $\dot{H}_{subl}$, respectively, are directly removed from the frozen core. The sublimation results in a shrinking of the frozen core region. While the vapor moves through the pores of the dried layer to the surfaces, the dried matter is removed from the frozen core region and is added to dried layer region, which is expressed by the exchange flows $\dot{m}_{fc\rightarrow dl}$ and $\dot{H}_{fc\rightarrow dl}$. The temperature of the frozen core and the dried layer region are calculated by:

$$T_{fc} = T_0 + \frac{h_{fc} - h_{fc,0}}{c_{p,fc}}, \tag{5}$$

$$T_{dl} = T_0 + \frac{h_{dl} - h_{dl,0}}{c_{p,dl}}, \tag{6}$$

with the specific enthalpy $h = H/m$, the specific heat capacity $c_p$ and the reference temperature $T_0$.

The infinitesimal particle surface is modeled by a quasi-stationary enthalpy balance, which takes into account the incoming radiative heat transfer and the outgoing conductive heat transfer from the surface into the dried layer as follows:

$$\dot{Q}_{rad} - \dot{Q}_{p\rightarrow dl} = \epsilon A_p \sigma (T_\infty^4 - T_p^4) - k_{dl} \cdot 4\pi \cdot (T_p - T_{dl}) \cdot \frac{r_p \cdot r_{dl}}{r_p - r_{dl}} = 0, \tag{7}$$

with the emissivity $\epsilon$, the Stefan–Boltzmann constant $\sigma$, the thermal conductivity of the dried layer $k_{dl}$ and the radius of the respective region $r$. The implicit equation is solved internally to calculate the temperature of the particle surface $T_p$.
The conductive heat flow between the dried layer and the frozen core is calculated by:

$$\dot{Q}_{dl\rightarrow fc} = k_{dl} \cdot 4\pi \cdot (T_{dl} - T_{fc}) \cdot \frac{r_{dl} \cdot r_{fc}}{r_{dl} - r_{fc}}. \tag{8}$$

The mass flow of dry material from the frozen core to the dried layer and the respective enthalpy flow can be described as a function of the vapor mass flow, which is released from the sublimation process as follows:

$$\dot{m}_{fc\rightarrow dl} = \dot{m}_{subl} \cdot \frac{w_w}{1 - w_w}, \tag{9}$$

$$\dot{H}_{fc\rightarrow dl} = \dot{m}_{fc\rightarrow dl} \cdot c_{p,dl} \cdot (T_{fc} - T_0). \tag{10}$$

where $w_w$ is the water weight fraction. The sublimation enthalpy flow is calculated using the equation

$$\dot{H}_{subl} = \dot{m}_{subl} \cdot \Delta h_{subl,w}, \tag{11}$$

with the specific enthalpy of sublimation $\Delta h_{subl,w}$.

**Calculation of the sublimation mass flow $\dot{m}_{subl}$**

The total sublimation mass flow $\dot{m}_{subl}$ depends on the vapor transport through a single pore and the number of pores which connect the frozen core with the environment. It is calculated by

$$\dot{m}_{subl} = \varepsilon_p \cdot 4\pi r_{fc}^2 \cdot \dot{m}''_{pore}, \tag{12}$$

where $\dot{m}''_{pore}$ is the vapor mass flux through the porous layer. For the calculation of the vapor mass flux, two different models are applied that are subsequently shown. The Knudsen number $Kn$ is used as a measure for the characteristic movement of particles in confined regions and can be calculated by

$$Kn = \frac{\lambda}{d_{pore}} = \frac{k_B T}{\sqrt{2}\pi\varsigma^2 p d_{pore}}, \tag{13}$$

where $\lambda$ is the mean free path, $k_B$ is the Boltzmann constant and $\varsigma$ is the molecular diameter. A Knudsen number above unity describes a mass transfer in pores, which is mainly influenced by molecular wall interactions. A Knudsen number $0.01 < Kn < 1$ is in the transition regime between viscous flow and Knudsen flow and for Knudsen numbers $Kn \ll 1$ a purely viscous flow can be assumed [17].

The Knudsen number is determined for all experiments. The values range between $23 < Kn < 84$ for the validation experiments, considering a pore size between 2.88 and 10.49 μm at −30 °C at a chamber pressure of 0.4 mbar.

*Mass transfer of vapor through the open pores by Knudsen Diffusion*

For the simple Knudsen flow approach in case of a transitional flow regime, the sublimation mass flow rate through the porous layer is calculated using Equations (14) and (15):

$$\dot{m}''_{pore,Kn} = -\frac{D_{Kn}}{RT}\frac{\partial p_w}{\partial r}, \tag{14}$$

$$D_{Kn} = \frac{4}{3} \cdot d_{pore}\frac{\epsilon_p}{\tau_p} \cdot \sqrt{\frac{RT}{2\pi M}}, \tag{15}$$

where $D_{Kn}$ is the Knudsen diffusion coefficient, $R$ is the universal gas constant, $M$ is the molecular weight and $\frac{\partial p_w}{\partial r}$ the gradient of the partial pressure in radial direction.

*Mass transfer of vapor through the open pores by Dusty gas Model*

The dusty gas model was shown to predict the mass transfer in catalysts [18–20], and was recently used for the modeling of spray freeze-dried coffee particles on trays [5]. The mass flux $\dot{m}''_{pore,Du}$ is provided by the following equation:

$$\dot{m}''_{pore,Du} = -\frac{M_w}{R_g \cdot T_d} \cdot (k_1 + k_2 \cdot p_{w,d})\frac{\partial p_{w,d}}{\partial r}, \tag{16}$$

where $k_1$ and $k_2$ are constants describing the bulk diffusivity and the self-diffusivity, respectively. These constants can be calculated by the following equations, where Equation (17) describes the Knudsen flow term and Equation (18) describes the viscous flow as follows:

$$k_1 = \frac{C_2 \cdot D^0_{w,in} \cdot K_w}{C_2 \cdot D^0_{w,in} + K_{mx} \cdot P}, \tag{17}$$

$$k_2 = \frac{K_w \cdot K_{in}}{C_2 \cdot D^0_{w,in} + K_{mx} \cdot P} + \frac{C_{01}}{\mu_{mx}}. \tag{18}$$

$K_w$ is the Knudsen diffusivity of water and $K_{in}$ is the Knudsen diffusivity of the inert gas, which are calculated by Equations (19) and (20). The Knudsen diffusivity of the binary mixture $K_{mx}$ can be calculated by its molar fractions $y_w$ and $y_{in}$, as shown in Equation (21):

$$K_w = C_1 \cdot \left(\frac{R_g \cdot T_I}{M_w}\right)^{0.5}, \tag{19}$$

$$K_{in} = C_1 \cdot \left(\frac{R_g \cdot T_I}{M_{in}}\right)^{0.5}, \tag{20}$$

$$K_{mx} = y_w \cdot K_w + y_{in} \cdot K_{in}. \tag{21}$$

The viscosity of the vapor phase, which is needed to calculate the viscous flow term, is calculated as function of the temperature:

$$\mu_{mx} = 18.4858 \cdot 10^{-7} \cdot \left(\frac{T_I^{1.5}}{T_I + 650}\right). \tag{22}$$

$C_{01}$, $C_1$ and $C_2$ are constants for the Darcy flow permeability, the Knudsen flow permeability and the bulk diffusivity. Liapis and Bruttini [10] propose to calculate these constants as a function of structural parameters such as the porosity $\varepsilon_p$, the tortuosity $\tau_p$ of the particle and the pore diameter $d_{pore}$:

$$C_1 = \frac{\varepsilon_p}{\tau_p} \cdot \frac{48.5 \cdot d_{pore}}{R_g^{0.5}}, \tag{23}$$

$$C_2 = \frac{\varepsilon_p}{\tau_p}. \tag{24}$$

The expression for the Darcy flow permeability is adjusted in comparison to the literature used, since in this work no particle bed but a single particle is investigated. Therefore, the equation is taken from Kast [17] who gives an expression for the Darcy flow permeability in pores:

$$C_{01} = \frac{\varepsilon_p}{\tau_p} \cdot \frac{d_{pore}^2}{32}. \tag{25}$$

The tortuosity $\tau_p$ is difficult to measure experimentally and is therefore assumed as $\tau_p = \sqrt{2}$, as it was done by other authors for soluble coffee [5,10]. The vapor pressure at the interface as a function of temperature was already proposed by Marti and Mauersberger [21]. It is calculated by:

$$p_w^{sat} = 10^{\frac{-2663.5}{T_{fc}} + 12.537}. \tag{26}$$

*Mass transfer of vapor for second drying stage modeled by first-order reaction kinetic*

The secondary drying is modeled by a first-order kinetic expression as already shown in previous attempts [5,10]. Therefore, the vapor flow is assumed to decrease with proceeding drying progress, which is characterized by the desorption constant $k_{des}$:

$$\dot{m}_{des} = -k_{des} \cdot c_{sw}. \tag{27}$$

## 2.2. Validation Experiments

For the validation of the model, freeze-drying experiments are carried out. For the experiments, the setup as described in [22] is used to structure the material. In addition to the described trials, the samples are gassed with nitrogen using a different overrun. The overrun is defined as the amount of gas $V_g$ in comparison to the amount of liquid $V_l$:

$$OR = \frac{V_g}{V_l} = \frac{\rho_l - \rho_f}{\rho_f} = \frac{V_{tot} \cdot \rho_l - m_f}{m_f}, \tag{28}$$

where $\rho_l$ and $\rho_f$ are the liquid and foam density, respectively, $V_{tot}$ is the total volume and $m_f$ corresponds to the foam mass. To change the internal porous structure, two parameters are changed: (1) the amount of overrun is varied between 50%, 100% and 150% to generate foams with varying porosity and (2) the freezing method is changed. For freezing, the slurry is distributed in a 10 mm layer on a circular aluminum tray with a diameter of 80 mm. Three methods are applied to achieve a change in ice crystal size and structure: Airblast freezing is used inside of a −60 °C freezer, using a fan (3212 JH4, EBM-papst, Germany), which is placed 120 mm above the sample. The second way of freezing is carried out by placing the sample on a steel plate inside the freezer at −60 °C, using conductive heat transfer on the bottom. For the third method, the sample is placed within of a polystyrene box with a wall thickness of 20 mm into the freezer to reduce the heat transfer and allow for lower freezing rates. The scraping speed inside the SSHE was held constant at 300 rpm and the temperature of the coolant was at −25 °C for all trials. After the freezing is completed, the frozen cake is milled and fractionated by sieving at −60 °C. Only particles between 2.24 mm and 2.80 mm are considered for the freeze-drying process. For the drying, the experimental setup is used as shown in [22]. A total of 1 g of the frozen granules is placed on a polymer mesh, with a mesh size of 0.5 mm, hanging from a microbalance (Martin Christ

Gefriertrocknungsanlagen, Osterode am Harz, Germany) to receive heat by radiation from the bottom and the top. The reduction in moisture is monitored over time with a resolution in time of 1 s and resolution in weight of 1 mg. The drying profile used is summarized in Table 1.

**Table 1.** Drying profile for single-layer drying to determine the freeze-drying kinetics of instant coffee.

| Step | $T_{start}$ (°C) | $T_{end}$ (°C) | $P_{chamber}$ (mbar) | t (min) |
|---|---|---|---|---|
| 1 | −30 | −30 | 1013 | 1 |
| 2 | −30 | 46 | 0.4 | 38 |
| 3 | 46 | 46 | 0.4 | 90 |
| 4 | 46 | 25 | 0.4 | 20 |

For the analysis of the dried coffee particles, the Sauter mean diameter is calculated from the particle size distribution (PSD) to confirm the homogeneity in particle size of all samples. The PSD is measured (CAMSIZER XT, Retsch GmbH, Haan, Germany) using the particle size based on the measured area $x_{area}$. Particles smaller than 1 mm are neglected for the Sauter mean diameter calculation, to avoid variance due to abrasion. The pore size distribution and the porosity of the samples are measured by mercury porosimetry (Pascal 140 + 440, Thermo Fisher Scientific Inc., Waltham, MA, USA). A total of 0.15 g of the sample are analyzed at pressures between 0.1 kPa and 400 MPa with an accuracy of 0.2%. The pressure ranges allow for the analysis of pore diameters of 3.8 to 116 µm and 0.0036 to 15 µm for the different units, respectively. For the determination of the closed pore porosity, a helium pycnometer (AccuPyc 1330, Micromeritics, Norcross, GA, USA) is used. For the precise method to calculate the pore diameter and the closed porosity, refer to [22]. Images of selected samples are taken by scanning electron microscopy to verify the obtained results for the pore structure (Gemini 1530, Carl Zeiss Meditec, Jena, Germany). The particles are sputter-coated with a 40 nm gold layer to avoid interferences during measurements. Magnifications of 25, 100, 400 and 1000 times are performed with an electron beam of 5 kV.

## 3. Results

### 3.1. Experimental Results

To validate the model, experimental data are generated. To compare the drying kinetics and to define the transition between the end of the primary drying and the beginning of secondary drying, the time, which is needed to remove 80% of the water, is referred to as primary drying time. The residual water is removed by desorption during secondary drying. Since primary drying and secondary drying are not strictly separated, a part of the bound water is desorbing during the primary drying step as already discussed by other authors [4,5,8,9].

The structural parameters are summarized in Table 2. For the characterization of the particle diameter, the Sauter mean diameter is chosen, considering the surface area with respect to the volume, since freeze-dried coffee particles are not spherical. The primary drying time $t_{80}$ and the averaged evaporation rate $\dot{m}_{80,mean}$ are used as measures for the description of the drying kinetics, where the latter is averaged over 15 s to overcome noice suppression.

**Table 2.** Drying time and mean evaporation rate of coffee granules as a function of Sauter mean diameter $d_{32,p}$ of the particle size distribution and structural parameters such as mean pore size $d_{50,pore}$ and open porosity $\epsilon$.

| $d_{32,p}$ (mm) | $d_{50,pore}$ (µm) | $\epsilon_p$ (-) | $t_{80}$ (min) | $\dot{m}_{80,mean}$ (kg/h) |
|---|---|---|---|---|
| 2.65 | 2.88 | 0.743 | $44.4 \pm 2.2$ | $2.36 \pm 0.20$ |
| 2.67 | 7.20 | 0.771 | $39.4 \pm 0.1$ | $3.03 \pm 0.00$ |
| 2.84 | 9.49 | 0.759 | $39.1 \pm 0.5$ | $3.05 \pm 0.04$ |
| 2.60 | 5.34 | 0.692 | $44.4 \pm 1.2$ | $3.56 \pm 0.10$ |
| 2.50 | 7.71 | 0.796 | $37.4 \pm 1.6$ | $2.47 \pm 0.11$ |

In Figure 2, the internal pore structure of the coffee granules is shown for various interconnecting pore sizes, measured by mercury porosimetry. Here, the SEM images proof the range of interconnecting pore diameters, which are connecting the gas pores. The analysis of the images with the software ImageJ shows that gas bubble size can be held constant for all experiments at $(23.1 \pm 4.5)$ µm.

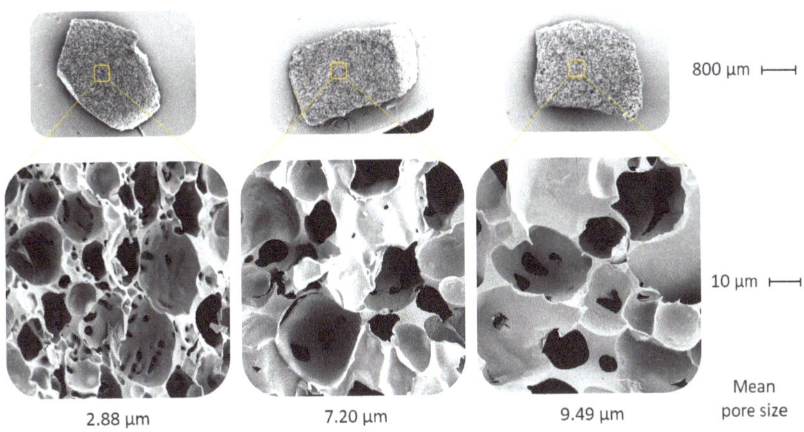


**Figure 2.** Scanning electron microscopy images of coffee granules showing the gas bubble size and the size of the interconnecting pores, left behind by ice crystals.

The results show that an increasing pore diameter leads to decreasing primary drying time. These results go along with the findings of Searles et al. [16], Hottot et al. [3] and our previous findings [22]. Mass transfer limitations seem to become more significant at small pore diameters of 3 µm or lower, but have less strong effects at higher pore diameters of 7 to 10 µm.

For the increased porosity, a decreasing trend in drying time can be observed, but the interpretation of experimental data need to be treated with care. At constant sample mass and particle diameter, but increasing overrun, the number of particles is increasing. As only a single layer is observed, the particle surface area available for radiation is increasing, which improves the heat transfer as well. In consequence, an increase of overrun leads to a reduction of the total water amount per particle and increases the drying rates. Therefore, the mean evaporation rate per particle is more meaningful compared to the drying time and shows a decreasing rate with increasing overrun, since heat transfer limitations become dominant.

Due to increased robustness the experiments are conducted using a single layer of material instead of single particle investigations. However, for a single layer of particles, not the complete surface area is available for the heat transfer due to the slightly overlapping of the particles. These effects turn out to be more significant for samples with increasing porosity and are corrected by a fitting parameter $f$. As shown in previous publications [23,24], vials drying on the edges of the plate dry faster compared to the center vials, since a larger part of the vial surface is exposed to a radiative source. Additionally, the shelves are not the only sources for radiative heat transfer, but also the other surfaces inside the freeze-dryer and their reflection of heat radiated by the front window. Therefore, besides the shelf temperature also an average environmental temperature needs to be defined to adapt the model to experimental conditions. These overlapping effects and the heat source are relevant for the heat transfer and are considered in the model approach as fitting parameters for the radiative heat transfer:

$$\dot{Q}_{rad} = f \epsilon A_p \sigma((g \cdot T_{sh}^4 + (1-g)T_{amb}^4) - T_p^4), \tag{29}$$

where $f$ is the fitting parameter to consider the reduction in surface area for the heat transfer and $g$ gives a ratio between heat transferred by the shelves in comparison to the heat transferred by other sources, where $T_{amb}$ is the ambient temperature of 20 °C.

### 3.2. Model Results

#### 3.2.1. Validation

Both model approaches are validated with the experimental data. All samples for the validation process have an initial coffee concentration of 50% $w/w$ and are foamed with nitrogen. The samples shown in Section 3.1 differ in their porosity and pore size. In Figure 3, the model results are compared with experimental data, varying the pore size diameters of the coffee granules between 2.88 µm and 9.49 µm, keeping the applied overrun constant at 100%.

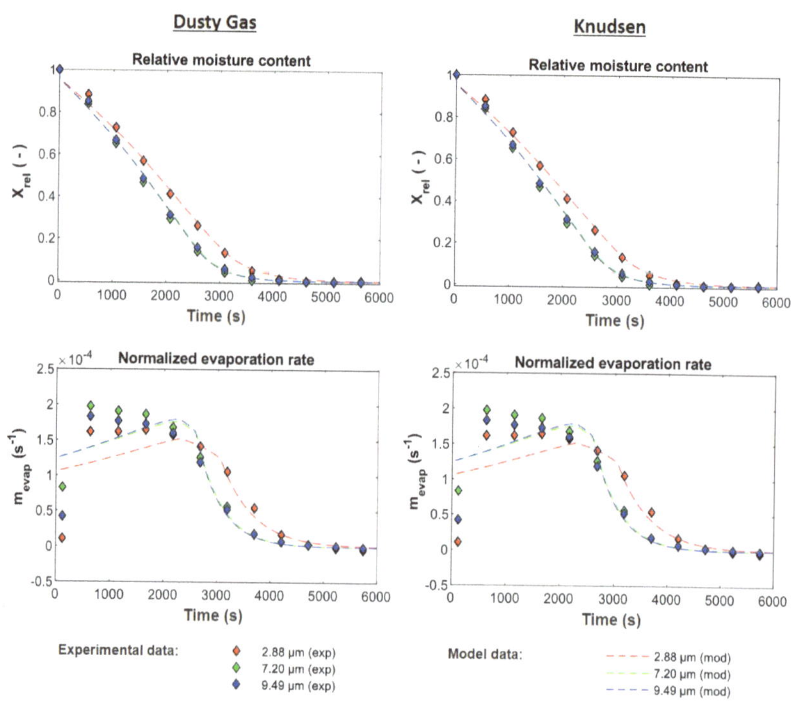


**Figure 3.** Validation of Knudsen model and dusty gas approach by means of experimental data during the freeze-drying of instant coffee granules by variation of pore size.

The total porosity (open and closed pore porosity) is comparable for all samples with 0.79 ± 0.02. The closed pore porosity for all samples ranges between 2 and 5%. The data show that the model can predict the loss of moisture over time, showing a slightly curved profile during primary drying, which is expected as shelf temperature is ramped up. The model and the experiment show that narrow pores of 2.88 µm decrease the drying rates significantly. Nevertheless, this limitation seems to decrease significantly for pore sizes between 7.2 µm and 9.5 µm, where the drying process gets limited rather by heat than mass transfer.

For the evaporation rate, the regions of primary and secondary drying can be observed in model and experiment. During primary drying a constant increase of the evaporation rate is observed, since shelf temperature increases in time. When shelf temperature is constant, a decreased rate is shown with time due to progressing heat and mass transfer limitations. During secondary drying a decay function can be observed, which approximates to an equilibrium value. While in the model primary drying and secondary drying are strictly separated, it can be observed that for the experiments the two mechanisms are overlapping, smoothening the curve in the transition region. Faster drying samples with usually increased pore diameter show a higher evaporation rate during primary drying (green and blue). During a later time-step between 2000 and 2500 s, the water content is decreasing significantly and the red curve shows the highest evaporation rate since primary drying is still not finished. The influence of the porosity on the drying kinetics are shown in Figure 4. The pore size is held constant in a range of $(6.50 \pm 1.18)$ µm.

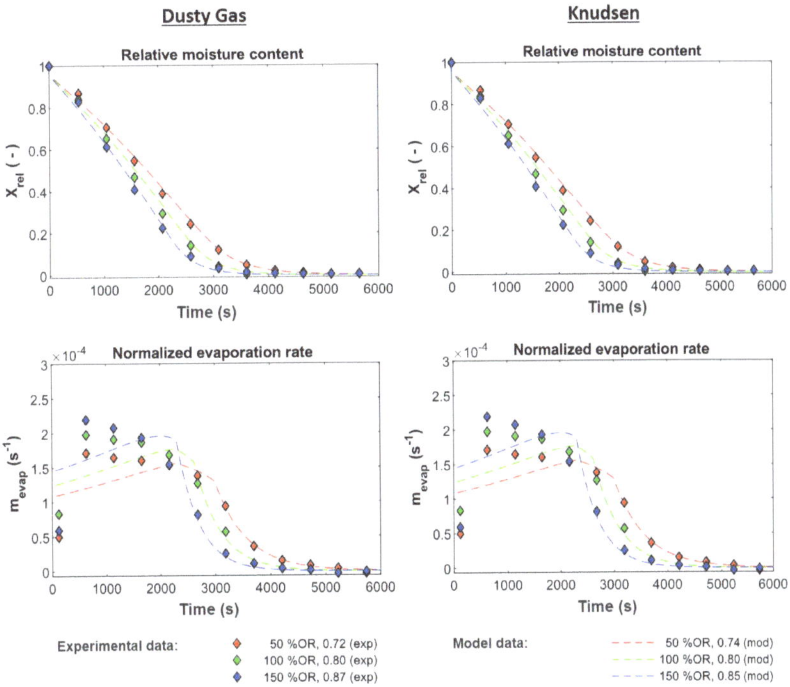


**Figure 4.** Validation of Knudsen model and dusty gas approach by means of experimental data during the freeze-drying of instant coffee granules by variation of overrun and the total porosity of the sample. Measured and calculated porosities for each sample are shown in the legend of each plot.

It shows that also for the varying porosity, the model fits the experimental data. With an increasing porosity, the drying time is decreasing and the vapor flow characterized by the normalized evaporation rate is increasing. The reason here is that for the experiments, the number of particles are increasing with increasing porosity and constant particle size and sample weight. Therefore, the heat transfer is increased. In the model, the volume of a particle is held constant and thus the surface area per particle stays constant. The amount of water is decreasing per particle. The normalization of the evaporation rate makes both approaches comparable and shows good trends for the primary drying, where low porosity shows the lowest normalized evaporation rate and high porosity shows the highest, as observed in the experiments. For secondary drying, all modeling approaches are in good agreement with the experimental data. In summary, the model shows similar trends in comparison to the experimental data for the Knudsen and the dusty gas approach, where only small deviations can be observed. Table 3 shows coeffcient of determination for both model approaches. A deviation both models is observed in the sixth decimal place. Both models show that an increasing pore diameter can decrease the limitations in mass transfer during primary drying and lead therefore to reduced drying times. An increase in porosity also shows a reduction in the drying time. However, this also leads to a reduction of absolute water content per volume of the particle on top of the improved pore space for vapor transport. The benefits and drawbacks of both parameters are discussed in detail in the sensitivity analysis.

**Table 3.** Coefficient of determination of the sample moisture evolution over time during freeze-drying, using the Knudsen approach $R^2_{Kn}$ and the dusty gas approach $R^2_{Du}$ in comparison to experimental data.

| $d_{32,p}$ (mm) | $d_{50,pore}$ (µm) | $\epsilon_p$ (-) | $R^2_{Kn}$ (-) | $R^2_{Du}$ (-) |
|---|---|---|---|---|
| 2.65 | 2.88 | 0.743 | 0.995649 | 0.995648 |
| 2.67 | 7.20 | 0.771 | 0.995247 | 0.995255 |
| 2.84 | 9.49 | 0.759 | 0.996340 | 0.996339 |
| 2.60 | 5.34 | 0.692 | 0.996624 | 0.996621 |
| 2.50 | 7.71 | 0.796 | 0.994986 | 0.994981 |

### 3.2.2. Sensitivity Analysis

For the sensitivity analysis, the model results are extrapolated and the influence of pore diameter and the porosity of the particle on the overall drying time and average sublimation rate is investigated. The pore diameter is varied from 0 to 20 µm and the porosity is varied from 0.6 to 0.9. In Figure 5, the primary drying time as a function of both structural parameters is shown.

The vapor flow through the porous dried layer is modeled with Knudsen diffusion and with the dusty gas model. For both simulations, the same fitting parameters are used. Both approaches come to similar results in prediction of the drying time, with marginal differences. An increase of the pore diameter or an increase of porosity decreases the primary drying time. As mentioned before a change in porosity shows several overlapping effects: An increase of porosity (1) decreases the amount of water which needs to be evaporated for a single particle, (2) increases the pore volume for the mass transfer and (3) reduces the conductivity of the dried layer and inhibits the heat transfer inside the particle towards the sublimation front. These effects are coupled and therefore complicate an interpretation for the optimal configuration.

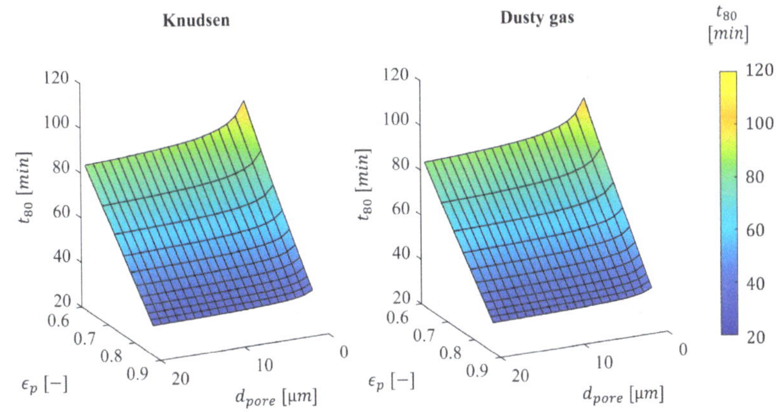


**Figure 5.** Primary drying time as a function of mean pore diameter and porosity, modeled by Knudsen diffusion and the dusty gas approach for the mass transfer processes through pores in the dried layer.

To understand the impact of the porosity on the heat and mass transfer mechanisms, the sublimation flow of water vapor needs to be considered to remove the influence of the decreasing water content, as shown in Figure 6. Here, only the dependency of the drying kinetics on pore volume and conductivity is considered.

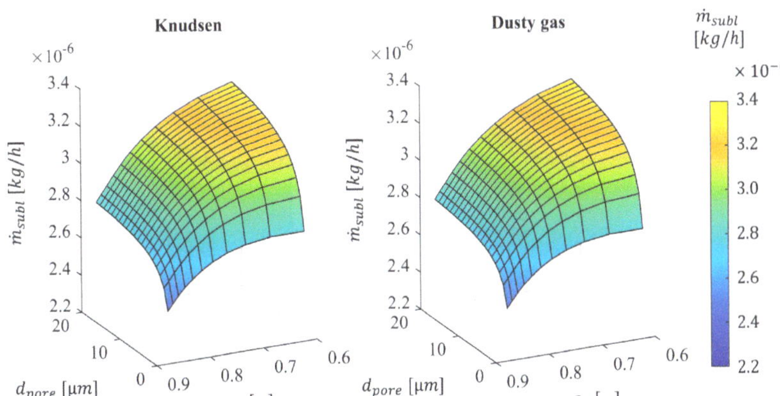


**Figure 6.** Averaged vapor mass flow during the freeze-drying process, modeled by Knudsen diffusion and the dusty gas approach as a function of mean pore diameter and porosity

Again both models show consistent results. It can be observed here that an increasing porosity of the sample reduces the vapor flow significantly. The optimum for both models is shown with 3.3 mg $h^{-1}$ at a pore size of 20 µm and an accessible porosity of 0.61 which corresponds to a sample where no gas was injected. With this configuration the optimal amount of vapor can be removed from the system during primary drying. It shows that the influence of porosity is increasing with increasing pore diameter since the freeze-drying process gets increasingly limited by the heat transfer. On the other hand, it shows that an increasing pore diameter has higher influences at low porosities of 60% in comparison to high porosities where the system is anyways limited by the heat transfer. At porosities of almost 0.9 a change in pore diameter from 10 to 20 µm does not significantly change the results, while a change from 1 to 10 µm has an impact. The highest limitations are observed at small pore diameters of 1 µm and high porosities of almost 0.9.

## 4. Conclusions

In this work, a mathematical model is presented to predict the freeze-drying time for bulky food materials, validated with experimental data. A Knudsen flow and a dusty gas model were used and compared for the mass transfer through the dried layer. Both model approaches are able to fit the experimental results in terms of moisture loss and sublimation rate over time during the freeze-drying process. The model shows room for improvement in the prediction of the transition regime between primary and secondary drying, where experiments have shown that a significant amount of water is already desorbed, while primary drying is still not completed. Nevertheless, this fact does not show any influence on the correct prediction of the drying kinetics of the model in comparison to the experimental approach. As it can be shown in the sensitivity analysis of this study, the pore size might be the appropriate tool to decrease the mass transfer limitations during freeze-drying. An increase of porosity also shows a reduction of mass transfer resistances, but on the expense of reduced heat transfer through the dried layer of the particle. Therefore, an increase of porosity leads to increasing drying time in the observed range of parameters at the given 50% $w/w$ coffee concentration. Furthermore, a change in porosity affects the density and fragility of the product which can impact the product quality.

Both model approaches, the Knudsen flow and the dusty gas approach, are comparable tools to model the freeze-drying of food granules in a range of pore sizes of 1 to 20 μm and porosities in the range of 0.6 to 0.9, which are typical for freeze-dried coffee products. In terms of complexity, an exclusive use of Knudsen flow is shown to be sufficient in this pore size range, shown to have a comparable coefficient of determination, where the dusty gas model may be of higher interest for bigger pore diameters or higher pressures, where other transport mechanisms such as viscous flow are dominating.

**Author Contributions:** Conceptualization, P.L., M.B., V.M., U.K., S.P.; methodology, P.L. and M.B.; software, P.L. and M.B.; validation, P.L. and M.B.; formal analysis, P.L. and M.B.; investigation, P.L.; resources, S.H.; data curation, P.L. and M.B.; writing—original draft preparation, P.L. and M.B.; writing—review and editing, P.L., M.B., V.M., U.K., S.P. and S.H.; visualization, P.L. and M.B.; supervision, V.M., S.P. and S.H.; project administration, S.H.; funding acquisition, S.H. All authors have read and agreed to the published version of the manuscript.

**Funding:** This research received no external funding

**Institutional Review Board Statement:** Not applicable.

**Informed Consent Statement:** Not applicable.

**Data Availability Statement:** Not applicable.

**Conflicts of Interest:** The authors declare no conflict of interest.

## Abbreviations

The following abbreviations are used in this manuscript:

| | |
|---|---|
| OR | Overrun |
| PSD | Particle size distribution |
| SEM | Scanning electron microscopy |
| SSHE | Scarped surface heat exchanger |

## References

1. Khalloufi, S.; Robert, J.L.; Ratti, C. Solid foods freeze-drying simulation and experimental data. *J. Food Process Eng.* **2005**, *28*, 107–132. [CrossRef]
2. Ratti, C. Hot air and freeze-drying of high-value foods: A review. *J. Food Eng.* **2001**, *49*, 311–319. [CrossRef]
3. Hottot, A.; Vessot, S.; Andrieu, J. A Direct Characterization Method of the Ice Morphology. Relationship Between Mean Crystals Size and Primary Drying Times of Freeze-Drying Processes. *Dry. Technol.* **2004**, *22*, 2009–2021. [CrossRef]

4. Pikal, M.J.; Cardon, S.; Bhugra, C.; Jameel, F.; Rambhatla, S.; Mascarenhas, W.J.; Akay, H.U. The nonsteady state modeling of freeze drying: In-process product temperature and moisture content mapping and pharmaceutical product quality applications. *Pharm. Dev. Technol.* **2005**, *10*, 17–32. [CrossRef] [PubMed]
5. Liapis, A.I.; Bruttini, R. A mathematical model for the spray freeze drying process: The drying of frozen particles in trays and in vials on trays. *Int. J. Heat Mass Transf.* **2009**, *52*, 100–111. [CrossRef]
6. Sebasti ao, I.B.; Bhatnagar, B.; Tchessalov, S.; Ohtake, S.; Plitzko, M.; Luy, B.; Alexeenko, A. Bulk Dynamic Spray Freeze-Drying Part 1: Modeling of Droplet Cooling and Phase Change. *J. Pharm. Sci.* **2019**, *108*, 2063–2074. [CrossRef]
7. Song, C.S.; Yeom, G.S. Experiment and numerical simulation of heat and mass transfer during a spray freeze-drying process of ovalbumin in a tray. *Heat Mass Transf.* **2009**, *46*, 39–51. [CrossRef]
8. Sadikoglu, H.; Liapis, A.I. Mathematical Modelling of the Primary and Secondary Drying Stages of Bulk Solution Freeze-Drying in Trays: Parameter Estimation and Model Discrimination by Comparison of Theoretical Results with Experimental Data. *Dry. Technol.* **1997**, *15*, 791–810. [CrossRef]
9. Sheehan, P.; Liapis, A.I. Modeling of the primary and secondary drying stages of the freeze drying of pharmaceutical products in vials: Numerical results obtained from the solution of a dynamic and spatially multi-dimensional lyophilization model for different operational policies. *Biotechnol. Bioeng.* **1998**, *60*, 712–728.:6<712::AID-BIT8>3.0.CO;2-4. [CrossRef]
10. Liapis, A.I.; Bruttini, R. Fundamentals of modeling and analysis of spray freeze drying: The drying of frozen pharmaceutical and food particles in packed beds. In Proceedings of the 17th International Drying Symposium, Magdeburg, Germany, 3–6 October 2010; pp. 71–80.
11. Her, J.Y.; Song, C.S.; Lee, S.J.; Lee, K.G. Preparation of kanamycin powder by an optimized spray freeze-drying method. *Powder Technol.* **2010**, *199*, 159–164. [CrossRef]
12. Warning, A.D.; Arquiza, J.; Datta, A.K. A multiphase porous medium transport model with distributed sublimation front to simulate vacuum freeze drying. *Food Bioprod. Process.* **2015**, *94*, 637–648. [CrossRef]
13. Gruber, S.; Vorhauer-Huget, N.; Foerst, P. In situ micro-computed tomography to study microstructure and sublimation front during freeze-drying. *Food Struct.* **2021**, *29*, 100213. [CrossRef]
14. Vorhauer-Huget, N.; Mannes, D.; Hilmer, M.; Gruber, S.; Strobl, M.; Tsotsas, E.; Foerst, P. Freeze-Drying with Structured Sublimation Fronts—Visualization with Neutron Imaging. *Processes* **2020**, *8*, 1091. [CrossRef]
15. Vorhauer, N.; Först, P.; Schuchmann, H.; Tsotsas, E. Pore network model of primary freeze drying. In Proceedings of the 21th International Drying Symposium, Valencia, Spain, 11–14 September 2018. [CrossRef]
16. Searles, J.A.; Carpenter, J.F.; Randolph, T.W. The ice nucleation temperature determines the primary drying rate of lyophilization for samples frozen on a temperature–controlled shelf. *J. Pharm. Sci.* **2001**, *90*, 860–871. [CrossRef] [PubMed]
17. Kast, W.; Hohenthanner, C.R. Mass transfer within the gas-phase of porous media. *Int. J. Heat Mass Transf.* **2000**, *43*, 807–823. [CrossRef]
18. Jackson, R. Transport in porous catalysts. In *Chemical Engineering Monographs*; Elsevier: Amsterdam, The Netherlands, 1977; Volume 4.
19. Mason, E.A.; Malinauskas, A.P.; Malinauskas, A.P. Gas transport in porous media: The dusty-gas model. In *Chemical Engineering Monographs*; Elsevier: Amsterdam, The Netherlands, 1983; Volume 17.
20. Gloor, P.J.; Crosser, O.K.; Liapis, A.I. Dusty-gas parameters of activated carbon absorbent particles. *Chem. Eng. Commun.* **1987**, *59*, 95–105. [CrossRef]
21. Marti, J.; Mauersberger, K. A survey and new measurements of ice vapor pressure at temperatures between 170 and 250 K. *Geophys. Res. Lett.* **1993**, *20*, 363–366. [CrossRef]
22. Levin, P.; Meunier, V.; Kessler, U.; Heinrich, S. Influence of Freezing Parameters on the Formation of Internal Porous Structure and Its Impact on Freeze-Drying Kinetics. *Processes* **2021**, *9*, 1273. [CrossRef]
23. Wegiel, L.A.; Ferris, S.J.; Nail, S.L. Experimental Aspects of Measuring the Vial Heat Transfer Coefficient in Pharmaceutical Freeze-Drying. *AAPS PharmSciTech* **2018**, *19*, 1810–1817. [CrossRef] [PubMed]
24. Rambhatla, S.; Pikal, M.J. Heat and mass transfer scale-up issues during freeze-drying, I: Atypical radiation and the edge vial effect. *AAPS PharmSciTech* **2003**, *4*, E14. [CrossRef] [PubMed]

processes

MDPI

Article

# Effect of Drying Methods and Processing Conditions on the Quality of *Curcuma longa* Powder

**Sandra M. Llano [1], Ana María Gómez [1] and Yudy Duarte-Correa [2,*]**

[1] INTAL Research Group, INTAL Foundation, Institute of Food Science and Technology, Itagüí 05542, Colombia; llanogils@gmail.com (S.M.L.); ana.gomez10@udea.edu.co (A.M.G.)
[2] BIOALI Research Group, Food Department, Faculty of Pharmaceutical and Food Sciences, Universidad de Antioquia, Medellin 050010, Colombia
* Correspondence: yudy.duarte@udea.edu.co

**Abstract:** Turmeric (*Curcuma longa*) is a spice that has been used for a long time in traditional medicine for its anti-inflammatory properties and recently used in the food industry for its dyeing and flavoring properties. This work studied the effect of different drying methods (convection oven drying, fluidized bed drying, and traditional solar drying) on the quality of *Curcuma longa* powder. The effect of UV radiation on turmeric powder using different packaging materials (glass, aluminum foil bag, and low-density polyethylene bag), was also studied. Subsequently, the fluidized bed drying method was used to evaluate the effect of drying temperature. The results show that convection and fluidized bed drying had no significant impact on turmeric quality. However, solar drying degraded curcuminoids by 36.5% and the ORAC value decreased by 14%. Regarding the packaging materials, the aluminum bag prevented the deterioration of 14% of the curcuminoids for the powder exposed to UV radiation. Finally, the effect of temperature on fluidized bed drying was evaluated at 50–80 °C, finding that there were no significant differences in the curcuminoid content and antioxidant capacity of turmeric powder. This implies that the range of temperature used in this study is appropriate for drying this material using fluidized bed drying, producing a turmeric powder with a high content of bioactive compounds, when compared to convection oven and solar drying. Therefore, the turmeric powder obtained in this way can be used as an active ingredient in the formulation of different kinds of foods and supplements.

**Keywords:** *Curcuma longa*; drying methods; antioxidant capacity; curcuminoids; packaging

**Citation:** Llano, S.M.; Gómez, A.M.; Duarte-Correa, Y. Effect of Drying Methods and Processing Conditions on the Quality of *Curcuma longa* Powder. *Processes* **2022**, *10*, 702. https://doi.org/10.3390/pr10040702

Academic Editor: Timothy Langrish

Received: 11 March 2022
Accepted: 2 April 2022
Published: 5 April 2022



## 1. Introduction

Turmeric root (*Curcuma longa*) is widely used for culinary, medicinal, and cosmetic purposes and as a dietary supplement. It is a rhizomatous small perennial plant belonging to the *Zingiberaceae* family, native to India. It is distributed throughout tropical and subtropical regions of the world and widely cultivated in Southeast Asia [1–3], where it is used as a natural coloring and as a flavoring agent, especially for curries, as well as for dyeing [4]. India and other Asian countries like Bangladesh, Pakistan, Sri Lanka, Taiwan, China, Myanmar, Indonesia, and Thailand are the lead growers of this crop [5]. However, turmeric is farmed in many warm regions of the world [6]. There are several reports of turmeric's health-promoting properties, such as antioxidant activity and anti-infectious, anti-inflammatory, anti-microbial, anti-viral, anti-carcinogenic, and anti-tumor properties [2,3,5,7,8], making it an appealing ingredient to develop functional foods. Globally, the demand for turmeric has grown due to its therapeutic functions and low toxicity [3]. Turmeric has become a relevant crop in Colombia, given its multiple uses and ease of processing into other products (e.g., powder and oleoresin). Its ease of processing promotes its use in traditional agriculture areas and as an alternative for illegal crops (United Nations Office on Drugs and Crime—UNODC, 2020). As of 2020, 50.2 hectares of turmeric were cultivated in Colombia and 755 tons were produced, with a yield of 15.44 ton/ha. Each ton was valued at USD 1610 (Ministerio de Agricultura de Colombia, 2020; Ministerio de Agricultura de Colombia, 2021) [9].

The major bioactive compounds in the turmeric rhizome are curcuminoids, phenolic acids, and flavonoids, which possess health-promoting properties [5]. These properties are mainly related to curcuminoids, a group of phenolic compounds comprising of curcumin, dimethoxy curcumin, and bisdemethoxycurcumin [2,10]. Curcuminoids are the main dyes of turmeric; they belong to a group of phenolic compounds called diarylheptanoids. Three main curcuminoids have been identified: curcumin—((1E,6E)-1,7-bis(4-hydroxy-3-methoxyphenyl)-1,6-heptadiene-3,5-dione), desmethoxycurcumin—((1E,6E)-1-(4-Hydroxy-3-methoxyphenyl)-7-(4-hydroxyphenyl)hepta-1,6-diene-3,5-dione), and bisdemethoxycurcumin—((1E,6E)-4-hydroxyph enyl)-7-(4-hydroxyphenyl)hepta-1,6-diene-3,5-dione). There are reports of curcuminoid's antioxidant [11], anti-inflammatory [12], anticancer [13]), and antimicrobial [14] properties. Additionally, they have shown to have a potential effect on neurodegenerative diseases and atherosclerosis [15–17]. Curcuminoids are not stable in the presence of light, especially in solutions. Some studies report that curcumin shows photodegradation when exposed to UV/Visible radiation, in both solution and on solid-state thin film. Several degradation products have been reported, including products from the cyclization of curcumin [18]. Additionally, certain processing conditions or exposure to UV radiation in storage could affect the stability and bioactivity of curcumin [19].

In the market, turmeric is generally found fresh or dried (powder). Currently, turmeric production is about 37,000 tons, valued at USD 40,160,000,000 [5]. Commercially, the powder is the most convenient presentation of turmeric [20] because it can be used directly as a spice and for turmeric oleoresin and oil manufacturing [21]. Therefore, the demand for dried turmeric is increasing in global markets [7]. Turmeric powder comes from the plant rhizomes, involving various post-harvest processes comprising of disinfection, drying, milling, and packing. These post-harvest treatments could affect the quality of turmeric powder [22] since they affect the content and stability of curcuminoids.

Drying or dehydration is the most widely used technique for turmeric preservation [5]. Convection oven drying, freeze drying, vacuum drying, solar drying, and osmotic drying are some drying techniques developed in the past and during recent years [23]. Solar drying is the traditional method used for obtaining turmeric powder. The climate conditions in tropical countries make solar energy practically attractive for food drying processes [24]. In this regard, solar drying is an adequate solution for developing countries that lack convectional energy resources but have a constant solar input throughout the year [25]. However, the choice of optimal processes would determine the physicochemical and functional properties of the final product. Some studies reported that solar drying decreases the overall content of curcuminoids [26]. Others have found that drying does not affect their content [26,27]. However, there are no records on the effect of fluidized bed drying on the curcuminoid content of turmeric powder.

On the other hand, packaging is another postharvest process that can significantly affect the quality of the final product. The packaging material is a critical factor that can protect and delay the formation of different compounds such as ((2Z,5E)-2-hidroxi-6-(4hidroxi-3-metoxifenil)-4-oxohexa-2,5-dienal, ferulic acid, and feruloilmetane, which result from the contact of turmeric with oxygen and light transmission through the packing [28]. The findings of previous studies have demonstrated that undesirable effects of gamma radiation on food products—such as softening, breathing, or lipid oxidation—are significantly diminished by modified atmosphere packaging [29]. Some authors evaluated the effect of gamma irradiation under various packaging atmospheres (air, $N_2$, and vacuum) on the physicochemical properties of turmeric powder [30]. However, there are no studies evaluating the effect of packaging materials on turmeric powder's properties, which is a crucial factor in food preservation. For instance, Fikreyesus et al. (2021) assessed the influence of packaging material on the physicochemical, microbial, and sensory properties of infant powder meal using paper, polyethylene, and polypropylene bags. They reported that powder meals are a moisture-sensitive product, and long-term storage using permeable packaging materials can lead to moisture absorption from the powder. On the other hand, Tripetch and Borompichaichartkul (2019) studied the effects of packaging materials (high-density polyethylene (HDPE) bag and jute sack) and storage time on the phenolic content, chlorogenic acid, and antioxidant capacity variations in arabica green coffee beans. The authors found that there were no significant differences in the content of phenolics in coffee beans during four months of storage for both types of packaging, but after one year of storage, the content of phenolics and antioxidant capacity of coffee beans stored in a jute sack were higher than those stored in an HDPE bag [31].

Therefore, the objective of this study was to evaluate the effect of drying methods, drying temperature, and packaging material on the curcuminoids' content, antioxidant capacity, and color, which are parameters directly related to the quality of turmeric powder.

## 2. Materials and Methods

### 2.1. Reagents

6-Hydroxy-2,5,7,8-tetramethylchroman-2-carboxylic acid (Trolox) was obtained from Biosynth Carbosynth (Berkshire, UK); 2,2′-Azobis(2-methylpropionamidine) dihydrochloride (AAPH) and gallic acid from Sigma-Aldrich (Saint Louis, MO, USA); fluorescein sodium from Chem-Impex International, Inc (Wood Dale, IL, USA); methyl-beta-cyclodextrin from ACROS Organics™ (Fisher Scientific GmbH, Schwerte, Germany); sodium carbonate and Folin-Ciocalteu's Reagent from ITW Reagents (PanReac-AppliChem, Barcelona, Spain); phosphate buffered saline (Dulbecco A) tablets from Oxoid Limited (Hampshire, England); curcumin from Dr. Ehrenstorfer GmbH (Augsburg, Germany); desmethoxycurcumin and bisdemethoxycurcumin from Chengdu Biopurify Phytochemicals Ltd. (Chengdu, China); and analysis-grade acetone, ethanol, methanol, acetic acid, hexane, and LC/MS-grade acetonitrile from Merck KGaA (Darmstadt, Germany). Additionally, dichloran glycerol agar 18% was obtained from Merck KGaA (Darmstadt, Germany).

### 2.2. Vegetal Material

Fresh turmeric (*Curcuma longa*) rhizomes were collected from local farmers in Uramita, Antioquia, Colombia (06°52′37.2″ N, 76°12′03.1″ O). This research used rhizomes with no mechanical affectations and free of damage caused by pathogens and insects. The rhizomes were washed to remove all physical contaminants. After washing, they were dried with a paper towel to remove any excess moisture and subsequently disinfected. The disinfection was carried out by immersion in a solution of organic acids at 0.5% *v/v* Biodes-Ultra (Cory Industries, Medellín, Colombia), as has been reported for other food matrices [32,33]. This process is necessary to destroy microorganisms that may be present in the rhizomes. Afterwards, they were stored under refrigerated conditions (0–5 °C) until further analysis.

### 2.3. Drying Experiments

Before drying, the rhizomes were sliced in a food processor (Sammic, Spain) to obtain slices of 5 mm thickness. Samples were processed in batches and were immediately submitted to the respective drying experiment to prevent enzymatic browning [26]. Three drying methods were studied: traditional solar drying (TSD), fluidized bed drying (FBD), and convection oven drying (COD). The drying experiments were conducted until constant weight, with an upper limit of 8% (wb) (ICONTEC, 2021) [34].

#### 2.3.1. Traditional Solar Drying (TSD)

For TSD, sliced rhizomes were spread into a uniform layer (1 cm of thickness) over a perforated aluminum tray (70 cm × 50 cm) and exposed to direct sunlight (from 8 am to 6 pm). The material was carefully homogenized by hand, using gloves (every hour, approximately). TSD conditions comprised of an average temperature of 24 ± 2 °C for product surface, and a relative humidity of 76 ± 0.5%. The drying was carried out until constant weight was achieved. The pyranometer of SIATA (Sistema de Alerta Temprana del valle de Aburrá) station was utilized to monitor the solar radiation intensity.

#### 2.3.2. Convection Oven Drying (COD)

The COD was performed in a Memmert model UF 260—Schutzart DIN 40050—IP20 oven (Schwabach, Germany). Samples were also spread into a 1 cm thick layer on a perforated aluminum tray (25 × 30 cm). COD was conducted at 50 °C and dried to a constant weight.

#### 2.3.3. Fluidized Bed Drying (FBD)

FBD was performed in a fluidized bed dryer (Vibrasec S.A.S., Medellín, Colombia). FBD conditions consisted of a 2400 rpm fan (equivalent to an air velocity of 1.0 m/s) and a temperature of 50 °C. Samples were dried to a constant weight.

### 2.4. *Turmeric Powder Production*

The dried turmeric slices obtained with different drying methods were milled to a powder (40 μm particle size) in a hammer mill (Vibrasec S.A.S., Medellín, Colombia) and then stored in aluminum sealable bags at 25 °C and 75% relative humidity conditions until analysis.

### 2.5. *Turmeric Powder Characterization*

#### 2.5.1. Moisture Content and Water Activity

The moisture contents of fresh turmeric and dried turmeric were determined according to AOAC methods, using a gravimetric method, drying at 105 °C in an air oven to constant weight. Water activity was measured at 25 °C using a dewpoint hygrometer (AQUALAB-PRE) (AOAC, 2000). Triplicate analyses of each sample were carried out (n = 3).

#### 2.5.2. Determination of Curcuminoid Content

Curcuminoid content was determined by HPLC-DAD UltiMate™ 3000 with software Chromeleon 7.2. (Dionex, Thermo scientific, Sunnyvale, CA, USA), using a reverse-phase Hypersil GOLD C18 column (100 mm × 2.1 mm ID, the particle size of 1.9 μm; 0.2 μm × 2.1 mm ID prefilter) (Thermo Scientific™, Waltham, MA, USA). The mobile phases were 0.3% acetic acid in water (A) and 0.3% acetic acid in acetonitrile (B). The gradient was isocratic (35% of A and 65% of B), the flow rate was 0.3 mL/min, the injection volume was 10 μL, and the detector was set at 420 nm. The column temperature was maintained at 30 °C.

The individual contents of curcuminoids were determined by UPLC-DAD, using a modified procedure based on Cheng et al. (2010) [35] for fresh rhizomes and turmeric powder. Curcuminoids were quantified from external standard calibration curves of curcumin, desmethoxycurcumin, and bisdemethoxycurcumin methanolic solutions. A total of 0.25 g ± 0.01 g of the turmeric powder were weighed in a centrifuge tube and added to 10 mL of hexane. This mixture was homogenized for 5 min and centrifuged at 4000× *g* rpm for 5 min. The supernatant was discarded, and the remaining hexane was evaporated inside a Resprep Quick-Replace Vacuum Manifold (Restek Corporation, Bellefonte, PA, USA). Then, the solid was reconstituted with 25 mL of ethanol, sonicated at room temperature for 5 min, homogenized for 5 min, and centrifuged at 4000× *g* rpm for 5 min. The resulting supernatant was diluted at 1:200 with the mobile phase. Triplicate analyses of each sample were carried out (n = 3).

#### 2.5.3. Determination of Antioxidant Capacity

Sample preparation was carried out according to the method previously described by Cao et al. [36]. Briefly, 0.20 g ± 0.01 g of turmeric powder was weighed in a centrifuge tube and added with 5 mL of hexane. This mixture was homogenized for 1 min, sonicated at room temperature for 10 min, and centrifuged at 4000× *g* rpm for 15 min. The supernatant was transferred to a screwed amber vial. The previous lipid extraction was conducted twice, and both supernatants were united in the same vial. The hexane of this extract was completely evaporated inside a Resprep Quick-Replace Vacuum Manifold (Restek Corporation, Bellefonte, PA, USA). Later, the residue was reconstituted with 2 mL of acetone, and this mixture was thoroughly mixed. This extract was reserved under freezing conditions (−20° C) until the ORAC-lipophilic assay. The lipophilic extract was diluted at 1:50 in 10 mM phosphate-buffered saline solution (PBS). Then, 1000 μL of the diluted sample was placed in a screwed vial, and 150 μL of the 7% methyl-beta-cyclodextrin solution (in water: acetone, 50:50) was added. This mixture was homogenized for 10 s.

For the hydrophilic extract, the hexane was evaporated inside a Resprep Quick-Replace Vacuum Manifold (Restek Corporation, Bellefonte, PA, USA). Then, 5 mL of extraction solvent (acetone: water: acetic acid, 70:29.5:0.5) were added to the dried powder. This mixture was homogenized for 1 min, sonicated at room temperature for 6 min, and centrifuged at 4000× $g$ rpm for 15 min. The supernatant was reserved under freezing conditions (−20 °C) until the ORAC-hydrophilic assay. The hydrophilic extract was diluted at 1:10 in 10 mM PBS.

The antioxidant capacity of fresh turmeric and turmeric powder was determined by the ORAC method, following Akter et al. [37]. The ORAC-lipophilic and ORAC-hydrophilic assays were performed on a multimode microplate reader Varioskan™ LUX equipped with a SkanIt RE 6.0.1. Software (Thermo Scientific™, Waltham, MA, USA). Trolox standards were prepared in 10 mM PBS (5–600 μM). The reagents were added to each of the wells of an ELISA black plate as follows: 150 μL of 1 μM fluorescein and 25 μL of the sample (either lipophilic, hydrophilic dilutions, blank, or standard). Samples were incubated in the dark at 37 °C for 30 min. Then, 25 μL of 250 mM AAPH was added to each well. Finally, the fluorescence was measured at 485 and 520 nm (in kinetic intervals of 30 s, for 1.5 h, and at a constant temperature of 37 °C). The ORAC-lipophilic and ORAC-hydrophilic results of the turmeric powder were extrapolated from the calibration curve as micromoles of Trolox equivalents per gram of turmeric powder (μmol TE/g). Quantifications were made in triplicate. The total ORAC value was obtained by adding the ORAC-lipophilic and the ORAC-hydrophilic results.

#### 2.5.4. Color Measurements

The color of rhizomes was determined with a Nanocolor UV/VIS Machery–Nagel set to D65 illuminant and a 10° observation angle. The CIELAB color space was used to determine the parameters: L* (black (0) to White (100)), a* (greenness (−) to redness (+)) and b* (blueness (−) to yellowness (+)). The mean values were obtained from triplicate readings. The total color difference (ΔE) was determined according to Equation (1) [30].

$$\Delta E = \sqrt{(\Delta l)^2 + (\Delta a)^2 + (\Delta b)^2} \tag{1}$$

These solutions were prepared by taking 25 mg of flour and 50 mL of ethanol. The mixture was stirred for 20 min at room temperature (25 ± 2 °C). The mixture was centrifuged at 3000× $g$ rpm at 20 °C. This solution was added using a quartz cell for reading [36].

### 2.6. *UV Radiation Effect Using Different Packaging Materials*

The three packing materials used were glass, aluminum foil bag, and low-density polyethylene (LDPE) bag. Also, unpackaged turmeric powder was used as a control. For these tests, small quantities (1 g) of powder were employed so that packaged material formed a thin sheet of powder inside the package (2 mm wide). The thin layers of the sample received radiation from a UV lamp. The radiation and power were 360 nm and 6 W, respectively.

The samples were exposed to radiation for 9 h at room temperature (25 ± 2 °C). Tracking of samples was conducted to analyze the content of curcuminoids and antioxidant capacity. Color assessment was also made. The assay was performed in triplicate.

### 2.7. Effect of Temperature of Drying

One of the three dehydration methods (TSD, FBD, COD) was selected to continue the research based on the curcuminoid content and antioxidant capacity (ORAC assay) results.

For the selected drying method (FBD), the effect of temperature (40, 50, 60, 70, and 80 °C) on the functional properties of turmeric powder (curcuminoid content and antioxidant capacity) was determined as well as the moisture content and water activity.

Finally, drying curves for each temperature were analyzed. Weight changes of turmeric during the drying process were monitored by taking samples at every 1 h interval, with a digital balance (0.001 g). Samples were dried to a constant weight.

A dimensionless moisture ratio (MR) was calculated from the drying curves using the following Equation (2):

$$MR_t = \frac{X_t - X_e}{X_o - X_e} \quad (2)$$

where $X_t$ is the moisture content at any time $t$ (g water/g dry basis), $X_e$ is the moisture content at the equilibrium (g water/g dry basis), and $X_o$ is the initial moisture content (g water/g dry basis). Values of $X_e$ are considered to be relatively small compared to $X_t$ or $X_o$. Thus, $(X_t - X_e)/(X_o - X_e)$ can be simplified to $(X_t/X_o)$ [37].

### 2.8. Statistical Analysis

Statistical analysis was performed using GraphPad Prism 5 software (GraphPad Software, Inc., San Diego, CA, USA) and Statgraphics Centurion XVII (Statpoint Technologies, Inc., Warrenton, VA, USA). The response variables were analyzed by an analysis of variance (ANOVA), and the means were compared using the Tukey test at a 95% significance level to determine the significant differences between samples.

## 3. Results

### 3.1. Drying Experiments Results

Table 1 shows the moisture content, water activity, and drying time results for each drying method and for fresh turmeric rhizomes. The drying was conducted until a constant weight was reached, obtaining different drying times for each treatment: TSD required 20 h, while COD and FBD needed 13 and 5 h, respectively. As expected, for all the dehydration methods, the moisture content (weight basis, wb) of the samples decreased, and the lowest value was obtained with the FBD process (7.5%). Additionally, the lowest $a_w$ was achieved with FBD, which ensured better stability since the values reported in Table 1 for TSD and COD were located in the region of fungal, yeast, and bacterial growth [36]. TSD was the drying method that required the longest time, since this occurs through incident solar radiation. Several authors have reported a longer drying time for the TSD than for hot air drying [38].

**Table 1.** Moisture content and water activity for fresh and dried turmeric.

| Drying | Moisture Content (% *w*/*w*) | Time (h) | $a_w$ |
|---|---|---|---|
| Fresh Turmeric | 83.2 ± 0.5 | — | 0.81 ± 0.003 |
| TSD | 7.9 ± 0.2 [a] | 20 | 0.08 ± 0.003 [a] |
| COD | 7.6 ± 0.1 [a] | 13 | 0.07 ± 0.003 [a] |
| FBD | 7.5 ± 0.2 [a] | 8 | 0.05 ± 0.003 [b] |

Data are expressed as mean ± standard deviation (n = 3). Common letter in the same column indicates that they are not significantly different ($p < 0.05$).

#### 3.1.1. Curcuminoid Content

The results show that the three drying methods, TSD, FBD and COD, affected the total curcuminoid content of the rhizomes. The kinetics of degradation of total curcuminoids for the three drying methods is shown in Figure 1.

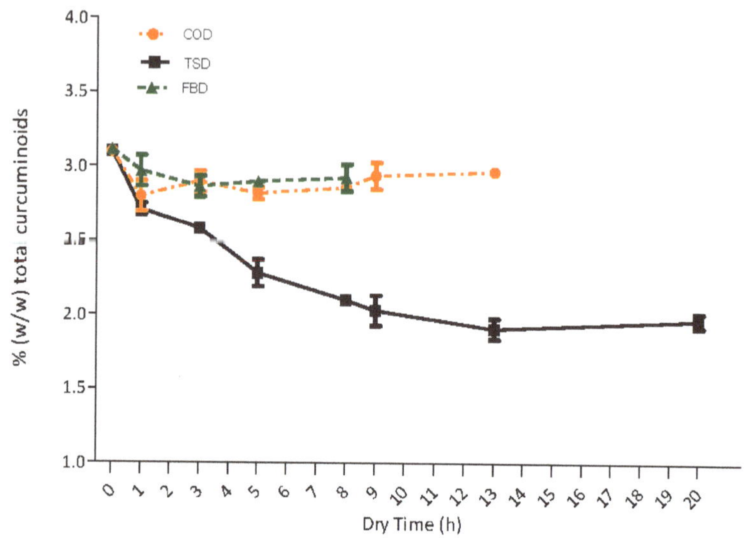


**Figure 1.** Degradation kinetics plot of total curcuminoids content, for convection oven drying (COD), solar drying (TSD), and fluidized bed drying (FBD).

TSD had a significant effect on the curcuminoid content. TSD decreased the curcuminoid content by 36.5%, in contrast to the fresh sample. This behavior was constant throughout the drying process (see Figure 1), since drying occurs through incident solar radiation (5.408 $kWhm^{-2}$). TSD, due to the low and inconsistent temperatures, takes longer to dry, which generates losses in the quality of the product [39]. On the other hand, COD and FBD presented a slight decrease in the curcuminoid content, resulting in a reduction of approximately 2% and 1.9%, respectively.

Some studies report that curcumin is stable up to temperatures of 190 °C, according to differential scanning calorimetry analysis (DSC) [5]. In our study, all the dryings were performed between 24 and 50 °C, thus, within this temperature range, curcuminoids are non-thermolabile. However, for TSD, the rhizomes were exposed to UV radiation. Previous studies on turmeric oleoresin have reported photodegradation processes of curcuminoids (curcumin, desmethoxycurcumin, and bisdemethoxycurcumin) [40]. Other authors have found that curcumin undergoes photodegradation in solution and solid form [41]. The latter could explain the progressive reduction in total curcuminoid content from the TSD samples.

Prathapan et al. (2009) measured the total curcuminoid content of turmeric at the end of the solar drying process. They observed a reduction in the concentration of curcuminoids [26]; however, they did not measure the curcuminoid content throughout the drying process as we did in the present study. We obtained a higher curcumin reduction rate (36.5%) for TSD than that reported in the Prathapan et al. study (18%) [26]. These discrepancies could be related to turmeric crop origin, the edaphoclimatic conditions of the soil, and the postharvest treatments before drying. Given that TSD decreases the concentration of curcumin, desmethoxycurcumin, and bisdemethoxycurcumin [7], the decline of these metabolites impacts the physicochemical and functional properties of turmeric powder (*Curcuma longa*) quality.

As can be seen in Figure 1, the other drying methods (COD and FBD) preserved the quality of the turmeric rhizomes, i.e., curcuminoids content and antioxidant capacity. For instance, FBD did not affect the content of curcuminoids and the antioxidant capacity of turmeric at the end of the process and presented a shorter drying time when compared to COD. FBD is a technique used in the food industry for various types of products; it has some advantages over TSD and COD: shorter drying periods, color retention, and preservation of bioactive compounds [19]. Other benefits include greater thermal efficiency, low costs, and easy control [42]. Therefore, the quality of the turmeric powder is best preserved with this drying method.

### 3.1.2. Antioxidant Capacity

The results for antioxidant capacity (ORAC) are shown in Figure 2. TSD had a significant effect on the antioxidant capacity of the rhizomes since the antioxidant capacity had reduced by 15% at the end of the drying (20 h). COD did not present a significant decrease in the ORAC value (1.8%). The reduction for FBD was only 0.2%.

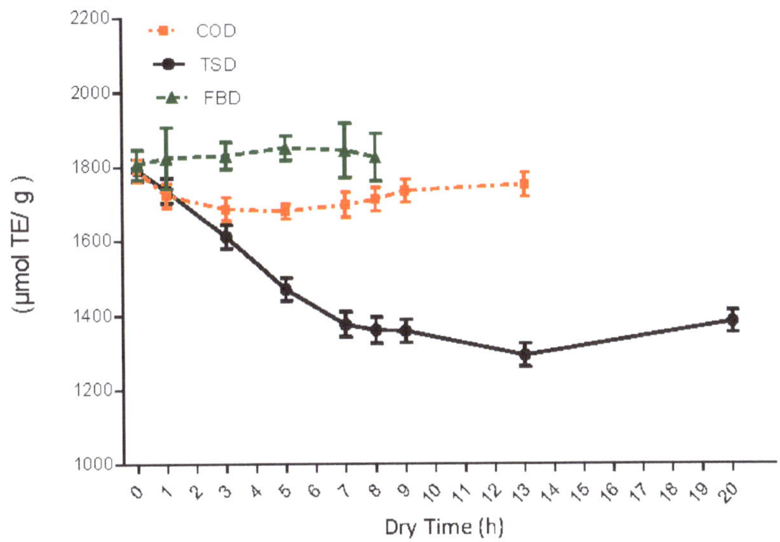


**Figure 2.** Effect drying treatments in curcuma longa rhizomes. Variation of antioxidant capacity (ORAC) of rhizomes during TSD, COD, and FBD.

There is strong evidence correlating antioxidant capacity with a higher content of curcuminoids since these bioactive compounds present antioxidant properties [12,43]. However, Sing et al. (2010) reported other active substances with antioxidant capacity in turmeric, such as ar-turmerone and alpha-turmerone, the main constituents of essential oil and oleoresin [44]. The authors suggest that these compounds probably exert synergistic or additive effects to the total antioxidant capacity because this reduction in antioxidant capacity is not equivalent to the same content reduction present in curcuminoids. In addition, according to Chumroenphat et al. (2021), curcumin is degraded and transformed into other phenolic compounds, particularly vanillin and ferulic acid, during the drying processes [41]. This degradation could be responsible for the final increase in the antioxidant capacity of TSD.

The photodegradation analysis showed a decrease in scavenging activities against DPPH radical [45], which may be associated with curcuminoid degradation. The decline in the concentration of curcumin, desmethoxycurcumin, and bisdemethoxycurcumin directly affects the antioxidant capacity since these substances have antioxidant properties following the hydrogen atom transfer (HAT) and single electron transfer (SET) mechanisms [46] and are the main bioactive compounds of turmeric.

FBD presented the best performance in drying time, moisture content, water activity, and preservation of the antioxidant capacity and total curcuminoid content of the turmeric. Hence, it was selected as the method to continue on in the study, evaluating some processing conditions. The study variable selected was temperature, i.e., the effect of fluidized bed drying temperature on the curcuminoid content and antioxidant capacity.

#### 3.1.3. Color

Color is often related to the quality of turmeric. Table 2 shows the results for color parameters (L*, a*, and b*) of the turmeric dyed solutions. TSD showed significant differences in b* (yellow/blue coordinate) and L*(lightness). The color difference (ΔE) was used to quantify the potential color change in turmeric samples processed by different drying techniques concerning fresh turmeric. Significant differences in ΔE were observed for all of the drying methods. However, TDS had the greatest effect on total color change, which signifies the highest color difference between the fresh and dried samples. The latter could be due to a decrease in curcuminoid content, which is affected by photodegradation [7]. Moreover, oxidation, thermal degradation, and glycosylation could also explain the previous results [39]. Ray et al. (2022) evaluated the effect of different drying methods (e.g., solar and convective drying) and found the highest color change in convective drying at 60 °C, explaining that intense browning reactions (Maillard), oxidation, and change in the surface structure are the principal causes of degradation of the color [47].

**Table 2.** Effect of drying treatments on the CIElab parameters.

| Drying | Color Analysis | | | |
|---|---|---|---|---|
| | L | a | B | ΔE |
| TSD | 90.4± 0.4 [a] | −10.7 ± 0.2 [a] | 123.9 ± 0.1 [a] | 6,30 ± 0.03 [a] |
| COD | 91.6 ± 0.1 [b] | −11.0± 0.1 [a] | 124.9 ± 0.3 [b] | 5,1 ± 0.1 [b] |
| FBD | 92.2 ± 0.3 [b] | −10.9 ± 0.2 [a] | 125.9 ± 0.3 [c] | 4,24 ± 0.02 [c] |

Data are expressed as mean ± standard deviation (n = 3). Common letters in the same column indicates that they are not significantly different ($p < 0.05$).

The parameter b* was statistically different for the three treatments, with its highest value in FBD. This solution had a more yellow hue. On the other hand, TSD had the lowest b* value. The latter is because curcuminoids give the characteristic yellow tone to the rhizomes [2], and the TSD sample also showed lower curcuminoid content. The drying method had no effects on the parameter a* (red/green coordinate). Komonsing et al. (2022) showed that the light exposure caused color fading in turmeric slices [7]. In TSD, the sun (light radiation) affects turmeric rhizomes' color, affecting the quality of the powder. Since higher lightness (L*) values are preferable in the case of dried food products [48], the FBD process lead to the best quality in color.

### 3.2. *Effect of Drying Temperature on Curcuminoids Content, Antioxidant Capacity and Drying Time*

Following the aforementioned results, fluidized bed drying shortened the drying process of turmeric rhizomes, as compared to convective oven drying and solar drying. Moreover, FBD led to the slightest variation in total curcuminoids content and antioxidant activity with better color preservation and the lowest water activity. Hence, FBD was selected to continue on in the study to evaluate the effect of drying temperature.

The effect of temperature on FBD in the drying curves of turmeric rhizomes is shown in Figure 3. For all temperatures, we observed that the MR value decreased rapidly after the first hour of drying, as observed in Figure 3, being more evident for the temperature of 80 °C. Sharma et al. evidenced a similar trend for hot air drying and direct solar drying methods [41]. In our study, the drying time to reach a final moisture content of less than 8% was 8, 8, 6, and 5 h for 50, 60, 70, and 80 °C, respectively. According to these results, with a temperature of 80 °C, the drying time was reduced by 37.5% as compared to drying at 50 °C. Several researchers have previously reported on the dependence of drying time on temperature. This behavior is due to higher temperature increasing the system's enthalpy, which increases the transfer of mass and energy, accelerating the migration of water [49].

Chan and Kuo (2018) evaluated different temperatures (80, 100, 120 °C) in a fluidized bed dryer on wheat germ, finding that the efficiency of dehydration was better at a higher set temperature (120 °C). Additionally, shorter drying times might be recommended for a hot and humid environment such as the summer season in Asia [50].

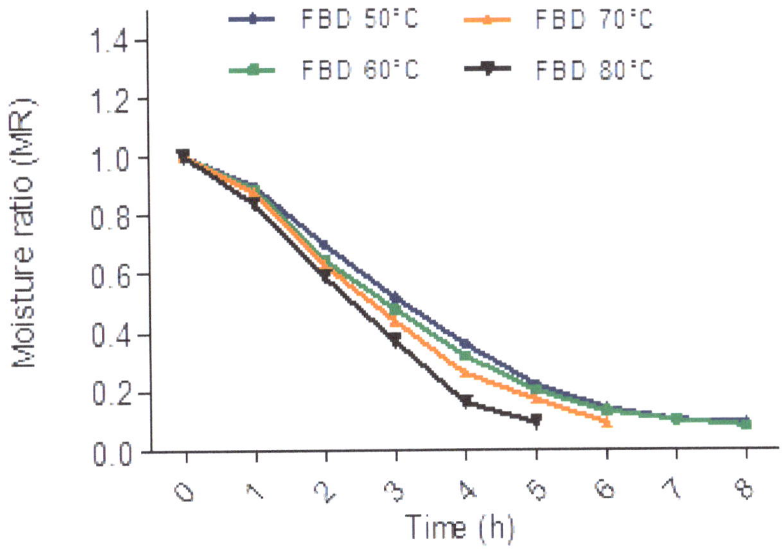


**Figure 3.** Fluidized bed drying kinetics at different temperatures.

Regarding the effect of temperature on curcuminoid content and antioxidant capacity, the results are shown in Table 3. Statistical analysis shows there were no significant differences ($p > 0.05$) in the curcumin, desmethoxycurcumin, and bisdemethoxycurcumin concentration between the dried products obtained under different drying conditions (50 °C, 60 °C, 70 °C and 80 °C).

**Table 3.** Effects of drying temperature of FBD on curcuminoid content and antioxidant capacity in turmeric powder and fresh turmeric.

| Temperature (°C) | Curcumin (% *w/w*) | Demethoxy-Curcumin (% *w/w*) | Bisdemethoxy-Curcumin (% *w/w*) | Total Curcuminoids (% *w/w*) | ORAC (μmol TE/g) | $a_w$ | Moisture Content (% *w/w*) |
|---|---|---|---|---|---|---|---|
| Fresh | 2.55 ± 0.35 [a] | 1.09 ± 0.14 [a] | 0.66 ± 0.09 [a] | 3.53 ± 0.73 [a] | 1805.45 ± 39.99 [a] | 0.85± 0.01 [a] | 83.2 ± 0.5 [a] |
| 50 | 2.73 ± 0.38 [a] | 1.36 ± 0.18 [a] | 0.87 ± 0.11 [a] | 3.15 ± 0.64 [a] | 1854.35 ± 81.88 [a] | 0.051 ± 0.005 [b] | 7.5 ± 0.2 [b] |
| 60 | 2.39 ± 0.33 [a] | 1.01 ± 0.16 [a] | 0.83 ±0.11 [a] | 3.43 ± 0.72 [a] | 1851.48 ± 71.98 [a] | 0.050 ± 0.004 [b] | 7.4 ± 0.2 [b] |
| 70 | 2.70 ± 0.38 [a] | 0.92 ± 0.15 [a] | 0.76 ± 0.099 [a] | 3.33 ± 0.71 [a] | 1826.19 ± 63.91 [a] | 0.042 ± 0.005 [b] | 7.4 ± 0.3 [b] |
| 80 | 2.56 ± 0.35 [a] | 1.15 ± 0.21 [a] | 0.85 ± 0.119 [a] | 4.054 ± 0.78 [a] | 1733.04 ± 35.94 [a] | 0.041 ± 0.005 [b] | 6.7 ± 0.2 [b] |

Data are expressed as mean ± standard deviation (n = 3). Common letter in the same column indicates that they are not significantly different ($p < 0.05$).

Previous studies evaluated the degradation of curcuminoids at different temperatures with light exposure, showing that light radiation can accelerate the drying process but detriments the curcuminoid content [7]. Although curcumin is stable above 100 °C in thermogravimetric (TGA) studies [41], these correspond to isolated curcumin. On the contrary, there are no reports of the stability of curcuminoids (curcumin, desmethoxy-curcumin, and bisdemethoxycurcumin) in turmeric powder obtained from FBD. In our study, FBD-dried rhizomes did not receive light radiation. Thus, we could observe that the functional properties of the turmeric powder prevailed at all the evaluated temperatures.

The results from HPLC-DAD show that curcumin was the most abundant curcuminoid in the turmeric powder, followed by bisdemethoxycurcumin and desmethoxycurcumin, which is consistent with a previous report [51]. Additionally, the concentration of the curcuminoids was stable at the different drying temperatures without showing significant degradation. Prathapan et al. (2009) found no significant differences in curcuminoid content when compared with fresh and thermally treated material (70, 80, 90, and 100 °C for 30 min) [26]. However, they also observed that solar drying significantly decreased the concentration of the curcuminoids [26].

On the other hand, from Table 3, it can be seen that there were no significant differences in antioxidant capacity, as assessed by ORAC method, in samples dried at 50, 60, 70, and 80 °C and fresh turmeric, thus preserving the functional properties of turmeric. Another study evaluated the effect of different drying temperatures (40, 50, 60, 70, and 80 °C) on the antioxidant capacity through DPPH, ABTS, FRAP, and TPC methods [7]. They also reported no significant changes in antioxidant capacity with temperature variation. The various types of antioxidants in turmeric also contribute to the overall antioxidant capacities and could explain the previous statement. Phenolic compounds of turmeric include gallic acid, curcumin, ferulic acid, epicatechin, catechin, cinnamic acid, protocatechuic acid, chlorogenic acid, rutin, genistein, and coumarin [7]. Therefore, although the antioxidant capacity of the flour is mainly related to the content of curcuminoids [12,46], these other substances also contribute to the total antioxidant capacity of turmeric flour.

For moisture content and Aw, we determined that higher drying temperatures led to lower moisture content and water activity of the turmeric powder (see Table 3). This has also been reported for other food matrices [38,52,53]. Additionally, this decrease helps reduce the risk of microbiological contamination [54], making the product more stable.

According to the information presented above, FBD at temperatures from 50 °C to 80 °C did not affect the quality of the turmeric powder but affected the drying time, obtaining a shorter drying time at 80 °C. This reduction in time can be significant since, in small processing plants that have drying equipment of this type (which is the case of several drying plants in Colombia), it allows for the continuation of the grinding and packing process in less time, without altering the final quality of the product.

### 3.3. UV Radiation Effect Using Different Packaging Materials

We studied the effect of UV radiation on the quality of turmeric powder, packaged in different materials: glass, low-density polyethylene, and an aluminum foil bag (see Figure 4). The content of curcuminoids decreased by 14% for the unpackaged powder (control) after 1.3 h of exposure. For glass-packaged powder, the concentration of the curcuminoids declined by 9.6%, while for LDPE and aluminum foil, the decline was 8% and 3.2%, respectively.

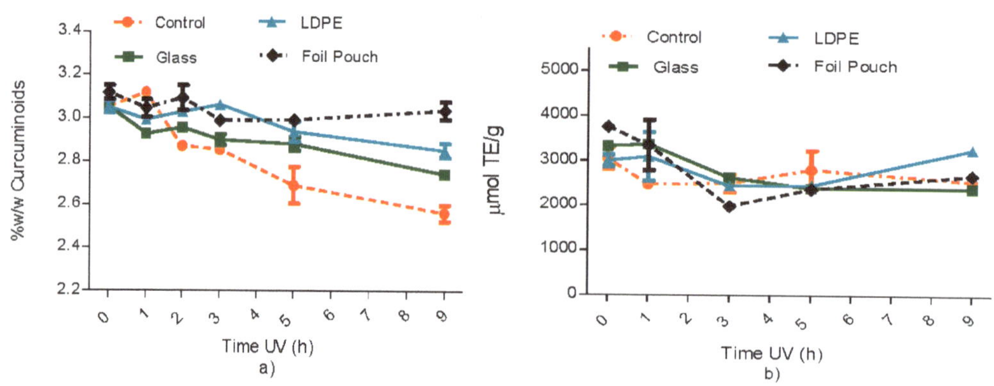


**Figure 4.** Effect of UV radiation on curcuminoid content using different packaging materials for turmeric powder. (**a**) Variation of curcuminoid content vs. time. (**b**) variation of the antioxidant capacity of the powder vs. time.

The antioxidant capacity of turmeric flour after exposure to UV radiation did not change significantly with time or with the packaging material. The color analysis of the UV-radiated turmeric powder showed an effect on the coordinate a*. Unpackaged powder presented a greener hue, while packaged samples showed no significant difference in color parameters, as shown in Table 4. The comparison of the total color change between the packaging materials did not show a significant difference between them.

**Table 4.** UV radiation effect on the CIElab parameters of turmeric powder.

| Packing Material | Color Analysis L | A | B | ΔE |
|---|---|---|---|---|
| Control | 96.3 ± 1.6 [a] | −5.0 ± 0.1 [a] | 119.9 ± 0.3 [a] | 7.86 ± 0.06 [a] |
| Glass | 93.5 ± 0.9 [b] | −3.9 ± 0.2 [b] | 119.4 ± 1.2 [a] | 8.9 ± 0.9 [a] |
| LDPE | 94.0 ± 0.7 [a,b] | −3.8 ± 0.1 [b] | 120.0 ± 0.8 [a] | 8.4 ± 0.6 [a] |
| Foil Pouch | 93.8 ± 0.7 [a,b] | −3.9 ± 0.1 [b] | 118.9 ± 0.4 [a] | 9.3 ± 0.3 [a] |

Data are expressed as mean ± standard deviation (n = 3). Common letters in the same column indicates that they are not significantly different ($p < 0.05$).

All the packaged powders were similar in terms of a*. Direct exposure to UV radiation can lead to photodegradation reactions that favor oxidation processes that generate compounds increasing the green hue.

Photoreactions affect the nutritional composition (i.e., fats, vitamins), the color, and the concentration of bioactive compounds in several foods, such as curcuminoids in turmeric. Therefore, these products might have a shorter shelf life [55]. Metalized bags, such as aluminum foil, have a high barrier against light [55]. In the present study, the aluminum foil protected the turmeric powder and prevented the degradation of curcuminoids and the decline of the antioxidant capacity.

## 4. Conclusions

The findings of this study show the impact of different drying methods (convection oven drying, fluidized bed drying, and traditional solar drying) on the quality of turmeric powder, as well as the impact of material packaging. The results show that the drying method affected the total curcuminoids content, the antioxidant capacity, and the color of the turmeric. Traditional solar drying reduced the quality of the dried product by 36.5% for curcuminoid content and 15% for ORAC antioxidant capacity while fluidized bed drying led to a reduction of 1.9% and 0.2% on curcuminoids and antioxidant capacity, respectively. Therefore, by choosing a suitable drying method, the quality of the turmeric powder could be preserved. From the three evaluated methods, FBD was the most convenient since it preserved the curcuminoid content, antioxidant capacity, and color of the turmeric with the lowest water activity. Additionally, its drying time was the shortest.

For FBD, temperature variations from 50 to 80 °C did not affect the content of curcuminoids and the antioxidant capacity. Therefore, drying can be performed at 80 °C, which results in the lowest drying time with a similar powder quality.

Finally, the effective protection over curcuminoids content was presented by a foil pouch, followed by LDPE and glass packaging. Thus, the aluminum bag for turmeric powder as a packaging material is suggested as the best option to protect the curcuminoids from photodegradation during storage.

These results are valuable for the food industry, turmeric growers, and the spices/supplements manufacturers to produce a high-quality product that can be commercialized and used as an active ingredient in many formulations.

**Author Contributions:** Conceptualization S.M.L. and Y.D.-C.; data curation and methodology, S.M.L. and A.M.G.; formal analysis, S.M.L. and Y.D.-C.; funding acquisition S.M.L.; investigation, S.M.L. and A.M.G.; supervision Y.D.-C.; writing—original draft, S.M.L. and A.M.G.; writing—review and editing Y.D.-C. and S.M.L. All authors have read and agreed to the published version of the manuscript.

**Funding:** This research was funded by Gobernación de Antioquia, Secretaría de Agricultura y Desarrollo Rural and Minciencias (Contract 4600007658-779).

**Institutional Review Board Statement:** Not applicable.

**Informed Consent Statement:** Not applicable.

**Data Availability Statement:** Data are contained within the article.

**Acknowledgments:** Authors wish to express their thankfulness to Gobernación de Antioquia and Ministerio de Ciencia Tecnología e Innovacion (Minciencias) for providing financial support for the present study. Corporación Secadora de Occidente for providing vegetal material.

**Conflicts of Interest:** The authors declare no conflict of interest.

## References

1. Govindarajan, V.S.; Stahl, W.H. Turmeric—Cheamistry, technology, and quality. *Crit. Rev. Food Sci. Nutr.* **1980**, *12*, 199–301. [CrossRef] [PubMed]
2. Martins, R.; Pereira, S.V.; Siqueira, S.; Salomão, W.F.; Freitas, L.A.P. Curcuminoid content and antioxidant activity in spray dried microparticles containing turmeric extract. *Food Res. Int.* **2011**, *50*, 657–663. [CrossRef]
3. Oh, S.H.; Jang, C.S. Development and Validation of a Real-Time PCR Based Assay to Detect Adulteration with Corn in Commercial Turmeric Powder Products. *Foods* **2020**, *9*, 882. [CrossRef] [PubMed]
4. Khodabakhshian, R.; Bayati, M.R.; Emadi, B. An evaluation of IR spectroscopy for authentication of adulterated turmeric powder using pattern recognition. *Food Chem.* **2021**, *364*, 130406. [CrossRef] [PubMed]
5. Chumroenphat, T.; Somboonwatthanakul, I.; Saensouk, S.; Siriamornpun, S. Changes in curcuminoids and chemical components of turmeric (*Curcuma longa* L.) under freeze-drying and low-temperature drying methods. *Food Chem.* **2020**, *339*, 128121. [CrossRef]
6. Kandiannan, K.; Chandaragiri, K.; Sankaran, N.; Balasubramanian, T.; Kailasam, C. Crop–weather model for turmeric yield forecasting for Coimbatore district, Tamil Nadu, India. *Agric. For. Meteorol.* **2002**, *112*, 133–137. [CrossRef]
7. Komonsing, N.; Khuwijitjaru, P.; Nagle, M.; Müller, J.; Mahayothee, B. Effect of drying temperature together with light on drying characteristics and bioactive compounds in turmeric slice. *J. Food Eng.* **2021**, *317*, 110695. [CrossRef]
8. Kilic, S.; Oz, E.; Oz, F. Effect of turmeric on the reduction of heterocyclic aromatic amines and quality of chicken meatballs. *Food Control* **2021**, *128*, 108189. [CrossRef]
9. MinAgricultura. Reporte: Importaciones del Sector Agropecuario. Valor y Volumen, por Producto, por Cadena, por Partida y País de Origen. Cúrcuma. Agronet, Ministerio de Agricultura de Colombia. 2021. Available online: https://www.agronet.gov.co/estadistica/Paginas/home.aspx?cod=24 (accessed on 26 March 2022).
10. Ali, Z.; Saleem, M.; Atta, B.; Khan, S.S.; Hammad, G. Determination of curcuminoid content in turmeric using fluorescence spectroscopy. *Spectrochim. Acta Part A Mol. Biomol. Spectrosc.* **2019**, *213*, 192–198. [CrossRef]
11. Jayaprakasha, G.K.; Jagan Mohan Rao, L.; Sakariah, K.K. Antioxidant activities of curcumin, demethoxycurcumin and bis-demethoxycurcumin. *Food Chem.* **2006**, *98*, 720–724. [CrossRef]
12. Jurenka, J.S. Anti-inflammatory properties of curcumin, a major constituent of Curcuma longa: A review of preclinical and clinical research. *Altern. Med. Rev. A J. Clin. Ther.* **2009**, *14*, 141–153.
13. López-Lázaro, M. Anticancer and carcinogenic properties of curcumin: Considerations for its clinical development as a cancer chemopreventive and chemotherapeutic agent. *Mol. Nutr. Food Res.* **2008**, *52*, S103–S127. [CrossRef] [PubMed]
14. Moghadamtousi, S.Z.; Kadir, H.A.; Hassandarvish, P.; Tajik, H.; Abubakar, S.; Zandi, K. A Review on Antibacterial, Antiviral, and Antifungal Activity of Curcumin. *BioMed Res. Int.* **2014**, *2014*, 186864. [CrossRef]
15. Mythri, R.B.; Srinivas Bharath, M.M. Curcumin: A potential neuroprotective agent in Parkinson's disease. *Curr. Pharm. Des.* **2012**, *18*, 91–99. [CrossRef] [PubMed]
16. Olszanecki, R.; Jawień, J.; Gajda, M.; Mateuszuk, L.; Gebska, A.; Korabiowska, M.; Chłopicki, S. Effect of curcumin on a therosclerosis in apo e/ldlr–double knockout mice. *J. Physiol. Pharm.* **2005**, *65*, 376–380.
17. Karlsen, J. Studies on curcumin and curcuminoids—5. Alkaline Degradation of Curcumin. *Z. Für Lebensm. Unters. Und Forsch.* **1985**, *180*, 132–134. [CrossRef]
18. Tønnesen, H.H.; Karlsen, J.; van Henegouwen, G.B. Studies on curcumin and curcuminoids VIII. Photochemical stability of curcumin. *Z. Lebensm. Unters. Forsch.* **1986**, *183*, 116–122. [CrossRef] [PubMed]
19. Lee, B.H.; Choi, H.A.; Kim, M.-R.; Hong, J. Changes in chemical stability and bioactivities of curcumin by ultraviolet radiation. *Food Sci. Biotechnol.* **2013**, *22*, 279–282. [CrossRef]
20. Hirun, S.; Utama-Ang, N.; Roach, P.D. Turmeric (*Curcuma longa* L.) drying: An optimization approach using microwave-vacuum drying. *J. Food Sci. Technol.* **2012**, *51*, 2127–2133. [CrossRef]
21. Madan, M.S. Turmeric. In *Turmeric: The Genus Curcuma*; Ravindran, K.S.P.N., Nirmal Babu, K., Eds.; CRC Press: Boca Raton, FL, USA, 2007; pp. 369–408.
22. Jose, K.P.; Joy, C.M. Solar tunnel drying of turmeric (*Curcuma longa* linn. syn. c. domestica val.) for quality improvement. *J. Food Process. Preserv.* **2009**, *33*, 121–135. [CrossRef]

23. An, K.; Zhao, D.; Wang, Z.; Wu, J.; Xu, Y.; Xiao, G. Comparison of different drying methods on Chinese ginger (*Zingiber officinale* Roscoe): Changes in volatiles, chemical profile, antioxidant properties, and microstructure. *Food Chem.* **2016**, *197*, 1292–1300. [CrossRef] [PubMed]
24. Nukulwar, M.R.; Tungikar, V.B. Drying kinetics and thermal analysis of turmeric blanching and drying using solar thermal system. *Sustain. Energy Technol. Assess.* **2021**, *45*, 101120. [CrossRef]
25. Borah, A.; Hazarika, K.; Khayer, S. Drying kinetics of whole and sliced turmeric rhizomes (*Curcuma longa* L.) in a solar conduction dryer. *Inf. Process. Agric.* **2015**, *2*, 85–92. [CrossRef]
26. Prathapan, A.; Lukhman, M.; Arumughan, C.; Sundaresan, A.; Raghu, K.G. Effect of heat treatment on curcuminoid, colour value and total polyphenols of fresh turmeric rhizome. *Int. J. Food Sci. Technol.* **2009**, *44*, 1438–1444. [CrossRef]
27. Cousins, M.; Adelberg, J.; Chen, F.; Rieck, J. Antioxidant capacity of fresh and dried rhizomes from four clones of turmeric (*Curcuma longa* L.) grown in vitro. *Ind. Crop. Prod.* **2007**, *25*, 129–135. [CrossRef]
28. Pérez-Vicente, A.; Serrano, P.; Abellán, P.; García-Viguera, C. Influence of packaging material on pomegranate juice colour and bioactive compounds, during storage. *J. Sci. Food Agric.* **2004**, *84*, 639–644. [CrossRef]
29. Gunes, G.; Ozturk, A.; Yilmaz, N.; Ozcelik, B. Maintenance of Safety and Quality of Refrigerated Ready-to-Cook Seasoned Ground Beef Product (Meatball) by Combining Gamma Irradiation with Modified Atmosphere Packaging. *J. Food Sci.* **2011**, *76*, M413–M420. [CrossRef]
30. Esmaeili, S.; Barzegar, M.; Sahari, M.A.; Berengi-Ardestani, S. Effect of gamma irradiation under various atmospheres of packaging on the microbial and physicochemical properties of turmeric powder. *Radiat. Phys. Chem.* **2018**, *148*, 60–67. [CrossRef]
31. Tripetch, P.; Borompichaichartkul, C. Effect of packaging materials and storage time on changes of colour, phenolic content, chlorogenic acid and antioxidant activity in arabica green coffee beans (*Coffea arabica* L. cv. Catimor). *J. Stored Prod. Res.* **2019**, *84*, 101510. [CrossRef]
32. Deng, L.-Z.; Mujumdar, A.S.; Pan, Z.; Vidyarthi, S.K.; Xu, J.; Zielinska, M.; Xiao, H.-W. Emerging chemical and physical disinfection technologies of fruits and vegetables: A comprehensive review. *Crit. Rev. Food Sci. Nutr.* **2019**, *60*, 2481–2508. [CrossRef]
33. Joshi, K.; Mahendran, R.; Alagusundaram, K.; Norton, T.; Tiwari, B. Novel disinfectants for fresh produce. *Trends Food Sci. Technol.* **2013**, *34*, 54–61. [CrossRef]
34. Instituto Colombiano de Normas Técnicas y Certificación, ICONTEC. Norma técnica Colombiana NTC 4423:1998 (reaprobada 2021). Food Industry. Spices and Condiments. Available online: https://tienda.icontec.org/ (accessed on 20 January 2022).
35. Cheng, J.; Weijun, K.; Yun, L.; Jiabo, W.; Haitao, W.; Qingmiao, L.; Xiaohe, X. Development and validation of UPLC method for quality control of Curcuma longa Linn.: Fast simultaneous quantitation of three curcuminoids. *J. Pharm. Biomed. Anal.* **2010**, *53*, 43–49. [CrossRef] [PubMed]
36. Cao, G.; Alessio, H.M.; Cutler, R.G. Oxygen-radical absorbance capacity assay for antioxidants. *Free Radic. Biol. Med.* **1993**, *14*, 303–311. [CrossRef]
37. Akter, J.; Hossain, M.A.; Takara, K.; Islam, M.Z.; Hou, D.-X. Antioxidant activity of different species and varieties of turmeric (*Curcuma* spp.): Isolation of active compounds. *Comp. Biochem. Physiol. Part C Toxicol. Pharmacol.* **2018**, *215*, 9–17. [CrossRef] [PubMed]
38. Duarte, Y.; Chaux, A.; Lopez, N.; Largo, E.; Ramírez, C.; Nuñez, H.; Simpson, R.; Vega, O. Effects of Blanching and Hot Air Drying Conditions on the Physicochemical and Technological Properties of Yellow Passion Fruit (*Passiflora edulis* Var. *Flavicarpa*) by-Products. *J. Food Process Eng.* **2016**, *40*, e12425. [CrossRef]
39. Tapia, M.S.; Alzamora, S.M.; Chirife, J. *Effects of Water Activity (aw) on Microbial Stability: As a Hurdle in Food Preservation. Water Activity in Foods: Fundamentals and Applications*, 2nd ed.; Wiley-Blackwell: Hoboken, NJ, USA, 2020; pp. 239–271. [CrossRef]
40. Sharma, S.; Dhalsamant, K.; Tripathy, P.P.; Manepally, R.K. Quality analysis and drying characteristics of turmeric (*Curcuma longa* L.) dried by hot air and direct solar dryers. *LWT* **2020**, *138*, 110687. [CrossRef]
41. Chen, Z.; Xia, Y.; Liao, S.; Huang, Y.; Li, Y.; He, Y.; Tong, Z.; Li, B. Thermal degradation kinetics study of curcumin with nonlinear methods. *Food Chem.* **2014**, *155*, 81–86. [CrossRef]
42. Na Jung, Y.; Hong, J. Changes in chemical properties and bioactivities of turmeric pigments by photo-degradation. *AIMS Agric. Food* **2021**, *6*, 754–767. [CrossRef]
43. Thamkaew, G.; Sjöholm, I.; Galindo, F.G. A review of drying methods for improving the quality of dried herbs. *Crit. Rev. Food Sci. Nutr.* **2020**, *61*, 1763–1786. [CrossRef]
44. Parlak, N. Fluidized bed drying characteristics and modeling of ginger (*Zingiber officinale*) slices. *Heat Mass Transf.* **2014**, *51*, 1085–1095. [CrossRef]
45. Maheshwari, R.K.; Singh, A.K.; Gaddipati, J.; Srimal, R.C. Multiple biological activities of curcumin: A short review. *Life Sci.* **2006**, *78*, 2081–2087. [CrossRef] [PubMed]
46. Singh, G.; Kapoor, I.; Singh, P.; de Heluani, C.S.; de Lampasona, M.P.; Catalan, C.A. Comparative study of chemical composition and antioxidant activity of fresh and dry rhizomes of turmeric (*Curcuma longa* Linn.). *Food Chem. Toxicol.* **2010**, *48*, 1026–1031. [CrossRef] [PubMed]
47. Llano, S.; Gómez, S.; Londoño, J.; Restrepo, A. Antioxidant activity of curcuminoids. *Phys. Chem. Chem. Phys.* **2019**, *21*, 3752–3760. [CrossRef] [PubMed]

48. Ray, A.; Mohanty, S.; Jena, S.; Sahoo, A.; Acharya, L.; Panda, P.C.; Sial, P.; Duraisamy, P.; Nayak, S. Drying methods affects physicochemical characteristics, essential oil yield and volatile composition of turmeric (*Curcuma longa* L.). *J. Appl. Res. Med. Aromat. Plants* **2021**, *26*, 100357. [CrossRef]
49. Osorio-Arias, J.; Delgado-Arias, S.; Cano, L.; Zapata, S.; Quintero, M.; Nuñez, H.; Ramírez, C.; Simpson, R.; Vega-Castro, O. Sustainable Management and Valorization of Spent Coffee Grounds Through the Optimization of Thin Layer Hot Air-Drying Process. *Waste Biomass Valorization* **2019**, *11*, 5015–5026. [CrossRef]
50. Chan, D.-S.; Kuo, M.-I. Wheat Germ Drying with Different Time-Temperature Combinations in a Fluidized Bed Dryer. *Processes* **2018**, *6*, 245. [CrossRef]
51. Dei Cas, M.; Ghidoni, R. Dietary Curcumin: Correlation between Bioavailability and Health Potential. *Nutrients* **2019**, *11*, 2147. [CrossRef]
52. Akoy, E.O.M. Experimental characterization and modeling of thin-layer drying of mango slices. *Int. Food Res. J.* **2014**, *21*, 1911–1917.
53. Royen, M.J.; Noori, A.W.; Haydary, J. Experimental Study and Mathematical Modeling of Convective Thin-Layer Drying of Apple Slices. *Processes* **2020**, *8*, 1562. [CrossRef]
54. Nyangena, D.N.; Mutungi, C.; Imathiu, S.; Kinyuru, J.; Affognon, H.; Ekesi, S.; Nakimbugwe, D.; Fiaboe, K.K.M. Effects of Traditional Processing Techniques on the Nutritional and Microbiological Quality of Four Edible Insect Species Used for Food and Feed in East Africa. *Foods* **2020**, *9*, 574. [CrossRef]
55. Bishop, C.A.; Mount, E.M. Vacuum Metallizing for Flexible Packaging. In *Multilayer Flexible Packaging*, 2nd ed.; William Andrew: Norwich, NY, USA, 2016; pp. 235–255. [CrossRef]

processes

MDPI

Article

# Drying Biomass with a High Water Content—The Influence of the Final Degree of Drying on the Sizing of Indirect Dryers

Jan Havlík [1,*], Tomáš Dlouhý [1] and Ján Piteľ [2]

1 Department of Energy Engineering, Faculty of Mechanical Engineering, Czech Technical University in Prague, Technicka 4, 16607 Prague, Czech Republic; tomas.dlouhy@fs.cvut.cz
2 Department of Industrial Engineering and Informatics, Faculty of Manufacturing Technologies, Technical University of Kosice, Bayerova 1, 080 01 Prešov, Slovakia; jan.pitel@tuke.sk
* Correspondence: jan.havlik@fs.cvut.cz

**Abstract:** This article deals with the influence of the final drying degree of moist biomass used as fuel in a power or CHP plant on indirect dryer sizing. For a description of the drying process, experiments with wet bark containing approx. 50 wt% of water were carried out in a laboratory indirect dryer. A new parameter called drying effectivity was introduced, whose size varies according to the degree of biomass being dried. Its maximum value corresponds to the optimal biomass drying, when the relative size of the indirect dryer to evaporate the required mass of water from the biomass would be smallest. Based on the experimentally determined drying characteristics of wet bark, the optimal drying of 13 wt% of water content was evaluated. If the bark was dried to a lower water content, the required relative size and price of the dryer would increase. Similarly, drying a bark with water content above 31 wt% is not very advantageous because drying effectivity continues to increase rapidly at this stage, and the required relative size of the dryer therefore decreases.

**Keywords:** biomass; drying; indirect dryer; drying rate; sizing

**Citation:** Havlík, J.; Dlouhý, T.; Piteľ, J. Drying Biomass with a High Water Content—The Influence of the Final Degree of Drying on the Sizing of Indirect Dryers. *Processes* **2022**, *10*, 739. https://doi.org/10.3390/pr10040739

Academic Editor: Timothy Langrish

Received: 11 March 2022
Accepted: 8 April 2022
Published: 11 April 2022



## 1. Introduction

The use of bioenergy as a renewable energy source for heat and power generation continues to increase in importance as coal combustion reduces. In 2019, renewable energy accounted for 19.7% of the energy consumed in the EU-27 [1], and bioenergy, as the most widely used source, covered 57% of this share [2]. Bioenergy contributes to the EU's energy security, as most is derived from local biomass production [3]. Biomass comes mainly from three sources: ligneous biomass, agriculture biomass and waste biomass, where ligneous biomass comprises the largest portion at approximately 70%. Examples are bark, saw dust or wood chips obtained from wood industries and forest management operations [2].

More and more small-scale biomass power plants have recently been constructed. The increasing use of biomass as a renewable source for energy purposes depletes the capacity of traditional high-quality forms such as dry wood, straw, etc. Shortages and the rising price of these fuels increases interest in the use of low-grade biomass, whose utilization is often complicated by high water content, and which was previously considered a waste material with disposal issues [4]. Furthermore, the quality of raw biofuels is often degraded by a higher water content, which reduces their calorific value and limits the efficiency of their application.

Nowadays, it is advantageous to use these materials despite their poorer operating parameters, or higher demands on pretreatment technology, should these factors be offset by the low cost of such materials [4]. Actual pieces of these kinds of biomass often possess non-uniform shapes and sizes and a high water content, which is affected by their origin and storage conditions [5]. Therefore, drying and dimensional unification are a suitable way to prepare optimal properties for the further energetic use of these materials [6,7].

Drying, as one of the suitable pretreatments of biomass for burning, improves its heating value and combustion efficiency, while reducing the required capacity of the boiler auxiliary equipment, and often lowering emissions and improving boiler operation [8,9]. Fuel pre-drying could significantly decrease the exergy loss in the combustion process. For equal heat exchange process in the boiler, the exergy efficiency could be increased by percentage points in comparison to conventional direct biomass usage in a power plant [10]. Moreover, pre-drying of biomass reduces the mass flow rate of fuel needed for the desired boiler output, which has an impact on fuel transport and storage, as well as the flow rate of combustion air and flue gas and the boiler size [8].

The aim of most industrial drying processes is to obtain a solid product of a desired quality at a minimum cost and a maximum throughput and to maintain these conditions consistently. Good quality implies that the product corresponds to a number of physical, chemical and biological parameters, each within specified limits [11]. The requirements for biomass drying before energy use are usually somewhat different. Since it is necessary to dry significant mass flows of biomass, it is usually not required to pay close attention to compliance with the exact conditions of the drying process, e.g., even a significant deviation from the optimal temperature or drying time does not necessarily mean excessive degradation of the product, similar to its not completely homogeneous final water content.

Modern drying systems could supply products with a high quality; however, these systems are often not only expensive, but also largely energy intensive, e.g., drying is the most energy- and time-consuming stage of the wood preparation process. Therefore, the main goal should be to manage the drying of significant mass and volume flows of material with the lowest energy intensity because thermal drying requires a large amount of power that accounts for up to 15% of all industrial energy consumption [6,11]. The most commonly used conventional dryers often operate at low thermal efficiency, typically between 25% and 50%, yet may also fall as low as 10% [11]. Therefore, it is crucial to find the optimum design and application of the drying process [6]. Different systems and drying methods always have to be evaluated individually for specific conditions and assignments [6]. These facts, along with an environmental perspective, force researchers to search for more effective ways of drying, such as modifications to the process, modernizing equipment, and waste heat recovery [12,13].

In general, industrial drying technologies can be divided into two basic categories—direct (convective) drying, where the drying medium is in direct contact with the material being dried, and indirect (conductive, contact) drying, where the heat for drying is supplied through a conductive wall separating the heated and drying spaces, so that the drying medium is not in direct contact with the material being treated [8,14].

Efforts to reduce the heat consumption for biomass drying before combustion leads to the application of indirect drying, which utilizes an external heat source [15]. No air or flue gas need to be used to support the drying process, which eliminates heat loss at the outlet of the dryer [8]. Therefore, this method possesses low heat consumption, which is close to the value of the internal latent heat of the evaporated water [15]. Another advantage is the gain of the waste vapor leaving the indirect dryer with a minimum of impurities, which may be easily utilized for external use, e.g., by suitable connection of indirect biomass dryer, the efficiency of a biomass fired steam power station or CHP plant can be significantly increased [16]. The integration of drying into the energy cycle enables a significant reduction in fuel consumption or replacement of high-quality kinds of biomass by low-grade sorts [16–18]. In addition, if the material contains dust particles, which is typically a case of low-grade biomass, there is a risk of fire. However, an indirect drying system for woody biomass, where the dried material is in direct contact with evaporated steam, is considered safe in the comparison with conventional air drying [19].

Ligneous biofuels may possess a high initial water content of 50 wt% to 65 wt% [20]. Factors such as storage duration, season, assortment and rain protection have a statistically significant influence on water content of woody biomass [21,22]. The water content of fresh wood chips is approximately 50 wt%, but this could vary between 66 wt% and 36 wt%

according to storage conditions [21]. For efficient combustion, it would be appropriate to reduce the biomass moisture to a level of 10 wt% to 15 wt% [4]. At these water content levels, all the free water is dried and drying of the bound water occurs, which is controlled by the diffusivity of water vapor in the material; thus, the drying rate significantly reduces [23,24]. The requirement for a higher degree of pre-drying prolongs the drying time, which significantly increases the required dryer dimensions and energy consumption [14]. Water contents significantly lower than 18 wt% are known to increase the risk of dust explosions, as dust clouds may be ignited if oxygen is present [20]. In conventional drying systems (e.g., air drying), the water content of wooden material is commonly reduced from 50 wt% to 25 wt% [25].

Contact drying of biomass is not yet widely used. The basic rules for their design and the optimization of their operation are still lacking. The aim of this article is to determine the required size of a contact dryer for the drying of biomass in relation to the required water content before combustion in the boiler. The required dryer size is designed according to experimentally determined drying rates of wet bark sourced from outdoor storage and containing about 50 wt% of water.

## 2. Materials and Methods

### 2.1. Tested Material

The tested material was a wet bark (see Figure 1) that was stored outdoors as a representative of waste material with high water content. This material was produced during wood processing by crushing the bark of spruce and pine trees, so it had a very inhomogeneous composition, and the particles varied considerably in shape and ranged in size from one millimeter to a few centimeters. The thickness of the particles ranged mostly from 3 to 5 mm. The bulk density was 330 $kg/m^3$.

**Figure 1.** Wet bark before and after drying.

Fuel composition gained from laboratory analysis is given in Table 1. The water content in the raw biomass was around 50 wt%. As the wet bark was stored outdoors, its water content could vary over time. Therefore, the water content of the bark was determined by laboratory analysis at the beginning of each drying test.

### 2.2. Experimental Dryer

The experimental dryer (see Figure 2) was designed as an indirect dryer in a paddle drum configuration. The dryer is electrically heated, thus allowing for a straightforward evaluation of energy consumption and temperature control with numerous settings to simulate various heat conditions, e.g., by hot water or low-pressure steam. It is presumed that the heat transfer in the drying process is not dependent on the way the heat is supplied because the thermal resistance between the heating medium, the conducting wall and the thermal resistance across the conducting wall are negligible compared to the resistance between the wall and the surface of the drying material and the resistance across its layer.

Assuming sufficient material mixing, it is possible to compare different concepts of indirect dryers, regardless of their method of heating.

**Table 1.** Results of ultimate and proximate analyses of bark and calorific value of its combustible.

| | | |
|---|---|---|
| C daf | wt%, daf | 51.75 |
| H daf | wt%, daf | 5.97 |
| O daf | wt%, daf | 41.82 |
| S daf | wt%, daf | 0.03 |
| N daf | wt%, daf | 0.43 |
| Ash db | wt%, db | 2.8 |
| W | wt% | ~50 |
| HHV daf | kJ/kg | 20,550 |

db: dry basis; daf: dry ash-free basis; W: water content; HHV: higher heating value.

**Figure 2.** Experimental paddle dryer located on a tensometric scale.

The drum of the experimental dryer has an inner diameter of 0.256 m and a length of 1 m. It works in a batch mode. The material inside the drum is mixed by paddle wheel stirrer with a rotation speed of 18 rpm. The drum volume is 0.051 $m^3$, and the heated surface area of the drum is 0.81 $m^2$.

The dryer is placed on a tensometric scale, which monitors the decrease in weight of the material during the drying process with accuracy of 0.002 kg. Five thermocouples are embedded in the wall of the dryer to measure the temperature of the heating surface. Two other thermocouples measure the temperature inside the dryer and the temperature of the exhaust vapor at the dryer outlet, which flows out into the surrounding by pipes in the upper part of the dryer due to small overpressure developing inside the dryer by the generated vapor.

### 2.3. Experiment Evaluation

At the beginning of each measurement, the initial weight of the batch and the water content in the bark were determined. The variation in the water content of the bark during drying, along with the drying rate, were continuously evaluated according to the weight loss of the dryer, which corresponds to the amount of evaporated water. The drying process could then be indicated by a drying curve defining the dependence of the actual water content in the bark on the drying time, as is usual.

### 2.4. Determination of Drying Rate

Due to the continuous record of weight loss of the dried bark, it was possible to evaluate the change in drying rate during each experiment. The drying rate *DR* expresses the weight of dried water from the material in 1 min and its instantaneous value changes during drying.

### 2.5. Determination of the Required Dryer Size

For the design of dryers where the material is dried from the initial water content $W_0$ to the final water content $W_1$, it is necessary to use the mean value of the drying rate for the residence time of the material in the dryer. The mean drying rate $\overline{DR}$ can be derived from the integral for instantaneous drying rate as a function of water content $W$ in the drying interval between the water contents of $W_0$ and $W_1$.

$$\overline{DR} = \frac{1}{(W_1 - W_0)} \int_{W_0}^{W_1} DR_{(W)} dW \tag{1}$$

The square evaporation capacity $EC$ defines the mass flow of evaporated water from dried material per 1 $m^2$ of dryer-heated area and drying time. The evaporation capacity can be derived from the mean drying rate in a given drying interval related to the heated surface of the dryer $A_{dryer}$.

$$EC = \frac{\overline{DR}}{A_{dryer}} \tag{2}$$

The heated area of dryer $A_{dryer}$ required for the drying of a mass flow of bark $M_{fuel}$ can be calculated based on the experimentally determined evaporation capacity corresponding to the analyzed drying interval from $W_0$ to $W_1$ and the mass flow of evaporated water $M_{\Delta W}$.

$$A_{dryer} = \frac{M_{\Delta W}}{EC} = \frac{\Delta W \cdot M_{fuel}}{EC} \tag{3}$$

The mass of water evaporated from drying of 1 kg of bark $\Delta W$ is determined by the following equation.

$$\Delta W = \frac{W_0 - W_1}{1 - W_1} \tag{4}$$

## 3. Results and Discussion

### 3.1. Results of Drying Tests

To describe the drying process in a wide range of operating conditions, the effects of different heating temperatures and dryer filling ratios (expressed as the volume of dried material to the dryer volume) on the drying characteristics were tested. In Figure 3, drying curves that describe the drying process for the various dryer heating temperatures (HT) and dryer filling ratios (FR) are shown.

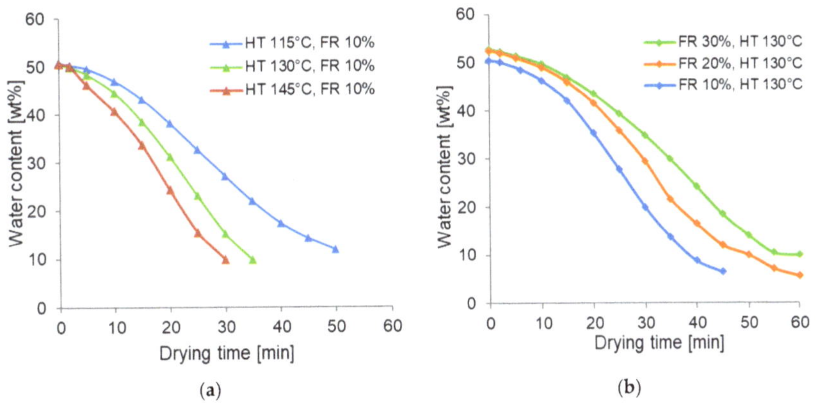


**Figure 3.** Comparison of drying curves for various (**a**) heating temperatures (HT) and (**b**) dryer filling ratios (FR).

The tested material was taken from the outdoor storage, and the input water content varied slightly depending on weather conditions.

If the filling ratio increases, more material is being dried, and the drying time increases. At the same time, the drying rate increases due to the larger contact area of the heated surface with a dried material.

### 3.2. Evaluation of Drying Rates

The course of drying rates was obtained by evaluating the weight of the dried material as a function of the drying time. To eliminate the fluctuation of the instantaneous drying rate, its value was calculated as an average over a period of 5 min. The courses of instantaneous drying rates for drying of wet bark from 50 wt% to 10 wt% water content under different conditions are shown in Figure 4. The drying tests differed in the heating temperature HT from 115 °C to 145 °C and in the dryer volume filling ratio of dryer FR from 10% to 30%.

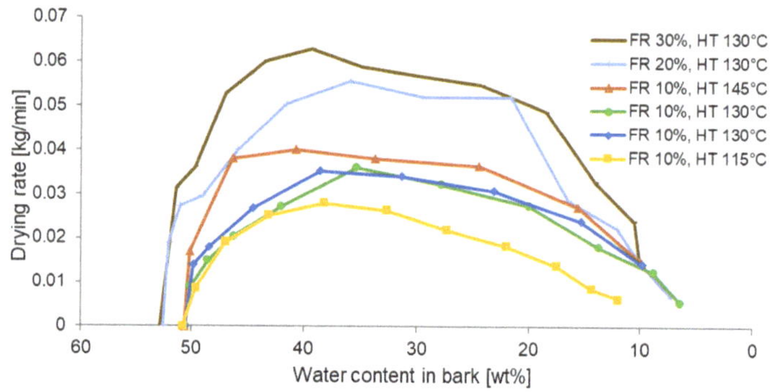


**Figure 4.** The course of drying rates for various conditions.

The change in drying intensity during the drying process corresponds to the three periods described in [14]. The lower initial drying rate increases steeply to its maximum value, which is related to the heating of the material to the drying temperature. Then the drying rate begins to decrease gradually and, in the final stage, when the material is relatively dry and only a small amount of vapor is released, the drying rate is at its lowest. The final drying phase is therefore not very intense and has little benefit in improving the conditions for burning the bark.

According to [14], the value of the overall heat transfer coefficients between the heating surface and the drying material for this type of indirect dryer can be assumed to be around 50 W/m$^2$K. This value of the overall heat transfer coefficient can be derived from experimental values using the main drying rate for atmospheric drying by multiplying by the latent heat of evaporated water and dividing by the dryer surface and the temperature difference between the heated surface and material being dried. The values determined in this way range from 40 to 67 W/m$^2$K, and they are in agreement with the value given in [14].

The process of indirect drying has been experimentally investigated in previous research works; however, these experiments were performed with laboratory material as sand or ceramic particles with a spherical shape and small dimensions. Some works focused on the research of the heat transfer coefficient (HTC) between the heated wall and bed of material in an indirect drum dryer. The value of the HTC for quartz sand with a mean particle diameter of 0.2 mm and mixing with a rotational speed of 6 rmp achieved 300 W/m$^2$K [26], and for glass beads with a diameter of 2 mm it achieved 100 W/m$^2$K [27]. For a rotational speed of 20 rmp and silica sand with a diameter of 1 mm, the HTC was 200 W/m$^2$K, and it was 80 W/m$^2$K for glass beads with a diameter of 4 mm [28]. These results show a strong dependence of the HTC on particle size. For larger particle sizes, they correspond with the experimental results given above.

The similarity of the course of measurements FR 10% and HT 130 °C (blue and green lines) confirms the repeatability of the experiment.

### 3.3. Effect of Final Degree of Drying on the Size of the Dryer

Based on the procedure described above, the evaporation capacities for various final degree of drying were determined according to Equations (1) and (2). The square evaporation capacity evaluated from the experimental data (see Figure 5) was calculated according to the mean drying rate in the drying interval from $W_0 = 50$ wt% to $W_1 = \langle 50, 10\rangle$ wt% divided by the heated surface of the experimental dryer $A_{exp}$.

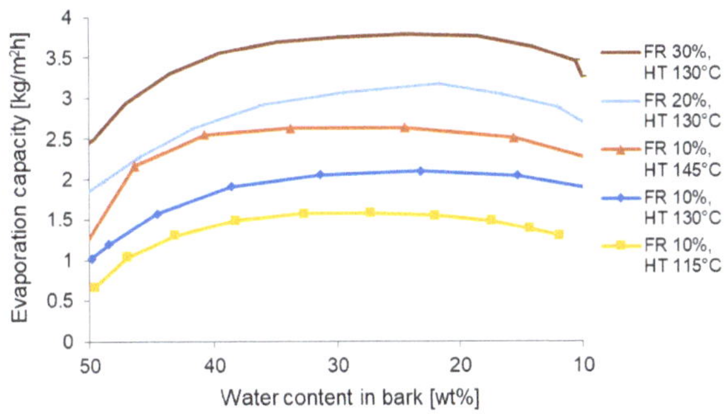


**Figure 5.** Square evaporation capacities for drying to various final water content and under different process conditions.

For comparison with other studies, only results for other materials are available. In [29], the drying rates of granulated porous aluminum silicate particles (diameter of 4.3 mm) were measured in a disc dryer with indirect heating with magnetic stirrer. The shapes of drying rates correspond with Figure 4. The values of the evaporation capacity (EC) for a rotational speed of 60 rpm ranged from 1 to 3 kg/m$^2$h for atmospheric drying. The EC for hygroscopic peat in a tray dryer [30] with HT of 130 °C and a mixer speed of 40 rpm for particles with diameters from 0.75 mm to 6 mm achieved a value from 8 to 17 kg/m$^2$h. Agitated contact drying of sewage sludge in a continuous paddle dryer was performed in [31], and the EC achieved the values up to 18 kg/m$^2$h for HT of 160 °C and a stirrer speed of 42 rpm. The results in [32] for grained ceramic material (AlSi) with a diameter of 1.1 mm pertain to the paddle dryer after vacuum conditions with a revolution speed of 30 rpm. For the temperature difference between the saturation temperature corresponding with vacuum pressure and the heating temperature from 23 to 79 °C, the EC ranged from 5 to 16 kg/m$^2$h. Achieving higher values of the EC for smaller particles is again evident here. Furthermore, good agreement with the experimental results is obtained assuming larger particle sizes of the bark.

The required size of the dryer heating area in the dependence on final water content for 1 kg per hour of dried bark evaluated from experiments under various drying conditions is shown in Figure 6.

After the first drying phase, where the heating of the material takes place at the same time, the drying process begins to be stable, and the size of the dryer increases linearly up to the final water content about 20 wt%. With further reduction of the water content, the required dryness size begins to increase progressively.

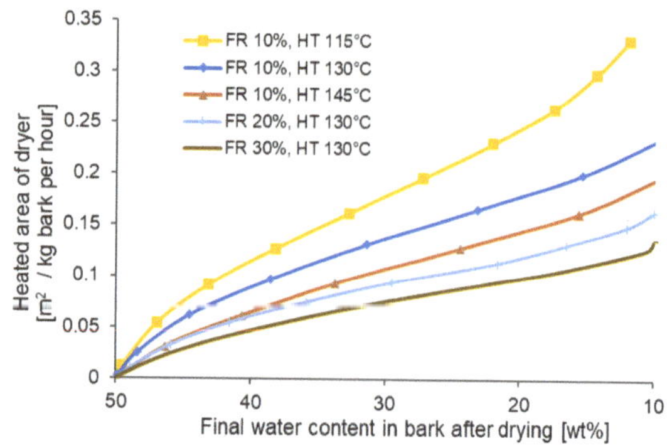


**Figure 6.** Required heating surface of the dryer for drying 1 kg of bark with a water content of 50 wt% to various final water contents placed under different processing conditions.

The required size of the dryer decreases both with increasing temperature of the heating medium and with increasing filling of the dryer. It can be said that filling has a greater effect. For indirect dryers, the main thermal resistance in the heat transfer process is between the wall and the surface of the bed of the drying material and the resistance across the bed of the drying material [14]. Assuming sufficient material mixing, there should also be similarity for different concepts and operating mode (batch or continuous) of indirect dryers [33].

### 3.4. The Drying Process Effectivity

Relative expression of the analyzed parameters was introduced for the general evaluation of the drying process effectivity.

The relative drying rate represents the ratio of the mean drying rate $\overline{DR}$ in given drying intervals from $W_0 = 50$ wt% to $W_1 = \langle 50,\ 10\rangle$ wt% up to the maximum value of the mean drying rate $\overline{DR}_{max}$ within intervals from $W_0 = 50$ wt% to $W_1 = 10$ wt% determined by each experiment. In other words, $\overline{DR}_{max}$ corresponds to the period in Figure 4, at which the largest amount of water is dried between the inlet water content $W_0 = 50$ wt% and the required final water content $W_1$.

$$\text{relative } DR = \frac{\overline{DR}}{\overline{DR}_{max}} \tag{5}$$

Analogously, the relative mass of evaporated water $\Delta W$ represents the ratio of the mass of water evaporated from the fuel in a given drying interval to the total mass of water evaporated $\Delta W_{max}$ when drying the bark from 50 wt% to 10 wt% of water content.

$$\text{relative } \Delta W = \frac{\Delta W}{\Delta W_{max}} \tag{6}$$

In this way, the values of each analyzed parameter were divided by their maximum values, and the obtained relative expression was then plotted as a function of the relative mass of evaporated water.

The dependence of the relative drying rates and the relative mass of evaporated water on the final water content in the bark is shown in Figure 7. The curves of the relative drying rates for all the drying experiments were plotted as a linear increase in the relative mass of evaporated water. Their highest values lie in the middle of the drying interval. With a higher value of the relative drying rate, a smaller size of dryer is needed to evaporate the same amount of water from the fuel.

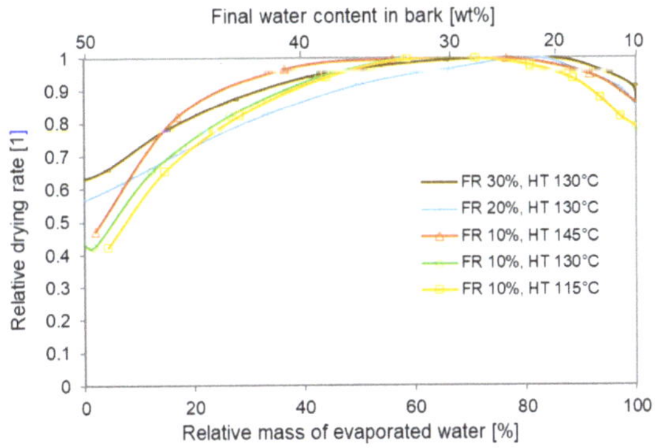


**Figure 7.** The dependence of the relative drying rates on the relative mass of evaporated water for various conditions in the range of 50 to 10 wt% of water content in bark.

The drying effectivity $DE$ is introduced as the product of the relative mass of evaporated water and the relative drying rate.

$$DE = \text{relative } \Delta W \cdot \text{relative } DR \tag{7}$$

The curves of individual drying rates in Figure 7 have a very similar course; in absolute values they differ slightly in the initial and final stages. Thus, it is possible to use the average relative drying rate to express the drying effectivity and to calculate its generalized dependence on the relative mass of evaporated water during drying, independent of the process conditions. The result is shown in Figure 8. The differences in the course of individual drying effectivities for different drying conditions are evident in a water content of 20 wt%; nevertheless, the maximum values of the courses are located in the similar values of final water contents.

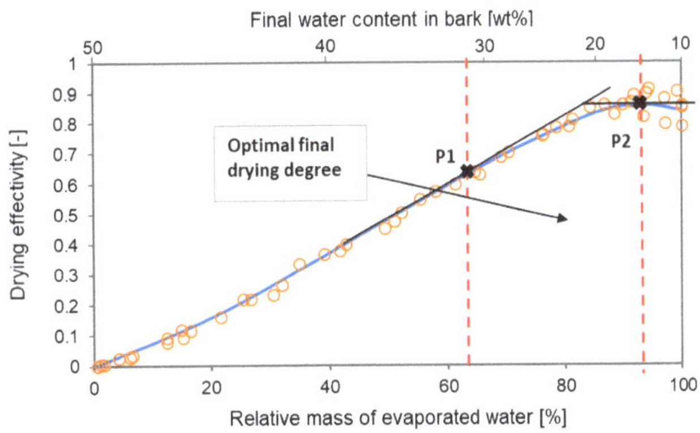


**Figure 8.** Dependence of the drying process effectivity in a range of 50 wt% to 10 wt%.

The drying effectivity steeply increases to a value of output water content around 31 wt%, from when the gradient of the curve begins to decrease (point P1 expressed by the tangent on the curve in Figure 8), and then from the value around 13 wt%, it starts to decrease (point P2 expressed as a maximal value of the curve in Figure 8).

Based on the obtained results, it is possible to determine the optimal final water content in the dried fuel, when the drying process would be at its most effective and when the relative size of the dryer for evaporating the required mass of water would be the smallest. According to the obtained experimental data, the optimal water content in the bark dried in the contact dryer is 13 wt%. If the bark was dried to a lower water content, the required relative size and price of the dryer would increase. Similarly, drying to a bark water content above 31 wt% is not very advantageous because the drying effectivity continues to grow rapidly at this stage and the required relative size of the dryer decreases.

## 4. Conclusions

Drying of the moist biomass before combustion is advantageous in order to improve combustion and increase boiler efficiency. A higher degree of drying brings a greater benefit but requires a larger and more expensive dryer.

The effect of the final drying degree of wet biomass on the required size of the indirect dryer was analyzed. To determine the rate of the drying process, experiments with wet bark containing approximately 50 wt% of water were carried out in a laboratory indirect dryer. The term drying effectivity was introduced for the general evaluation of the full course of the drying process. Based on drying experiments, the dependence of drying effectivity on the relative mass of evaporated water corresponding to the final biomass drying degree was determined. Depending on the biomass drying degree, the drying efficiency first increases, reaches its maximum value and then begins to fall. In this way, it is possible to determine the optimal drying of biomass, when the drying effectivity will be at its maximum and when the relative size of the dryer to evaporate the required amount of water would be the smallest. Based on the experimentally determined drying characteristics of the wet bark, the optimal drying at 13 wt% of water content was evaluated. It can also be concluded that drying above 31 wt% of water content in the bark proves fairly disadvantageous due to the fact that, at this stage, the drying effectivity continues to increase rapidly and the required relative size of the dryer decreases.

The results of the analysis show that it is not advantageous to dry the material to too low a water content in the dried material, because the required size of the dryer grows progressively, and the benefit obtained is small. The described procedure can be used to determine the optimal degree of drying for any material with regard to the effective use of the considered indirect dryer.

Future work will focus on the integration of the dryer into the biomass power plant and the evaluation of the optimal drying of biomass before combustion for the operation of the whole energy system. If the dryer is integrated into the energy plant, the benefits of drying for other equipment, especially for the efficiency and operation of the boiler, must be considered, and it is necessary to include the economic aspect in the evaluation. Only in this way will it be possible to determine the economically optimal drying degree of the fuel before combustion.

**Author Contributions:** Conceptualization, J.H., T.D. and J.P.; methodology, J.H. and T.D.; formal analysis, J.H.; investigation, J.H.; data curation, J.H.; writing—original draft preparation, J.H.; writing—review and editing, T.D. and J.P.; supervision, T.D. and J.P.; funding acquisition, T.D. All authors have read and agreed to the published version of the manuscript.

**Funding:** This work was supported by the Ministry of Education, Youth and Sports under OP RDE grant number CZ.02.1.01/0.0/0.0/16-019/0000753 "Research Centre for Low-Carbon Energy Technologies".

**Institutional Review Board Statement:** Not applicable.

**Informed Consent Statement:** Not applicable.

**Data Availability Statement:** Not applicable.

**Acknowledgments:** The article was also written as a result of the successful solving of the Project of the Slovak Research and Development Agency under the contract No. APVV-15-0602.

**Conflicts of Interest:** The authors declare no conflict of interest.

## References


1. Eurostat. Renewable Energy Statistics. 2020. Available online: https://ec.europa.eu/eurostat/statistics-explained/index.php?title=Renewable_energy_statistics (accessed on 15 November 2021).
2. Bioenergy Europe. Available online: https://bioenergyeurope.org/article/301-bioenergy-europe-s-new-brochure-we-are-renewable-energy-champions.html (accessed on 12 November 2021).
3. Scarlat, N.; Dallemand, J.; Taylor, N.; Banja, M. *Brief on Biomass for Energy in the European Union*; Publications Office of the European Union: Luxembourg, 2019. [CrossRef]
4. Ross, C. *Biomass Drying and Dewatering for Clean Heat & Power*; NorthWest CHP Application Center: Olympia, WA, USA, 2008.
5. Jirjis, R. Storage and drying of wood fuel. *Biomass Bioenergy* **1995**, *9*, 181–190. [CrossRef]
6. Murugan, P.; Dhanushkodi, S.; Sudhakar, K.; Wilson, V.H. Industrial and Small-Scale Biomass Dryers: An Overview. *Energy Eng.* **2021**, *118*, 435–446. [CrossRef]
7. Kung, K.S.; Ghoniem, A.F. Multi-scale analysis of drying thermally thick biomass for bioenergy applications. *Energy* **2019**, *187*, 15989. [CrossRef]
8. Wade, A. *Report on Biomass Drying Technology*; National Renewable Energy Laboratory: Golden, CO, USA, 1998.
9. Van Loo, S.; Koppejan, J. *The Handbook of Biomass Combustion and Co-Firing*, 1st ed.; Earthscan: Sterling, VA, USA, 2008. [CrossRef]
10. Liu, M.; Zhang, X.; Han, X.; Li, G.; Yan, J. Using pre-drying technology to improve the exergetic efficiency of bioenergy utilization process with combustion: A case study of a power plant. *Appl. Therm. Eng.* **2017**, *127*, 1416–1426. [CrossRef]
11. Gebreegziabher, T.; Oyedun, A.; Hui, C. Optimum biomass drying for combustion—A modeling approach. *Energy* **2013**, *53*, 67–73. [CrossRef]
12. Liu, Y.; Aziz, M.; Kansha, Y.; Bhattacharya, S.; Tsutsumi, A. Application of the self-heat recuperation technology for energy saving in biomass drying system. *Fuel Process. Technol.* **2014**, *117*, 66–74. [CrossRef]
13. Wallin, E.; Fornell, R.; Räftegård, O.; Walfridson, T.; Benson, J. Design and Integration of Heat Recovery in Combination with Solar and Biomass-based Heating in a Drying Plant. *Chem. Eng. Trans.* **2020**, *81*, 1387–1392. [CrossRef]
14. Mujumdar, A. *Handbook of Industrial Drying*, 3rd ed.; CRC/Taylor & Francis: Boca Raton, FL, USA, 2007. [CrossRef]
15. Fagernäs, L.; Brammer, J.; Wilén, C.; Lauer, M.; Verhoeff, F. Drying of biomass for second generation synfuel production. *Biomass Bioenergy* **2010**, *34*, 1267–1277. [CrossRef]
16. Havlik, J.; Dlouhy, T. Integration of Biomass Indirect Dryers into Energy Systems. *J. Chem. Eng. Jpn.* **2017**, *50*, 792–798. [CrossRef]
17. Wang, J.; Salman, C.; Wang, B.; Li, H.; Thorin, E. Integrating sludge drying in biomass fueled CHP plants. *Energy Ecol. Environ.* **2021**, *6*, 1–12. [CrossRef]
18. McIlveen-Wright, D.; Huang, Y.; Rezvani, S.; Redpath, D.; Anderson, M.; Dave, A.; Hewitt, N. A technical and economic analysis of three large scale biomass combustion plants in the UK. *Appl. Energy* **2013**, *112*, 396–404. [CrossRef]
19. Adamski, R.; Siuta, D.; Kukfisz, B.; Mitkowski, P.T.; Szaferski, W. Influence of process parameters in superheated steam drying on fire and explosion parameters of woody biomass. *Fuel Process. Technol.* **2021**, *211*, 106597. [CrossRef]
20. Strömberg, B. *Fuel Handbook*; Värmeforsk Service AB: Stockholm, Sweden, 2006.
21. Hofmann, N.; Mendel, T.; Schulmeyer, F.; Kuptz, D.; Borchert, H.; Hartmann, H. Drying effects and dry matter losses during seasonal storage of spruce wood chips under practical conditions. *Biomass Bioenergy* **2018**, *111*, 196–205. [CrossRef]
22. Leoni, E.; Mancini, M.; Aminti, G.; Picchi, G. Wood Fuel Procurement to Bioenergy Facilities: Analysis of Moisture Content Variability and Optimal Sampling Strategy. *Processes* **2021**, *9*, 359. [CrossRef]
23. Holmberg, A.; Wadsö, L.; Stenström, S. Water vapor sorption and diffusivity in bark. *Dry. Technol.* **2015**, *34*, 150–160. [CrossRef]
24. He, X.; Wang, L. Experimental Determination and Modeling of Drying Process of Woody Biomass. *IOP Conf. Ser. Earth Environ. Sci.* **2020**, *552*, 012016. [CrossRef]
25. Vandenbroek, R. Biomass combustion for power generation. *Biomass Bioenergy* **1996**, *11*, 271–281. [CrossRef]
26. Herz, F.; Mitov, I.; Specht, E.; Stanev, R. Experimental study of the contact heat transfer coefficient between the covered wall and solid bed in rotary drums. *Chem. Eng. Sci.* **2012**, *82*, 312–318. [CrossRef]
27. Herz, F.; Mitov, I.; Specht, E.; Stanev, R. Influence of operational parameters and material properties on the contact heat transfer in rotary kilns. *Int. J. Heat Mass Transf.* **2012**, *55*, 7941–7948. [CrossRef]
28. Lybaert, P. Wall-particles heat transfer in rotating heat exchangers. *Int. J. Heat Mass Transf.* **1987**, *30*, 1663–1672. [CrossRef]
29. Tsotsas, E.; Schlunder, E.U. Contact Drying of Mechanically Agitated Particulate Material in the Presence of Inert Gas. *Chem. Eng. Process. Process Intensif.* **1986**, *20*, 277–285. [CrossRef]
30. Esotsas, E.; Metzger, T.; Gnielinski, V.; Schlünder, E.U. Drying of Solid Materials. In *Ullmann's Encyclopedia of Industrial Chemistry*; Wiley-VCH Verlag GmbH & Co. KGaA: Weinheim, Germany, 2010; pp. 11–13. [CrossRef]
31. Chen, S.; Wang, F.; Milhé, M.; Arlabosse, P.; Liang, F. Experimental and theoretical research on agitated contact drying of sewage sludge in a continuous paddle dryer. *Dry. Technol.* **2016**, *34*, 1979–1990. [CrossRef]
32. Schlünder, E.U.; Mollekopf, N. Vacuum contact drying of free flowing mechanically agitated particulate material. *Chem. Eng. Process. Process Intensif.* **1984**, *18*, 93–111. [CrossRef]
33. Havlík, J.; Dlouhý, T. Indirect Dryers for Biomass Drying—Comparison of Experimental Characteristics for Drum and Rotary Configurations. *ChemEngineering* **2020**, *4*, 18. [CrossRef]

processes

MDPI

Article

# Hierarchical Exploration of Drying Patterns Formed in Drops Containing Lysozyme, PBS, and Liquid Crystals

**Anusuya Pal [1,2,*], Amalesh Gope [3,*] and Germano S. Iannacchione [1,*]**

1 Order-Disorder Phenomena Laboratory, Department of Physics, Worcester Polytechnic Institute, Worcester, MA 01609, USA
2 Department of Physics, University of Warwick, Coventry CV4 7AL, UK
3 Department of Linguistics and Language Technology, Tezpur University, Tezpur, Assam 784028, India
* Correspondence: apal@wpi.edu (A.P.); amalesh@tezu.ac.in (A.G.); gsiannac@wpi.edu (G.S.I.)

**Abstract:** Biological systems, by nature, are highly complex. These systems exhibit diverse hierarchical spatial and temporal features when driven far from equilibrium. The generated features are susceptible to the initial conditions that largely depend on vast parameter space. Extracting information on their properties and behavior thus becomes far too complex. This work seeks to examine the drying kinetics of the drops containing a globular protein (lysozyme (Lys)), phosphate buffer saline (PBS), and thermotropic liquid crystal (LCs). The drying evolution and the morphological crack patterns of these drops are examined using high-resolution microscopy, textural image analysis, and statistical methods. This study observes that the textural parameters can identify the (i) phase separation of the salts present in the PBS and (ii) the LCs' birefringence during the drying evolution. This birefringence activities of the LCs slow down when the initial PBS concentration is increased from 0.25 to 1× despite using a fixed volume of LCs. To comprehend such a surprising effect, the combinations of (i) Lys+PBS and (ii) PBS+LCs are thoroughly examined. A phase diagram is established as a function of initial concentrations of Lys and PBS. The scanning electron microscopic images of Lys+PBS reveal that the tuning between lysozyme and salt concentrations in PBS plays a significant role in determining the morphological patterns. The Lys drops with and without LCs exhibit two distinct regions: the peripheral ring ("coffee-ring") and the central ones. This phase-separated ring formation indicates that the film containing Lys and salts might have formed on top of these LCs in the central region, which reduces the optical response (birefringence) of LCs. A physical mechanism is proposed in this paper to anticipate the redistributions of LCs in a multi-component system such as Lys+PBS+LCs.

**Keywords:** drying; drop; lysozyme; salts; liquid crystals; patterns; texture

**Citation:** Pal, A.; Gope, A.; Iannacchione, G.S. Hierarchical Exploration of Drying Patterns Formed in Drops Containing Lysozyme, PBS, and Liquid Crystals. *Processes* **2022**, *10*, 955. https://doi.org/10.3390/pr10050955

Academic Editor: Jan Havlík

Received: 10 March 2022
Accepted: 29 April 2022
Published: 11 May 2022



## 1. Introduction

Biological systems, by nature, are highly complex. The biosystems' emerging patterns are far more complicated than any other soft system. These features are hard to predict and susceptible, since many biological activities are interconnected and largely depend on vast parameter space. The mesoscopic or macroscopic resulting patterns are primarily sensitive to the initial conditions, including the preparation of the samples, perturbing fields, and so on [1]. A systematic and consistent approach is recommended to explore these systems that require altering one parameter at a time. The complex patterns associated with these bio-systems are due to the local self-assembling interactions between the constituent particles [2]. The transportation of a system from one state (initial bio-colloidal fluid) to another (dried film) through a non-equilibrium (drying) process requires (i) exchanging energy and matter to get acquainted with its micro-environment, and (ii) adjusting the self-assembling interactions between the constituent particles [3].

The research findings of the drying drop community have exhibited numerous studies in the salt-colloidal systems [4–14]. However, the number of findings drops off when the biological systems in the presence of external salts are concerned. Two globular protein samples, lysozyme and bovine serum albumin (BSA), are predominantly investigated to understand how the patterns are affected when the initial salt concentration is varied [15–22]. Gorr et al. [18] examined lysozyme protein at various concentrations of NaCl. The study by Gorr et al. reported the presence of three distinct regions in the drops with NaCl. The first one is formed in the peripheral ring, where most lysozyme is present. The second one forms different salt structures that occupy the secondary ring area (observed adjacent to the ring), and the final one is observed in the central regions. A similar observation is also reported in BSA-saline protein drying drops by Yakhno [15]. The study by Yakhno concludes that the salt crystals are phase-separated by forming different zones from homogeneous protein film near the periphery to the salt crystals in the central region. Pathak et al. [23] investigated the effects of multiple salts ($MgCl_2$ and KCl) on the BSA patterns. This study reveals that the crystal structures depend on the initial tuning ratio of these salts. Furthermore, a few recent studies examined the formation of these protein-saline drying drops at different elevated substrate temperatures. The findings of these experiments confirmed that the final morphological patterns are mainly dependent on the environmental conditions (higher vs. lower drying rate) [24–26].

New and more sophisticated image processing techniques are developed with the increasing demands of examining the complex images of drying stages. Pattern recognition tools, such as k-means clustering and the k-nearest neighbor algorithm, were applied by Gorr et al. [27] to differentiate the lysozyme-NaCl deposits based on the salts' initial concentration. Carreón et al. [21] applied the first-order statistics (FOS), and the gray level co-occurrence matrix (GLCM) specifying the textural image properties. The study explored the evolution of the final state of drying BSA-lysozyme films in NaCl salts' presence using the FOS and GLCM techniques. Pal et al. [28] also examined BSA drops at different initial concentrations of phosphate buffer saline (PBS). The statistical analysis incorporated in that work showed that these GLCM parameters' horizontal and vertical orientations have a non-significant effect when the pixel displacement is $\leq 1$. The pixel distribution is explored at different PBS initial conditions and regions, such as rim and non-rim regions. The study concluded that the BSA–BSA interactions are dominant over the BSA-saline interactions in the rim regions and vice versa.

On the other hand, many researchers also initiated simplification of these systems by reducing the system's complexity and preparing these protein samples in de-ionized water (without adding any external salts) [17,22,29]. In recent years, Pal et al. provided a new perspective while examining the protein droplets and explored the physics of the drying drops containing optically active particles such as thermotropic liquid crystals [30–32]. The complexity of the multi-component system studied by Pal et al. is more or less similar to the protein-saline drops, since both protein and liquid crystals are non-volatile. The detailed findings along these lines imply that the morphological patterns are altered during the drying process when a fixed volume of liquid crystals is added to different globular proteins (lysozyme, BSA, and myoglobin) [32]. The liquid crystals are distributed randomly in the light-weighted proteins (myoglobin and lysozyme). In contrast, these liquid crystals form umbilical defect structures in heavily weighted proteins such as BSA.

Despite the intense research on protein-saline drying drops, to the best of our knowledge, no systematic study is being performed to understand multiple salts' effects on various concentrated lysozyme protein solutions. It would also be interesting to extend the work from protein+liquid crystals+water to protein+liquid crystals+PBS. Since the liquid crystals (LCs) have polar head groups, and these salts have multiple ionic charges, it would not be surprising to get complex morphological patterns. However, the question here is what is the best possible way to explore the physics of such a complex system? The current paper systematically examines the drying drop consisting of (i) PBS, (ii) LCs+PBS, (iii) lysozyme+PBS, (iv) lysozyme+PBS+LCs, at different initial concentrations. High-resolution (bright-field, cross-field, and scanning electron) microscopy, image-processing tools, and statistical methods are incorporated in our experiments and analysis to explore their drying evolution and the morphological patterns. In hindsight, this paper attempts to address a few fundamental questions: (i) what are the effects of multiple salts on lysozyme drops with and without LCs? (ii) Do the three regions (as reported in [18]) in lysozyme-saline drops emerge at every initial concentration? (iii) Does it behave uniformly as reported in the BSA drops? (iv) Is it possible to draw fundamental insights on the self-assembly of LCs, proteins, and salts using the proposed image analysis techniques? If so, how important could these insights be?

## 2. Materials and Experimental Methods

The materials used in this study include a lyophilized form of hen-egg white lysozyme, PBS (phosphate buffer saline), and thermotropic liquid crystal (5CB (4-Cyano-4′-pentyl biphenyl)). The lysozyme (Catalog number L6876) was purchased from Sigma-Aldrich, USA. The different concentrations of the PBS were prepared by diluting 1 (purchased from the Fisher BioReagents, USA (Catalog number BP24384)) into 0.75, 0.5, and 0.25×.

The lysozyme has a roughly ellipsoid shape. Its dimension is $3.0 \times 3.0 \times 4.5$ nm$^3$, with an aspect ratio of 1.5. Its molecular mass is ~14.3 kDa [22]. Lysozyme is made up of 129 amino acids. The isoelectric point of lysozyme is 11.1, which allows it to carry a net positive charge. The globular shape and stability of this protein are attributed to the disulfide bridges, hydrogen bonds, and hydrophobic interactions [22]. The 1× PBS solution contains 0.137 M (~8.0 mg mL$^{-1}$) NaCl, 0.002 M (~0.2 mg mL$^{-1}$) KCl, and 0.0119 M (~1.44 mg mL$^{-1}$ of $Na_2HPO_4$ and ~0.24 mg mL$^{-1}$ of $KH_2PO_4$) phosphates at a pH of ~7.4. The presence of the -cyno groups in 5CB makes it an optically active and polar thermotropic LC. This LC is ~2 nm long and ~0.5 nm in width, with an aspect ratio of 4. It undergoes a phase transition from a crystalline to a nematic phase at 24 °C and from the nematic LC phase to an isotropic phase at ~35 °C [3].

The various amounts, i.e., 100, 75, 50, 35, 25, and 10 mg of lysozyme, are weighed and mixed separately in 1 mL of these PBS solutions. The samples contained the initial concentrations of lysozyme ($\phi_{Lys} = 9.0, 6.9, 4.8, 3.3, 2.4,$ and 1.0 wt%) and the initial concentrations of the PBS ($\phi_{PBS} = 1, 0.75, 0.5, 0.25,$ and 0×). The $\phi_{PBS} = 0\times$ means the de-ionized water (Millipore, 18.2 MΩ·cm at ~25 °C). Finally, LC (Catalog number 328510, Sigma Aldrich, St. Louis, MO, USA), was heated above its transition temperature, and ~10 µL was added at the fixed lysozyme concentration ($\phi_{Lys} = 9.0$ wt%) as a third component to the different protein-saline drops.

A volume of ~1 µL sample solution is pipetted on a freshly cleaned coverslip (Catalog number 48366-045, VWR, Radnor, PA, USA) under ambient conditions (the room temperature of ~25 °C and the relative humidity of ~50%). The drying evolution is monitored every two seconds only for those samples, where the initial lysozyme concentration is kept fixed ($\phi_{Lys} = 9$ wt%), and the initial PBS concentrations ($\phi_{PBS}$) are varied from 1 to 0× with and without LC droplets. The clock started when the drops were deposited on the substrates. This paper displays the time as $t/t_d$, where the total drying time is denoted by $t_d$, and $t$ is the instantaneous time at which the respective images are captured. The images were captured under 5× magnification using bright-field and cross-polarized optical microscopy (Leitz, Wetzlar, Germany) configured in the transmission mode. An 8-bit digital camera (Model number MU300, Amscope, Irvine, CA, USA) was attached to the microscope to click the top-view images. All these experiments were repeated three times, and all the samples showed the highest reproducibility.

The textural image analysis is performed on the protein drops with and without LCs during the drying process. The oval tool of ImageJ is selected to capture the area of interest in such drops. The first-order statistical image parameters, such as mean gray values (I) and the standard deviation (SD), are extracted using ImageJ [33]. A detailed discussion on the image analysis process is available in [26]. A non-parametric Kruskal–Wallis test (an alternative to the parametric one-way ANOVA test) was preferred to examine possible significant interactions among the different initial PBS concentrations ($\phi_{PBS}$) in terms of these textural parameters (mean and SD). A similar procedure is also adopted for the samples containing LCs. In the Kruskal–Wallis test, the $\phi_{PBS}$ is the independent factor, with four levels (groups), 0.25, 0.5, 0.75 and 1×, whereas the mean and SD are the dependent variables. The significant level is kept as $p < 0.05$. A pair-wise comparison between all the $\phi_{PBS}$ is drawn using the Bonferroni test in R (Version 3.6.3) embedded with R studio (Version 1.2.1335, RStudio, Inc., Boston, MA, USA). For this, the function *pairwise.wilcox.test()* with the *p.adjust.method = "bonferroni"* is used. The *ggplot2* library and the function *geom_violin()* are used to draw the violin plots.

Scanning electron microscopy (JEOL-7000F, JEOL Inc., Peabody, MA, USA) is used for the samples by varying $\phi_{Lys}$ of 9.0, 4.8, 3.3, and 1.0 wt% at a fixed $\phi_{PBS}$ of 0.5× and $(\phi_{Lys}, \phi_{PBS}) = (9.0\text{ wt\%}, 0\times)$. For this, the ∼4 nm layer of gold nanoparticles is sputter-coated, and the images are captured at the accelerating voltage of 3 kV and the probe current of 5 mA.

## 3. Results

### 3.1. PBS Drying Drops with and without LC Droplets

The drying evolution of the drops at various initial concentrations of PBS ($\phi_{PBS}$) from 0 to 1× is imaged under bright-field optical microscopy; 0× indicates that the solvent is the de-ionized water. As expected, the drying evolution does not show any residue as the drop dries; however, the deposits' formation changes from 0.25 to 1× (see Figures S1–S3 in the supplementary section). The drying evolution of $\phi_{PBS} = 0.75\times$ is shown in Figure 1a–h. The timestamps ($t$) are calculated as the ratio of the instantaneous time and the total drying time ($t_d$). The water starts evaporating from the drop as soon as it is deposited on the substrate. The height of the drop decreases (see Figure 1a,b how the gray shade changes). The various angles of images and the reduction in drop size confirm that the drop is not pinned to the substrate (see the yellow circular dashed line in Figure 1a–h). The amount of water decreases as the evolution time progresses to such an extent (see Figure 1d) that the concentration of salts increases, and these are carried away with the flow (see Figure 1d–f). Finally, the salts crystallize and fall out of the solution. The final stage demonstrates the evaporation of the trapped water and the formation of a crystal cluster (see Figure 1f–h).

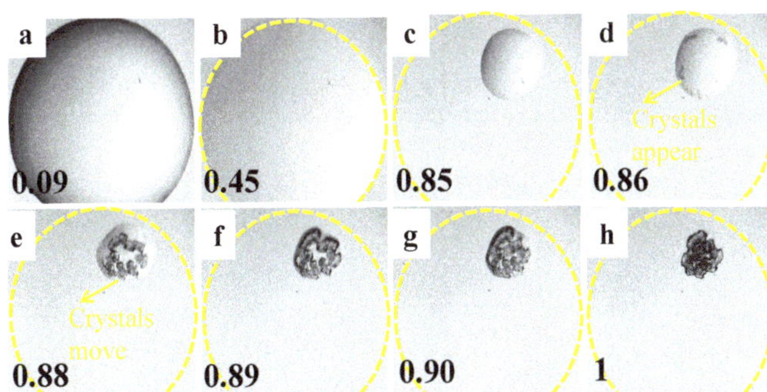


**Figure 1.** Drying evolution of the pattern formation in PBS droplet at the initial concentration ($\phi_{PBS}$) of 0.75× is displayed in (**a–h**). The timestamps are shown with respect to the total drying time ($t_d$) at the bottom left of each image. The yellow dashed circular line shows the drop lining as soon as the drop is deposited on the substrate. The white color in the right corner is a scale bar representing a length of 0.20 mm.

The images of drops containing LC droplets and PBS are captured during the drying process (see Figure 2a–l) using both bright-field and cross polarizing configurations of the optical microscopy. Capturing images consecutively at different configurations ensures the progression of these LC droplets in the PBS is tracked. The process also allows us to compare the drying evolution of the drops with and without LC droplets. The droplets are mostly carried away towards the periphery of the drop (see Figure 2a–d). A ring-like pattern forms that partially pins the drop on the substrate, unlike the PBS drop (see Figure 1a–h). The drop size does not decrease symmetrically with water evaporation, which can predominantly be seen in Figure 2e. We also observe that a few LC droplets fall out of the solution, i.e., they stay where they are. In contrast, some droplets are carried away with the water (see Figure 2e–h). The salt starts appearing later in the drying process, including the PBS drop (see Figure 2g–i). The LC droplets are observed to be deposited near these crystals as soon as the formation of the salt crystals starts (see Figure 2j–l).

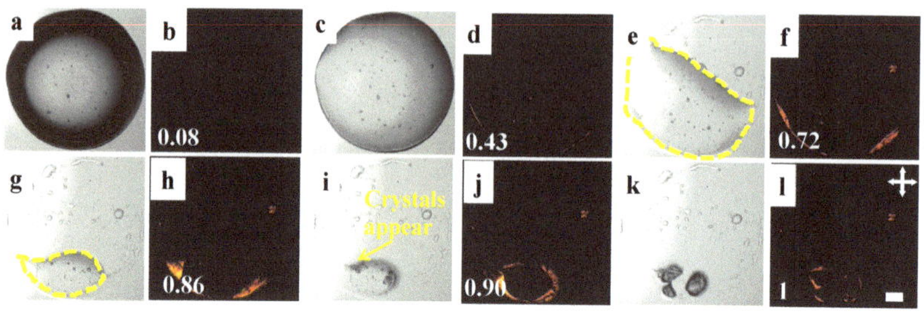


**Figure 2.** Time-lapse images of drying LC+PBS drop at an initial concentration ($\phi_{PBS}$) of 0.75×. The timestamps ($t$) are calculated as the ratio of the instantaneous time and the total drying time ($t_d$), shown at the bottom left of each image. The different stages of the drying process are imaged under bright-field (**a**,**c**,**e**,**g**,**i**,**k**) and crossed polarizing configurations (crossed double arrows). The LC droplets are shown with the bright patches in the crossed configurations (**b**,**d**,**f**,**h**,**j**,**l**). The white color in the right corner is a scale bar representing a length of 0.20 mm.

### 3.2. Lysozyme+PBS Drying Drops

#### 3.2.1. Morphological Patterns at Macro- to Micro-Scales

Figure 3 displays a morphological grid of the samples varying the initial concentrations of Lys ($\phi_{Lys}$) from 9.0 to 1.0 wt% (along the $Y$-axis) and the initial concentrations of the PBS ($\phi_{PBS}$) from 1 to 0.25× (along the $X$-axis). The $\phi_{PBS}$ = 0× embodies the lysozyme solution prepared in the de-ionized water. Though all these deposits show the "coffee-ring" effect [34], diverse patterns are observed for each $\phi_{Lys}$ and $\phi_{PBS}$. The lysozyme films show a mound-like structure when the solution is prepared without external salts. A dimple (or depression) is also noticed within this mound. The mound area gets wider as the $\phi_{Lys}$ increases. The random cracks are only observed in the peripheral ring at ($\phi_{Lys}$, $\phi_{PBS}$) = (1.0 wt%, 0×). However, these cracks spread throughout the film as the $\phi_{Lys}$ increases. The radial and orthoradial cracks promote well-connected (small and large) domains in these drops. Some fringes appear in the concentrated lysozyme samples at $\phi_{PBS}$ = 0×. Many domains in the ring are delaminated, which is predominantly observed at ($\phi_{Lys}$, $\phi_{PBS}$) = (9.0 wt%, 0×). The concentration dependence of these lysozyme drops in the salts' absence is thoroughly explained in [22]. Comparing these patterns reveals that the mound diminishes in the salts' presence. Remarkably, each drop reveals two distinct regions—(i) a peripheral ring ("coffee-ring" [34]) and (ii) a central region. Furthermore, the central regions also disclose two rings at ($\phi_{Lys}$, $\phi_{PBS}$) = (2.4–3.3 wt%, 0.5×) and (2.4 wt%, 0.25×). The presence of multiple rings is also observed in the highly concentrated lysozyme samples, i.e., ($\phi_{Lys}$, $\phi_{PBS}$) = (6.9–9.0 wt%, 0.5×) and (6.9–9.0 wt%, 0.25×). A unique trend is also noticed when $\phi_{Lys}$ is fixed and $\phi_{PBS}$ is varied. For example, at $\phi_{Lys}$ = 1.0 wt%, the ring width decreases with the increasing $\phi_{PBS}$. The central region becomes grainy, the texture becomes darker, and some thread-like structures appear in the central region. The samples at $\phi_{PBS}$ = 1× display a dark texture in the central region and a gray texture in the peripheral ring. Though many drops form the two distinct regions, various textures are observed in the central region. For example, the textures at ($\phi_{Lys}$, $\phi_{PBS}$) = (4.8 wt%, 0.25–0.5×) and (3.3 wt%, 0.25×) are different from ($\phi_{Lys}$, $\phi_{PBS}$) = (2.4–4.8 wt%, 1×). The random small cracks are observed at $\phi_{Lys}$ = 1.0 wt% whereas, the radial cracks predominantly appear in the peripheral ring as $\phi_{Lys}$ increases in the salts' presence. In contrast, no overall drift is found when these cracks intervene in the central region. For instance, the sample at ($\phi_{Lys}$, $\phi_{PBS}$) = (9.0 wt%, 1×) shows crack patterns, whereas other samples at $\phi_{PBS}$ = 1× do not.

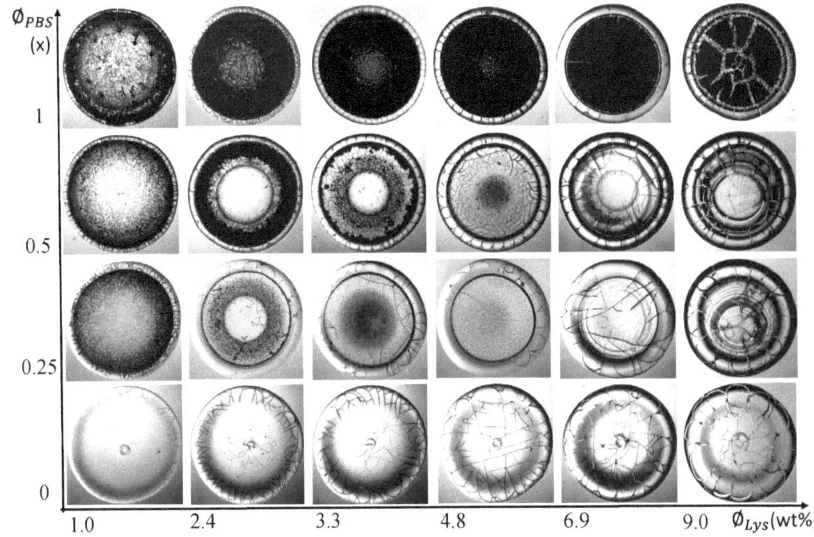


**Figure 3.** Morphological patterns observed in the drying drops of Lys are exhibited here. These patterns are detected for the various initial concentrations of Lys ($\phi_{Lys}$) dissolved in the initial concentrations of the phosphate buffer saline ($\phi_{PBS}$). The $\phi_{PBS}$ = 0× indicates that the solution is prepared in de-ionized water. The scale bar is of a length of 0.15 mm.

A few lysozyme drops, interestingly, follow different zones while moving from the drop's periphery to the central region. Similar zones are observed in other globular protein drops, viz., BSA [15,35]. A close observation of the samples from peripheral to central regions at ($\phi_{Lys}$, $\phi_{PBS}$) = (9.0 wt%, 0.5×) and ($\phi_{Lys}$, $\phi_{PBS}$) = (6.9 wt%, 0.5×) in Figure 3 display a zone of somewhat homogeneous lysozyme films (or glassy peripheral ring), then a zone of different lysozyme structures, gel-like structures, and the crystalline zone. Therefore, the lysozyme displays different patterns in different volumes of salts. It undergoes phase separation and forms different material properties; for instance, the lysozyme concentration is highest in the peripheral ring and systematically drops towards the central region. On the other hand, the salt concentration is almost null in the periphery but highest as we move towards the central regions. The lysozyme drops in the central region display different cracks in low salt concentrations at $\phi_{Lys} \geq 6.9$ wt%. It is to be noted that these optical images showcase these patterns globally (at macoscale). However, these images do not reveal any microstructural information.

Figure 4I–IV exhibits the microstructures of the various concentrated lysozyme samples at $\phi_{PBS}$ = 0.5×. The sample at ($\phi_{Lys}$, $\phi_{PBS}$) = (9.0 wt%, 0×) is displayed in Figure 4V. The different regions in the central and peripheral regions were emphasized in all these samples. The sample shows a uniform homogeneous texture without external salts (see Figure 4V). In contrast, a distinct texture is observed in the salts' presence (see Figure 4IV). Furthermore, a few non-uniform structures are also uncovered in the crack lines separating the periphery and the central regions. A comparison of the central and the peripheral regions confirms a smooth texture in the peripheral ring. However, some snowflake-like structures appear in the inner ring of the periphery at ($\phi_{Lys}$, $\phi_{PBS}$) = (3.3 wt%, 0.5×) (see Figure 4II). The crystal-like structures are spotted in the zone between the central and peripheral regions at ($\phi_{Lys}$, $\phi_{PBS}$) = (1.0 wt%, 0.5×) (see Figure 4I). These structures; however, do not seem to be too prominent as we move towards the central region of the film. The central region is mainly replaced with different forms of the dendrite structures; viz, long but thin structures at ($\phi_{Lys}$, $\phi_{PBS}$) = (3.3 wt%, 0.5×), but then again appears to be shorter but thicker structures at ($\phi_{Lys}$, $\phi_{PBS}$) = (9.0 wt%, 0.5×) in (see Figures 4II,IV). In these samples, a grainy amorphous layer mostly occupies the middle region (between the peripheral and central regions). On the other hand, it is hard to differentiate this layer between the middle and the central regions at ($\phi_{Lys}$, $\phi_{PBS}$) = (4.8 wt%, 0.5×) (see Figure 4III).

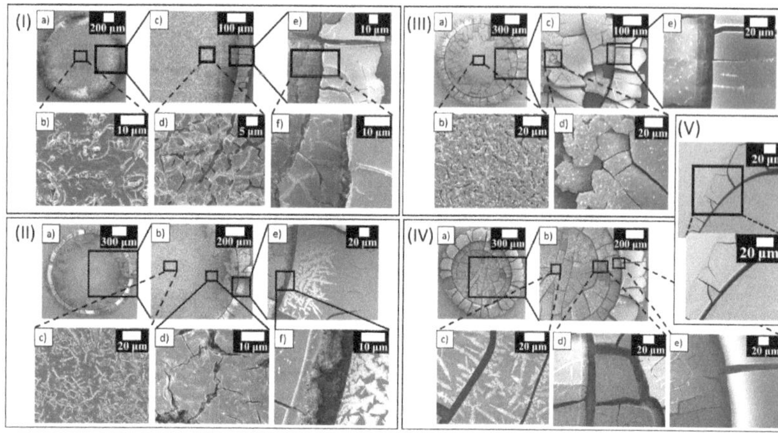

**Figure 4.** Microscopic images of the dried lysozyme samples are displayed in this figure. ($\phi_{Lys}$, $\phi_{PBS}$) = (1.0 wt%, 0.5×) in (**I**), ($\phi_{Lys}$, $\phi_{PBS}$) = (3.3 wt%, 0.5×) in (**II**), ($\phi_{Lys}$, $\phi_{PBS}$) = (4.8 wt%, 0.5×) in (**III**), ($\phi_{Lys}$, $\phi_{PBS}$) = (9.0 wt%, 0.5×) in (**IV**), and ($\phi_{Lys}$, $\phi_{PBS}$) = (9.0 wt%, 0×) in (**V**). Different length scales are shown as the scale bars in the upper-right corner of each image.

### 3.2.2. Qualitative and Quantitative Analysis of Drying Evolution

To understand how these distinct structures appear in different regions, we examine the drying evolution and dried morphology by keeping $\phi_{Lys}$ at 9.0 wt%, and only the $\phi_{PBS}$ is varied from 0.25 to 1×. Figure 5A–D describes the drying evolution of the lysozyme drops prepared with different PBS concentrations. The timestamps are calculated as the instantaneous time divided by the total drying time ($t_d$). A uniform gray texture with a dark peripheral band is observed in all the drops when the first image is captured of the drying process (see Figure 5A). The fluid front moves from the periphery to the central region with time progression (see Figure 5B). Surprisingly, the texture of the front movement changes in the salts' presence. Once the peripheral ring emerges, the grainy texture develops in the central region. A clear distinction is visible at the interface of the inner peripheral ring. At $\phi_{PBS}$ = 0.25×, the formation of the dark texture is not predominantly observed; however, the darkness increases as the $\phi_{PBS}$ rises. Simultaneously, the cracks propagate from the periphery toward the center. However, the propagation is not smooth, unlike $\phi_{PBS} = 0\times$, and the propagation is interrupted in the presence of the salts. The mound-like structure begins in the last stage of this fluid front movement, but the salts' presence diminishes its formation in the central region. The multiple rings are found as the front's radius gets smaller (see Figure 5B,C). After the visible drying process, the final morphological patterns are captured in Figure 5D.

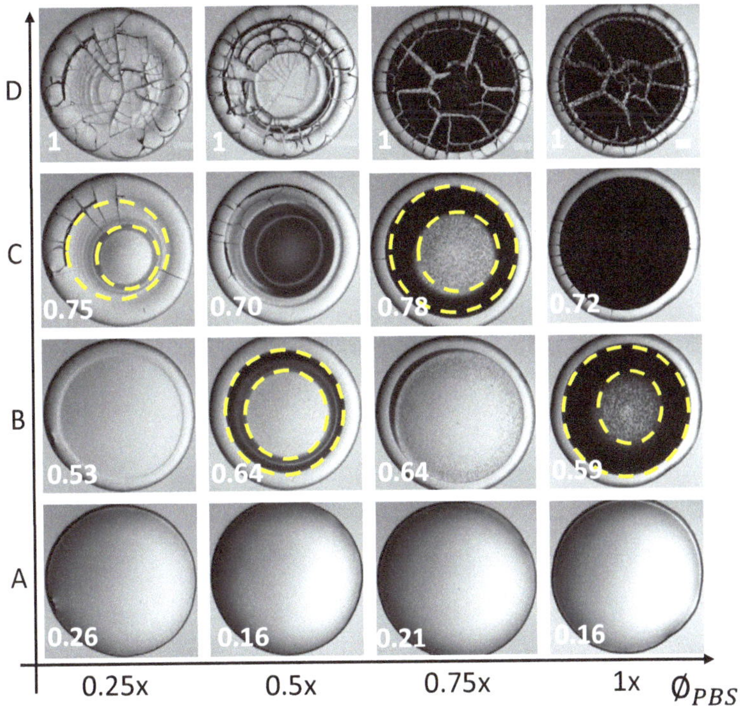


**Figure 5.** The time evolution of lysozyme drops ($\phi_{Lys}$ = 9 wt%) at various initial concentrations of salts in PBS ($\phi_{PBS}$) during the drying process is displayed in (**A–D**). The instantaneous time is divided by the total drying time ($t_d$) to calculate the timestamps, as shown in the bottom left of each image. The white rectangle represents a scale bar of 0.15 mm length in the top right.

Figure 6I–IV shows the quantitative analysis of the textural evolution during the drying process. The first-order statistical parameters, the mean gray values ($I$), and the standard deviation ($SD$) are exhibited as a function of the drying time (in seconds) at $\phi_{PBS}$ ranging from 0.25 to 1×. It is to be noted that these parameters describe the gray level distribution of the image's pixel intensity. The $I$ defines the averaged values, whereas the $SD$ illustrates the textural complexity. The $I$ stays nearly constant at the beginning of the drying process. It reduces and starts fluctuating (marked with a star in Figure 6I–IV). In contrast, the $SD$ decreases linearly until $t/t_d \sim 0.6$ and rapidly rises for $t/t_d \sim$ 0.6–0.7. It first decreases, and grows again (marked with a star in Figure 6I–IV). Finally, both the $I$ and $SD$ saturate in the last phase of the drying evolution. Interestingly, both the $I$ and $SD$ exhibit significant changes when $t/t_d \sim$ 0.6–0.9. The $I$ decreases, and the $SD$ increases when we prepare the drops by adding more PBS. For instance, the $I$ reduces from ~50 to ~20 a.u. at $\phi_{PBS} = 1\times$, whereas it only varies ~15 a.u. when $\phi_{PBS} = 0.25\times$. The images (see Figure 6I–IV) show that considerable variation in these textural parameters ($I$ and $SD$) occurs when the dark textured fluid front moves in the central region.

Figure 7I,II shows the textural parameters, mean, and standard deviation (SD) at different initial concentrations of PBS ($\phi_{PBS}$). The samples considered for the statistical analysis include the protein drops at a fixed initial concentration ($\phi_{Lys}$ = 9 wt%) that do not contain any liquid crystals (LCs). The $\phi_{PBS}$ is varied from 0.25 to 1×. The violin plots in Figure 7I,II combine the box plots with the density traces (viz., the peaks, valleys, lengths, etc.). This density trace is vertically plotted to both sides of the box plots. These plots display the statistical description in which a rectangle represents the second and third quartiles, and a line inside that box indicates the median value. The traces are symmetrically drawn; only the directions are reversed to observe the density magnitude in a better way. The box plots and the density traces allow us to get a quick and insightful comparison of the data distributions. However, these plots do not provide any information about their sample size [36]. Here, the median at all $\phi_{PBS}$ has been sifted for the textural parameters, mean, and the SD. The mean shows a bi-modal distribution for $\phi_{PBS}$ = 0.25 and 0.5×, and somewhat normal distribution is observed in the range of 0 to 20 (a.u.). However, a positive skewness is observed in the range of 40 to 60 (a.u.) in the $\phi_{PBS}$ = 0.75 and 1× (see Figure 7I). The SD reveals a uni-modal distribution at $\phi_{PBS}$ = 0.75×. The SD data, nonetheless, are skewed in the range of 10 to 15 a.u. (see Figure 7II). Both the textural parameters confirm a non-normal distribution. The skewness also confirms that the mean and the SD are not equally distributed throughout the drying process (see Figure 6I–IV). Therefore, a non-parametric Kruskal–Wallis test is performed. A detailed report of the pair-wise comparison is shown in Tables S1 and S2 of the supplementary section. It must be noted that the visual display of Figure 6I–IV also indicates possible differences in the textural parameters at each $\phi_{PBS}$. The proposed statistical test facilitates quantitatively distinguishing and confirming that the mean and SD are different at each $\phi_{PBS}$.

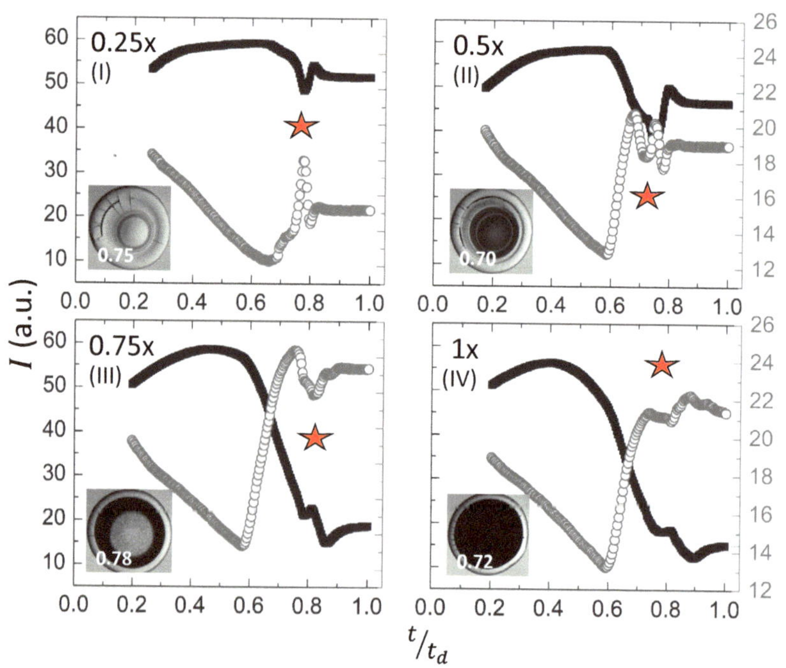


**Figure 6.** Textural analysis of the lysozyme drops ($\phi_{Lys}$ = 9 wt%) at various initial concentrations of salts in PBS ($\phi_{PBS}$) during the drying process is displayed in (**I–IV**). The $x$-axis displays $t/t_d$, where the total drying time is denoted by $t_d$, and $t$ is the instantaneous time captured by the respective image. The left-$Y$ axis (shown in back color squares) and the right-$Y$ axis (shown in gray color circles) in each graph describe the mean gray values ($I$) and the standard deviation ($SD$) in arbitrary units (a.u.), respectively. The star symbol indicates the fluctuations in these textural parameters.

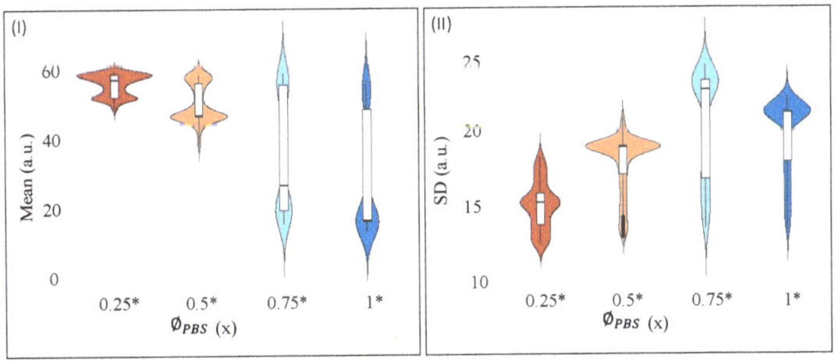


**Figure 7.** Violin plots for the first-order textural parameters—(**I**) represents the distribution of Mean, and (**II**) displays the distribution of Standard Deviation (SD) of the protein drops at different initial concentrations of PBS ($\phi_{PBS}$). The significant $\phi_{PBS}$ is marked with an asterisk (*). The protein drops have a fixed initial concentration of 9 wt% that do not contain any liquid crystals (LCs).

### 3.3. *Lysozyme+PBS Drying Drops with LC Droplets*

The drying evolution of the lysozyme+PBS drops with LC droplets at different $\phi_{PBS}$ of 0 to 1× under crossed polarizing configurations is shown in Figure 8A–D. The $\phi_{Lys} = 9.0$ wt%, and the volume of LC (i.e., ~10 µL) are kept constant, and only the $\phi_{PBS}$ is varied from 0 to 1×. The timestamps ($t$) are calculated by dividing the instantaneous time by the total drying time ($t_d$). Here, we are interested to know how the optical activities or the birefringence properties of these LC droplets are influenced due to the presence of the salts. The $\phi_{PBS} = 0\times$ denotes no external salt, whereas the initial salt concentration systematically increases from 0.25 to 1×.

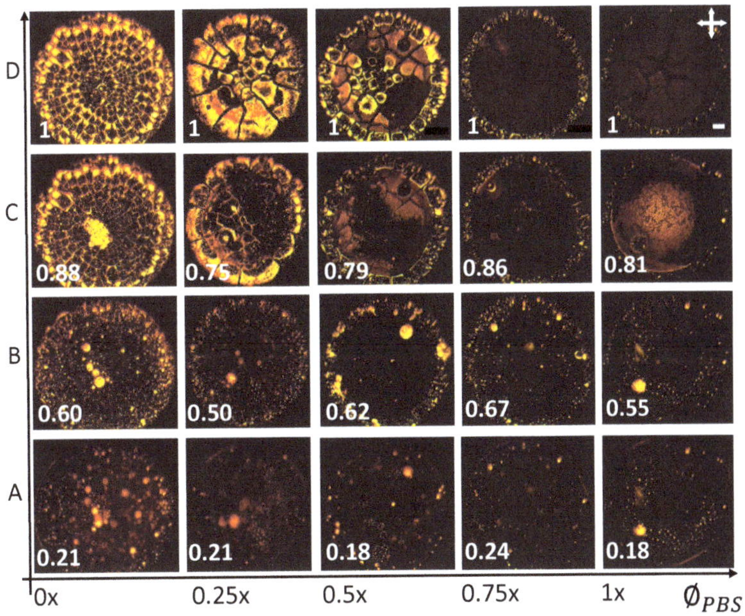


**Figure 8.** Time-lapse images of drying lysozyme+PBS drop with LC droplets at the initial concentration ($\phi_{PBS}$) of 0 to 1× are displayed in (**A–D**). The timestamps are shown at the bottom left of each image. The images are captured under crossed polarizing configuration (crossed double arrows). The white color in the right corner is a scale bar representing a length of 0.20 mm.

The LC droplets are carried away towards the periphery of the drop once this is deposited on the substrate, irrespective of the presence of the salts (see Figure 8A,B, $\phi_{PBS} = 0\times$). It is to be noted that the volume of the LC added in the drop is of equal quantity. Therefore, the image captured after depositing the drop on the substrate is expected to be similar. However, a slight variation is found as we increase the $\phi_{PBS}$. In the absence of the salts, these droplets follow the crack lines (see Figure 8C, $\phi_{PBS} = 0\times$). Subsequently, each cracked domain is partially filled with these droplets. Each domain has a central dark region surrounded by a bright region (see Figure 8D, $\phi_{PBS} = 0\times$). A detailed description of the drying evolution of lysozyme drop with LC droplets at $\phi_{PBS} = 0\times$ is available in [30]. On the other hand, the crack lines and the filling up of the domains are faintly observed at $\phi_{PBS} = 0.25\times$. Interestingly, this process diminishes as we increase $\phi_{PBS}$, and a new feature evolves. The central part of the drop changes from the dark to the bright grainy region (see Figure 8C). The bright regions get brighter at 0.5 and 0.75×, whereas these turn darker at $\phi_{PBS} = 1\times$ (see Figure 8D). Finally, two distinct regions, i.e., peripheral and central regions, can be identified in the presence of salts. The number of cracks decreases, increasing the size of the cracked domains as we keep on adding more salts into the system.

The visual observation is quantified using the textural image analysis of the first-order statistics (mean gray values (I in a.u.) and standard deviation (SD in a.u.)) (Figure 9I–IV at different $\phi_{PBS}$). The I and SD remain constant in the initial drying stage irrespective of the salt concentrations. However, their transient behavior at $t/t_d \sim 0.7$–$0.85$ is interesting. At $\phi_{PBS} = 0.25\times$, both SD and I rapidly increase and saturate. The I increases, whereas SD shows fluctuating behavior at $\phi_{PBS}$ of 0.5 to 0.75×. This increase in I, indeed, gets smaller as we increase $\phi_{PBS}$. For instance, it grows from ~15 to ~75 a.u. at $\phi_{PBS} = 0.25\times$ but only increases to ~25 a.u. at $\phi_{PBS} = 0.75\times$. The $\phi_{PBS} = 1\times$ shows different behavior from the rest. The I decreases again at $t/t_d \sim 0.8$ and eventually saturates. Overall, we can say that the birefringence (or the optical activity) of the LC droplets reduces as we increase the initial concentration of PBS.

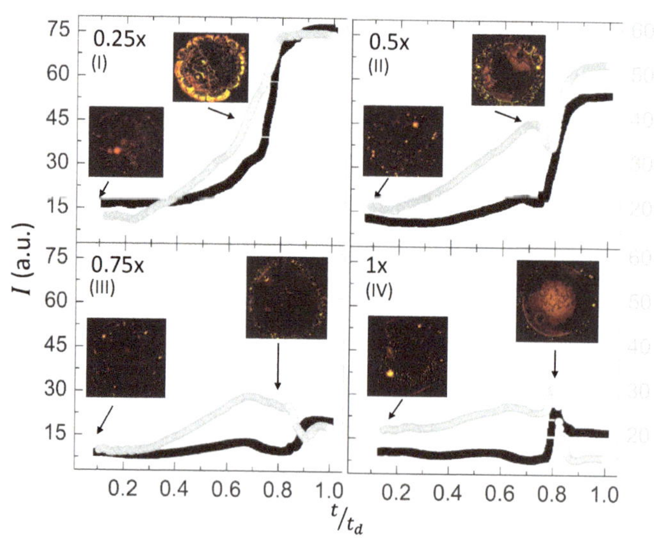


**Figure 9.** Textural analysis of the lysozyme drops ($\phi_{Lys} = 9$ wt%) with LC droplets at various initial concentrations of salts in PBS ($\phi_{PBS}$) during the drying process is displayed in (**I–IV**). The *x*-axis displays $t/t_d$, where the total drying time is denoted by $t_d$, and $t$ is the instantaneous time captured by the respective image. The left-*Y* (shown in back color squares) and the right-*Y* axes (shown in gray color circles) in each graph describe the mean gray values ($I$) and the standard deviation ($SD$) in arbitrary units (a.u.), respectively.

Comparing Figures 6 and 9I–IV, it is found that the textural analysis identifies different significant activities in the drop. Interestingly, this is captured for $t/t_d \sim 0.6$–0.85 during the drying process in the lysozyme drops with and without LC droplets.

Figure 10I,II displays the violin plots at different initial concentrations of PBS ($\phi_{PBS}$) for the textural parameters, mean and standard deviation (SD). The samples considered for the statistical analysis include the protein drops at a fixed initial concentration ($\phi_{Lys} = 9$ wt%), and contain liquid crystals (LCs). The $\phi_{PBS}$ is varied from 0.25 to 1×. Similar to Figure 7I,II, the median at all $\phi_{PBS}$ has sifted for both the textural parameters, mean and SD; however, it is not possible to make one-to-one comparisons between the protein drops with and without LCs (viz., between Figures 7 and 10), since the images without LCs are captured under bright-field configuration. In contrast, the images with LCs are captured under crossed polarizing configurations. However, both the textural parameters confirm a non-normal distribution. The skewness suggests that the mean and SD are not equally distributed throughout the drying process [also evident from Figure 9I–IV]. A detailed report of the pair-wise comparison is shown in Tables S3 and S4 of the supplementary section.

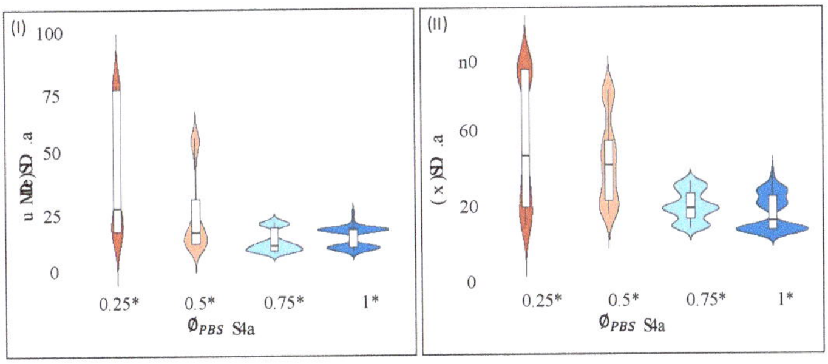

**Figure 10.** Violin plots for the first-order textural parameters. (**I**) represents the distribution of Mean, and (**II**) displays the distribution of Standard Deviation (SD) of the protein drops at different initial concentrations of PBS ($\phi_{PBS}$). The significant $\phi_{PBS}$ is marked with an asterisk [*]. The protein drops have a fixed initial concentration of 9 wt% and contain a fixed volume of liquid crystals (LCs).

## 4. Discussions

The height and the contact angle start reducing as soon as the drops are pipetted on the substrate. The non-uniform textural gradient in the optical images (for example, Figure 5A,B) indicates that the height of the drop initiates the decreasing process in the initial drying stage itself. These drops are of a spherical-cap shape and are partially wet (checked with a goniometer). The curvature of these drops induces the highest mass loss near the periphery compared to their central region. The drops get pinned to the substrate, and the lysozyme particles are transported through the outward capillary radial flow to compensate for this loss. The process leads to the popular "coffee-ring effect" [34] that is also observed in other bio-colloids as well [20,22]. Furthermore, most LC droplets are carried away towards the periphery of the lysozyme drops.

With the progression of the evolution time, the fluid front recedes from the periphery to the central region in the lysozyme drops. Concurrently, the contact angle reduces, contrary to the results reported in [19]. The deposits in the crack lines (see Figure 4) indicate a discontinuity in the crack lines at the ring interface (also evident in Figure 5B,C). It is to be noted that the evaporation of a significant amount of water initiates the formation of salt crystals at this time. Since the images were taken in the transmission mode, the thick film gives a dark texture. The dark textured front starts engulfing the central region (a similar phase transition phenomenon reported in [15]). A significant fluctuation in the textural evolution (see Figure 6I–IV) is also observed at this phase. The complexity (SD) increases as the salt crystals (inhomogeneities) appear.

The crack propagation of the pinned drops relieves the mechanical stress. The appearance of the salt crystals in different lysozyme concentrations affects the crack formation process (see Figure 3). It also alters the interaction between the lysozyme particles and changes their aggregation and precipitation processes (samples with and without adding external salts, evident from Figures 3 and 4). The images captured at high and low initial concentrations of PBS ($\phi_{PBS}$) at (i) fixed and (ii) different initial concentrations of lysozyme ($\phi_{Lys}$) reveal contrasting morphological characteristics. For example, the dried drops formed at $\phi_{Lys}$ = 1 wt%, ($\phi_{PBS}$) = 0.25×, exhibit a rough texture with block salt crystals in the central region. On the other hand, a very thin "coffee ring" and small needle-shaped structures appear in the peripheral region (see Figures 3 and 4(I,e)). In contrast, deposits with higher $\phi_{PBS}$ contain a thick "coffee ring" near the edge, and large rosette-like crystal structures appear in the central region (see Figure 4(IV,c)). These contrasting morphologies emerge from the different nucleation and super-saturation points during the drying process. Interestingly, this appearance of different structural morphology is not limited to the protein drops but is also reported to be observed in many polymer-saline drops [14].

This study also demonstrates that the three prominent regions (as reported in [18]) may or may not surf in the lysozyme drops when the salts are in the solution. This study, therefore, argues that the occurrence of three distinct regions should not be generalized for lysozyme+PBS drying drops. We further argue that the three distinct regions in the lysozyme drops depend on the relative initial concentrations of both lysozyme and salts. The only variation of the salt content might not provide us with a clear picture. Figures 3 and 4 show that the chemistry between multiple salts and lysozyme at various initial concentrations (both lysozyme and salts) is the crucial factor in determining morphological patterns. In this context, we established a phase diagram based on the initial concentrations of lysozyme ($\phi_{Lys}$) and PBS ($\phi_{PBS}$). Figure 11 exhibits the various phases that are colored uniquely. For example, the orange color describes a phase where the drops always show one ring that divides the central region from the periphery. Two representative images are shown in this phase to illustrate that the central region's grainy texture changes to black as we increase the $\phi_{PBS}$. However, we are ignoring this gradual textural change and counting all in one phase. Though the phase of white color also shows one ring, we used a different color to indicate that the texture of the central regions of both the phases is not the same. The phase of blue color suggests that the drops have multiple rings in the central region in addition to the peripheral ring (illustrated with the dashed red colored line in Figure 11). The green color indicates the presence of only two rings (one in the central region and another in the peripheral ring). The hatched lines for $\phi_{PBS} = 0\times$ indicate that the drops without adding any external salts are totally different from those with PBS. The presence of a mound-like structure (illustrated with the solid red colored line in Figure 11) in the central region confirms that the underlying physics of such drying-mediated patterns are different for the drops with and without salts. Different phases in Figure 11 also confirms that the hierarchical structures that are formed by the aggregation of the lysozyme and salts are not directly correlated with their initial concentrations. If so, we would have the same phase as we double their concentrations. For instance, that phase at ($\phi_{Lys}$, $\phi_{PBS}$) = (1.0 wt%, 0.25×) is not the same as the ($\phi_{Lys}$, $\phi_{PBS}$) = (2.4 wt%, 0.5×) or ($\phi_{Lys}$, $\phi_{PBS}$) = (2.4 wt%, 0.5×) is not the same phase as ($\phi_{Lys}$, $\phi_{PBS}$) = (4.8 wt%, 1×).

However, it is hard to predict how the adsorption process of various particles (lysozyme and salts) occurs during the drying process. The SEM images of lysozyme+PBS drops (see Figure 4I–IV) indicate that the lysozyme forms a film on the substrate during the initial drying stage. The salts are phase-separated during the mid-drying stage, forming two different regions. The peripheral region contains most lysozyme particles, whereas the central region contains lysozyme and salts. It is worthy of mentioning here that salt crystallization is a natural process formed due to the evaporation of a large volume of water from the system (similar to the salt formation in the sea beds). It can be assumed that the lysozyme particles in the film act as the nucleation points and initiate distinct types of salt crystallization. For instance, some salts crystallize to form dendrite-like structures on the protein films, whereas some appear as snowflakes, sword-like structures, etc. Some aggregated globules are also present in the crack lines (see Figure 12(AI,AII)). In this context, a plausible mechanism can be drawn in terms of the protein charges, affinity of salts, surface properties, etc. It can be assumed that the globule nature of these proteins is maintained. As a result, the hydrophobic residues of the proteins are buried inside the protein core, and both positively and negatively charged residues on the protein surface are exposed. It is to be noted that the coverslip (substrate) is negatively charged, and the overall charge of lysozyme is positive. Thus, the positively charged residues adsorb the substrate [22]. The probability of altering the overall interaction of lysozyme with the substrate is very low.

On the other hand, there is a high chance of different lysozyme–lysozyme and/or lysozyme–salts interactions. This is because the interactions between opposing surface charges and hydrophobic regions binding these globular proteins might be favored. Similarly, the interaction between the lysozyme particles and the solvent produces a charged layer (Stern layer of counter-ions) [37]. This layer comprises water molecules and/or different dissociating ions present in PBS (viz., $Na^{+}$, $K^{+}$, $Cl^{-}$). The oxygen and hydrogen atoms present in the water are likely to be attached to the protein surfaces' positively and negatively charged ions, respectively, due to their electrostatic interactions. The further progression of the drying process leads to the modification of the proteins' hydration shell. Finally, the evaporation of a significant amount of water from the drops facilitates the emergence of the super-saturation or nucleation points.

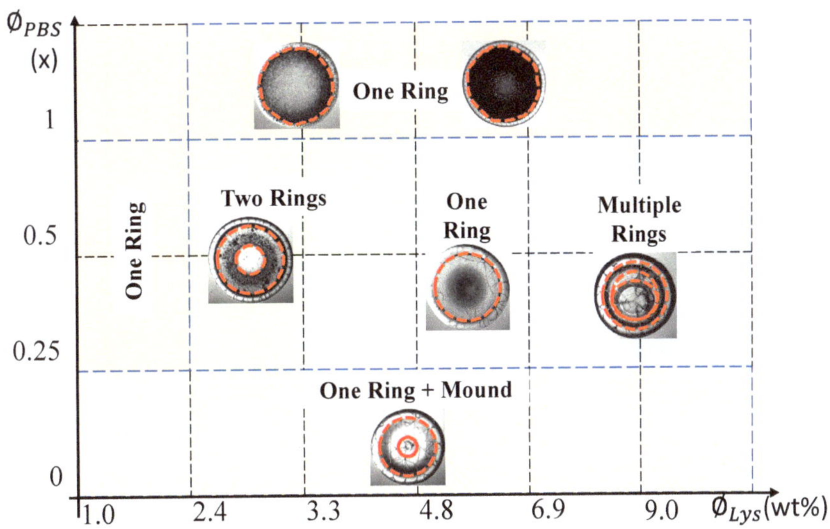


**Figure 11.** A phase diagram is established for the drying-mediated patterns observed in Figure 3. Different phases, indicated by distinct color shades—for example, one, two, and multiple rings in the drops—are presented as a function of various initial concentrations of Lys ($\phi_{Lys}$ along the $X$-axis) and the phosphate buffer saline ($\phi_{PBS}$ along the $Y$-axis). The hatched lines for $\phi_{Lys} = 0\times$ indicate that the drops without adding any external salts are different from the drops with PBS by the presence of the mound-like structure in the central region. The representative image in each phase is provided where the dashed red-colored lines mark the rings, and the solid line depicts the mound-like structure.

The addition of LC droplets in the lysozyme drops makes the scenario far more complicated. Let us look at the drops prepared in the de-ionized water ($\phi_{PBS} = 0\times$). The LCs fill the crack lines of the drop at random. Once the protein-cracked domains buckled up due to mechanical stress (see Figure 3, ($\phi_{Lys}, \phi_{PBS}$) = (9 wt%, 0×)), the randomly distributed LCs are sucked underneath these domains (see Figure 12(BIII,BV)). Accordingly, we can confirm that the bright regions observed under crossed polarized configurations are the randomly oriented LCs distributed underneath each lysozyme domain. The dark region in each domain corresponds to the attached protein layer that is not optically active.

A detailed discussion on this is available in [30]. It is to be noted that the aspect ratios of lysozyme and LC are 1.5 and 4, respectively. Therefore, it can be assumed that there is a probability of a size effect on the crack formation as reported in other multi-component systems [22,38]. The bright-field images of the lysozyme drops with and without LCs show that the cracks are more ordered in the presence of LCs, whereas a chaotic system is observed in their absence [31]. However, extracting the exact physical mechanism due to the difference in their sizes is not within the scope of this paper.

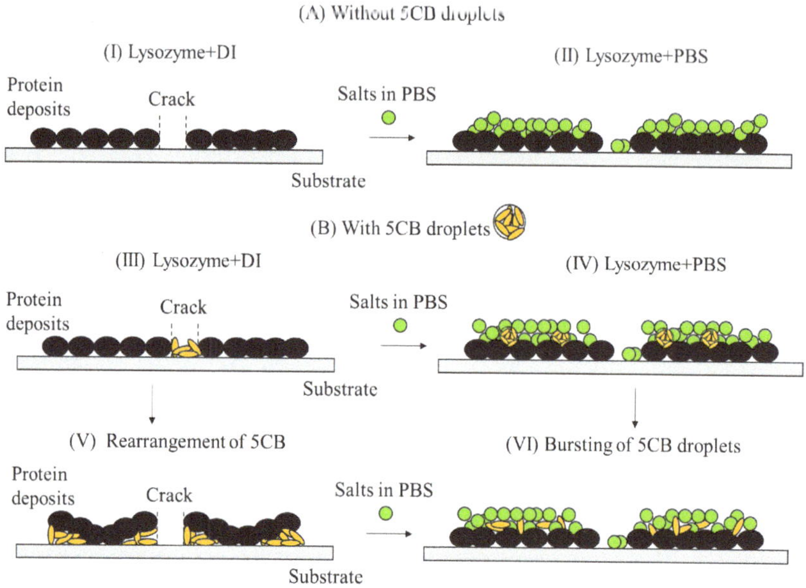


**Figure 12.** A schematic representation of lysozyme drops prepared in different concentrations of PBS ($\phi_{PBS}$) with and without LC droplets is shown in this figure. (**A**)(I,II) depicts the changes in the central region of the drop when we add concentrated $\phi_{PBS}$. (**B**)(III–VI) illustrates how the LC droplets affect the distribution of proteins, salts, and the final morphological patterns.

A proper explanation of the drying mechanism in the presence of salts, lysozyme, and LC droplets can be drawn from Figures 5–9. The LC droplets behave more or less like 0× at the lower salt concentration, $\phi_{PBS} = 0.25\times$ (see Figures 8 and 9I). The LC comprises organic molecules with (i) a rigid core of two phenyl groups and (ii) a side chain of a cyano group (-$CN^-$), making it polar. The cyano group in the LCs interacts with the dissociating positively charged ions of the PBS ($Na^+$, $K^+$, etc.) and the positively charged residues of the lysozyme. This interaction possibly impacts the packing of both lysozyme particles and LC droplets as more water evaporates during the drying process. Interestingly, we can see that the crack domains are larger than that of 0×, which is true for all the protein drops in the presence and absence of LCs. This indicates that the interactions between the salts–salts, salts–proteins, and proteins–proteins are rigid, and that they want to be together. It might also be the case that the presence of the salts has increased the film height, so the mechanical stress (via crack formation) acts differently from the drops without salts. This observation did not change when we added LCs into the system. This means LCs might not increase the film height but only affect the packing of the particles in the system. In other words, the LC droplets are somehow trapped in the layer between lysozyme particles and salts in the central regions. The evaporation of a further volume of water at the later drying stage ($t/t_d \sim 0.7$–$0.8$) propagates the bursting and random distribution of these droplets (see Figure 12(BIV,BVI)). Unlike 0×, there is an additional salt layer on top of the

LC distribution. In contrast, the peripheral region of the lysozyme+PBS+LC drop mostly contains the lysozyme particles, which drive the LCs to fill these cracked domains. Since the LCs do not have enough volume to fulfill the cracked domains, and the water leaves the system by that time, the LCs stay where they are. The whole process thus provides a plausible explanation of the reduced birefringence intensity (I under crossed polarizing configuration) when the $\phi_{PBS}$ was increased despite having a fixed volume of LCs (see Figure 9I–IV).

## 5. Conclusions

This paper reveals that the tuning between lysozyme and salts is essential in determining the morphological patterns. It rasters the parameter space by varying the initial concentrations of both lysozyme and PBS salts. The findings of this paper thoroughly explain the hierarchical complexity of the multi-component system consisting of (i) PBS, (ii) PBS+LCs, (iii) lysozyme+PBS, and (iv) lysozyme+PBS+LCs at different initial concentrations. The textual evolution indicates that the interactions between different lysozyme particles during the drying process are dependent on the amount of the salts present in PBS. It also shows that the occurrence of three distinct regions is not the general characteristic of lysozyme-saline droplets; rather, it depends on the relative initial concentrations of lysozyme and salts. Different phase-separated zones reported in the BSA drops are also observed in some lysozyme samples. This implies that the phase separation of salts, and lysozyme particles in different structural forms, might not be specific to the individual protein characteristics; instead, this phase separation is associated with all the globular proteins. Different pattern-recognized phases established by the phase diagram also confirm that the hierarchical structures formed by the aggregation of the lysozyme and salts are not directly correlated with their initial concentrations. This study proposes a plausible mechanism supporting the optical behavior (birefringence) of the LCs that relates the morphological patterns at the macroscale to the different interactions at the microscale. The experimental findings of this study anticipate the redistribution of these LCs. This study, however, does not advocate a particular experimental protocol or technique that can exactly explore the complex interactions at the interface, viz., protein–protein interactions, or the distribution of LCs, etc. Nonetheless, this study motivates researchers to develop in situ experimental set-ups to examine and explore the clear-cut complex simultaneous events during such non-equilibrium processes in the near future.

**Supplementary Materials:** The following supporting information can be downloaded at: https://www.mdpi.com/article/10.3390/pr10050955/s1, Figure S1. Drying evolution of PBS drop at the initial concentration ($\phi_{PBS}$) of (I) 0x and (II) 0.25x. Figure S2. Drying evolution of PBS drop at the initial concentration ($\phi_{PBS}$) of (I) 0.5x, and (II) 0.75x. The scale bar of length 0:15 mm. Figure S3. Drying evolution of PBS drop at the initial concentration ($\phi_{PBS}$) of 1x. Table S1. Pairwise comparison of mean among all PBS concentrations ($\phi_{PBS}$) for the drops at a fixed lysozyme initial concentration. These drops do not contain any liquid crystals (LCs). Table S2. Pairwise comparison of SD among all PBS concentrations ($\phi_{PBS}$) for the drops at a fixed lysozyme initial concentration. These drops do not contain any liquid crystals (LCs). Table S3. Pairwise comparison of mean among all PBS concentrations ($\phi_{PBS}$) for the drops at a fixed lysozyme initial concentration. These drops contain liquid crystals (LCs). Table S4. Pairwise comparison of SD among all PBS concentrations ($\phi_{PBS}$) for the drops at a fixed lysozyme initial concentration. These drops contain liquid crystals (LCs).

**Author Contributions:** All the authors were actively involved in preparing the final manuscript. The experiments were designed and performed by A.P. The interpretation of the results was carried out by A.P and A.G. The images were drawn, quantified, and computed by A.G. and A.P. G.S.I. supervised the work. A.G. also edited the final version of the manuscript. All authors have read and agreed to the published version of the manuscript.

**Funding:** This work is supported by the Department of Physics at WPI, USA, and the University of Warwick, UK. This study was partially funded by Leverhulme Trust (Grant No. RPG-2018-345).

**Institutional Review Board Statement:** Not applicable.

**Informed Consent Statement:** Not applicable.

**Data Availability Statement:** Not applicable.

**Conflicts of Interest:** The authors declare no conflicts of interest.

## References


1. Iannacchione, G.; Pal, A. Bio-colloidal Drying Droplets: Current Trends and Future Perspectives on Image Processing Applications. *Acad. Lett.* **2021**, 2. [CrossRef]
2. Rapis, E. Self-assembly of cluster protein films (allotropic nonequilibrium noncrystalline modification) during the process or their condensation. *Tech. Phys.* **2000**, *45*, 121–131. [CrossRef]
3. Pal, A. Self-Assembly and Morphological Patterns in Drying Droplets of Bio-Colloids. Ph.D Thesis, Worcester Polytechnic Institute, Worcester, MA, USA, 2021.
4. Kaya, D.; Belyi, V.A.; Muthukumar, M. Pattern formation in drying droplets of polyelectrolyte and salt. *J. Chem. Phys.* **2010**, *133*, 114905. [CrossRef]
5. Ma, W.; Wang, Y. Effect of Salt Concentration on the Pattern Formation of Colloidal Suspension. *Phys. Procedia* **2012**, *24*, 122–126. [CrossRef]
6. Dutta, T.; Giri, A.; Choudhury, M.D.; Tarafdar, S. Experiment and simulation of multifractal growth of crystalline NaCl aggregates in aqueous gelatin medium. *Colloids Surf. A Physicochem. Eng. Asp.* **2013**, *432*, 127–131. [CrossRef]
7. Choudhury, M.D.; Dutta, T.; Tarafdar, S. Growth kinetics of NaCl crystals in a drying drop of gelatin: Transition from faceted to dendritic growth. *Soft Matter* **2015**, *11*, 6938–6947. [CrossRef] [PubMed]
8. Roy, B.; Choudhuri, M.D.; Dutta, T.; Tarafdar, S. Multi-scale patterns formed by sodium sulphate in a drying droplet of gelatin. *Appl. Surf. Sci.* **2015**, *357*, 1000–1006. [CrossRef]
9. Shahidzadeh, N.; Schut, M.F.; Desarnaud, J.; Prat, M.; Bonn, D. Salt stains from evaporating droplets. *Sci. Rep.* **2015**, *5*, srep10335. [CrossRef]
10. Msambwa, Y.; Shackleford, A.S.; Ouali, F.F.; Fairhurst, D.J. Controlling and characterising the deposits from polymer droplets containing microparticles and salt. *Eur. Phys. J. E* **2016**, *39*, 1–8. [CrossRef]
11. Qazi, M.J.; Liefferink, R.W.; Schlegel, S.J.; Backus, E.H.; Bonn, D.; Shahidzadeh, N. Influence of Surfactants on Sodium Chloride Crystallization in Confinement. *Langmuir* **2017**, *33*, 4260–4268. [CrossRef]
12. Zhong, X.; Ren, J.; Duan, F. Wettability Effect on Evaporation Dynamics and Crystalline Patterns of Sessile Saline Droplets. *J. Phys. Chem. B* **2017**, *121*, 7924–7933. [CrossRef]
13. Morinaga, K.; Oikawa, N.; Kurita, R. Emergence of different crystal morphologies using the coffee ring effect. *Sci. Rep.* **2018**, *8*, 12503. [CrossRef] [PubMed]
14. Basu, N.; Mukherjee, R. Morphology modulation in evaporative drying mediated crystallization of sodium chloride solution droplet with surfactant. *Soft Matter* **2018**, *14*, 7883–7893. [CrossRef] [PubMed]
15. Yakhno, T. Salt-induced protein phase transitions in drying drops. *J. Colloid Interface Sci.* **2008**, *318*, 225–230. [CrossRef] [PubMed]
16. Annarelli, C.; Fornazero, J.; Bert, J.; Colombani, J. Crack patterns in drying protein solution drops. *Eur. Phys. J. E* **2001**, *5*, 599–603. [CrossRef]
17. Gorr, H.M.; Zueger, J.M.; Barnard, J.A. Lysozyme pattern formation in evaporating drops. *Langmuir* **2012**, *28*, 4039–4042. [CrossRef]
18. Gorr, H.M.; Zueger, J.M.; McAdams, D.R.; Barnard, J.A. Salt-induced pattern formation in evaporating droplets of lysozyme solutions. *Colloids Surf. B* **2013**, *103*, 59–66. [CrossRef]
19. Gorr, H.M. Lysozyme Pattern Formation in Evaporating Droplets. Ph.D Thesis, University of Pittsburgh, Pittsburgh, PA, USA, 2013.
20. Chen, G.; Mohamed, G.J. Complex protein patterns formation via salt-induced self-assembly and droplet evaporation. *Eur. Phys. J. E* **2010**, *33*, 19–26. [CrossRef]
21. Carreón, Y.J.; Ríos-Ramírez, M.; Moctezuma, R.; González-Gutiérrez, J. Texture analysis of protein deposits produced by droplet evaporation. *Sci. Rep.* **2018**, *8*, 9580. [CrossRef]
22. Pal, A.; Gope, A.; Athair, A.S.; Iannacchione, G.S. A comparative study of the drying evolution and dried morphology of two globular proteins in de-ionized water solutions. *RSC Adv.* **2020**, *10*, 16906–16916. [CrossRef]
23. Pathak, B.; Christy, J.; Sefiane, K.; Gozuacik, D. Complex pattern formation in solutions of protein and mixed salts using dehydrating sessile droplets. *Langmuir* **2020**, *36*, 9728–9737. [CrossRef] [PubMed]
24. Carreón, Y.J.; Ríos-Ramírez, M.; Vázquez-Vergara, P.; Salinas-Almaguer, S.; Cipriano-Urbano, I.; Briones-Aranda, A.; Díaz-Hernández, O.; Santos, G.J.E.; González-Gutiérrez, J. Effects of substrate temperature on patterns produced by dried droplets of proteins. *Colloids Surf. B Biointerfaces* **2021**, *203*, 111763. [CrossRef] [PubMed]
25. Efstratiou, M.; Christy, J.; Bonn, D.; Sefiane, K. The Effect of Substrate Temperature on the Evaporative Behaviour and Desiccation Patterns of Foetal Bovine Serum Drops. *Colloids Interfaces* **2021**, *5*, 43. [CrossRef]
26. Pal, A.; Gope, A.; Iannacchione, G. Temperature and Concentration Dependence of Human Whole Blood and Protein Drying Droplets. *Biomolecules* **2021**, *11*, 231. [CrossRef] [PubMed]

27. Gorr, H.M.; Xiong, Z.; Barnard, J.A. Pattern recognition for identification of lysozyme droplet solution chemistry. *Colloids Surf. B Biointerfaces* **2014**, *115*, 170–175. [CrossRef]
28. Pal, A.; Gope, A.; Iannacchione, G.S. Statistical Image Analysis of Drying Bovine Serum Albumin Droplets in Phosphate Buffered Saline. In *Biomedical Data Mining for Information Retrieval*; John Wiley & Sons, Ltd.: Hoboken, NJ, USA.
29. Carreón, Y.J.; González-Gutiérrez, J.; Pérez-Camacho, M.; Mercado-Uribe, H. Patterns produced by dried droplets of protein binary mixtures suspended in water. *Colloids Surf. B Biointerfaces* **2018**, *161*, 103–110. [CrossRef]
30. Pal, A.; Gope, A.; Kafle, R.; Iannacchione, G.S. Phase separation of a nematic liquid crystal in the self-assembly of lysozyme in a drying aqueous solution drop. *MRS Commun.* **2019**, *9*, 150–158. [CrossRef]
31. Pal, A.; Gope, A.; Iannacchione, G.S. A comparative study of the phase separation of a nematic liquid crystal in the self-assembling drying protein drops. *MRS Adv.* **2019**, *4*, 1309–1314. [CrossRef]
32. Pal, A.; Gope, A.; Iannacchione, G.S. Image-based analysis of patterns formed in drying drops. In Proceedings of the Pattern Recognition and Machine Intelligence, Tezpur, India, 17–20 December 2019; Deka, B., Maji, P., Mitra, S., Bhattacharyya, D.K., Bora, P.K., Pal, S.K., Eds.; Springer International Publishing: Cham, Switzerland, 2019; pp. 567–574.
33. Abràmoff, M.D.; Magalhães, P.J.; Ram, S.J. Image processing with ImageJ. *Biophotonics Int.* **2004**, *11*, 36–42.
34. Deegan, R.D.; Bakajin, O.; Dupont, T.F.; Huber, G.; Nagel, S.R.; Witten, T.A. Capillary flow as the cause of ring stains from dried liquid drops. *Nature* **1997**, *389*, 827. [CrossRef]
35. Efstratiou, M.; Christy, J.R.E.; Bonn, D.; Sefiane, K. Transition from Dendritic to Cell-like Crystalline Structures in Drying Droplets of Fetal Bovine Serum under the Influence of Temperature. *Langmuir* **2022**, *38*, 4321–4331. [CrossRef] [PubMed]
36. Hintze, J.L.; Nelson, R.D. Violin plots: A box plot-density trace synergism. *Am. Stat.* **1998**, *52*, 181–184.
37. Roberts, D.; Keeling, R.; Tracka, M.; Van Der Walle, C.; Uddin, S.; Warwicker, J.; Curtis, R. The role of electrostatics in protein–protein interactions of a monoclonal antibody. *Mol. Pharm.* **2014**, *11*, 2475–2489. [CrossRef] [PubMed]
38. Liu, W.; Midya, J.; Kappl, M.; Butt, H.J.; Nikoubashman, A. Segregation in drying binary colloidal droplets. *ACS Nano* **2019**, *13*, 4972–4979. [CrossRef] [PubMed]

processes

MDPI

*Article*

# Experimental Study on a New Combined Gas–Liquid Separator

**Lei Ji, Qin Zhao *, Huiming Deng, Lanyue Zhang and Wanquan Deng**

Key Laboratory of Fluid and Power Machinery, Ministry of Education, Xihua University, Chengdu 610039, China; 0119910030@mail.xhu.edu.cn (L.J.); denghuiming@stu.xhu.edu.cn (H.D.); 212019080700005@stu.xhu.edu.cn (L.Z.); 0119960034@mail.xhu.edu.cn (W.D.)

* Correspondence: zhaoqin@mail.xhu.edu.cn

**Abstract:** Gas–liquid separation at natural gas wellheads has always been a key technical problem in the fields of natural gas transportation and storage. Developing a gas–liquid separation device that is both universal and highly efficient is the current challenge. A new type of combined gas–liquid separation device was designed in this study, and the efficiency of the separator was studied using a laser Doppler anemometer and phase Doppler particle analyzer at a flow rate of 10–60 $Nm^3/h$. The results showed that the separation efficiency of the combined separator was above 95% at each experimental flow rate, verifying the strong applicability of the combined separator. Moreover, the separation efficiency was as high as 99% at the flow rates of 10 and 60 $Nm^3/h$, thereby realizing efficient separation. This study is significant to the development of gas–liquid separation devices.

**Keywords:** natural gas; gas–liquid separation device; laser Doppler anemometer; phase Doppler particle analyzer

**Citation:** Ji, L.; Zhao, Q.; Deng, H.; Zhang, L.; Deng, W. Experimental Study on a New Combined Gas–Liquid Separator. *Processes* **2022**, *10*, 1416. https://doi.org/10.3390/pr10071416

Academic Editor: Zhenmeng Peng

Received: 10 June 2022
Accepted: 19 July 2022
Published: 20 July 2022



## 1. Introduction

With the rapid development of the global economy and industries, the increasing demand for energy and worsening environmental issues have become major concerns. Natural gas has received attention as a clean and efficient high-quality energy source, and its development and application have made significant advances [1,2]. In the industrial context of increased demand for natural gas, transportation and storage concerns have gained significant attention.

The natural gas coming out of the wellhead contains saturated water vapor and a small amount of hydrocarbons. Inside the pipeline, some components of natural gas react with free water, which leads to the formation of acidic substances and hydrate crystallization; this, in turn, causes corrosion and pipeline blockages that reduce the gas transmission efficiency, thus increasing gas supply instability and economic losses [3,4]. Removal of free water before the natural gas enters the pipeline has become a key issue in the subsequent stages of natural gas transportation [5,6]. At present, the traditional natural gas dehydration methods used are the solid adsorption, solvent absorption and low temperature condensation methods [7].

The solid adsorption method mainly involves the use of molecular sieves, silica gel, alumina and other substances with good absorption capacity for water as adsorbents. When natural gas is in contact with such substances, the water in the gas is absorbed to achieve a dehydration effect. However, this method has disadvantages, such as the associated high cost and the requirement of a large space [8]. The solvent absorption method mainly involves the use of absorbents with a high absorption capacity for water, and dehydration equipment can be designed according to the absorbent capacity of the absorbents. Triethylene glycol is the most widely used absorbent for achieving the required dehydration effect. However, triethylene glycol dehydration equipment has several components, high complexity, high maintenance cost and risk of environmental pollution with improper operation [9]. In the low-temperature condensation method, the temperature of natural gas is reduced through throttling expansion and external refrigerant heat exchange such that the water in it condenses out, thereby achieving gas–liquid separation. However, throttling expansion refrigeration is only suitable for high-pressure aqueous natural gas, which also has limitations, while the external refrigerant heat exchange refrigeration method uses complex devices and consumes a large amount of energy [10]. Therefore, traditional natural gas dehydration methods are limited. With the continuous development of technologies, various researchers have optimized and improved the traditional dehydration methods of natural gas and gradually derived new natural gas dehydration methods, such as the ionic liquid natural gas dehydration method [11] and supersonic separator dehydration method [12]. These various methods have advantages and disadvantages, are not versatile and the separation mechanism of most separation equipment is not yet clear [13]. Therefore, the investigation, development and fabrication of efficient, low-resistance, universal natural gas dehydration methods and equipment is important in the field of natural gas transportation and storage. Compared with other gas-liquid separators, cyclone separators have the advantages of small size and long maintenance intervals; thus, they have become the focus of research. Matsubayashi et al. [14] studied an aero-liquid cyclone separator in a boiling water reactor and concluded that the diameter of the hub does not affect the efficiency of the separator; however, reducing the leaf placement angle will reduce the separation efficiency. Yu et al. [15] studied the effect of droplet particle size on the efficiency of a cyclone separator using Fluent software, and concluded that as the droplet particle size increases, the separator efficiency increases, and when the particle size is greater than 10 μm, the separation efficiency remains fixed. Han et al. [16] applied numerical calculations to study the flow field pressure, velocity and separation efficiency of a cyclone separator and concluded that with an increase in the blade envelope arc, the pressure, speed and separation efficiency of the internal flow field fluctuated greatly; they proposed that the blade envelope should be set in the range of 32–44 mm.

However, although several studies have been conducted on cyclone–liquid separators, the problem of low efficiency of a cyclone separator under small flow conditions is still not resolved. To improve the efficiency and applicability of the gas–liquid separator, this study optimized a design based on the cyclone separator. In this study, a combined natural gas dewatering device was designed, and the dewatering effect of the combined separator was measured using a laser Doppler anemometer (LDA) and phase Doppler particle analyzer (PDPA). This study has great research significance in the fields of natural gas transportation, storage and gas–liquid separation.

## 2. Experimental Principle and Process

The components of the combined separator designed in this study include the cyclone, steady flow, leaf grid and folding plate elements, as shown in Figure 1. The cyclone element is a preliminary separation component that relies on twisted blades to form a swirling flow of gas–liquid mixed fluid and the centrifugal action is applied to achieve the effect of gas–liquid separation [17]. The structure is similar to the internal flow field of a hydraulic mechanical transfer wheel chamber; after the fluid passes through the twisted blade, the flow state is twisted from the layer flow to turbulence, and the flow is extremely unstable [18,19]. This unstable flow can negatively impact the subsequent flow measurement if left untreated [20,21]. Therefore, a steady flow element is added downstream of the cyclone element to smooth the flow state of the strong turbulent fluid and have a certain separation effect on the droplets. Downstream of the steady flow element, the leaf grid and folding plate elements are installed sequentially and separated from the liquid contained in the natural gas to improve the separation efficiency according to the principle of inertial separation.

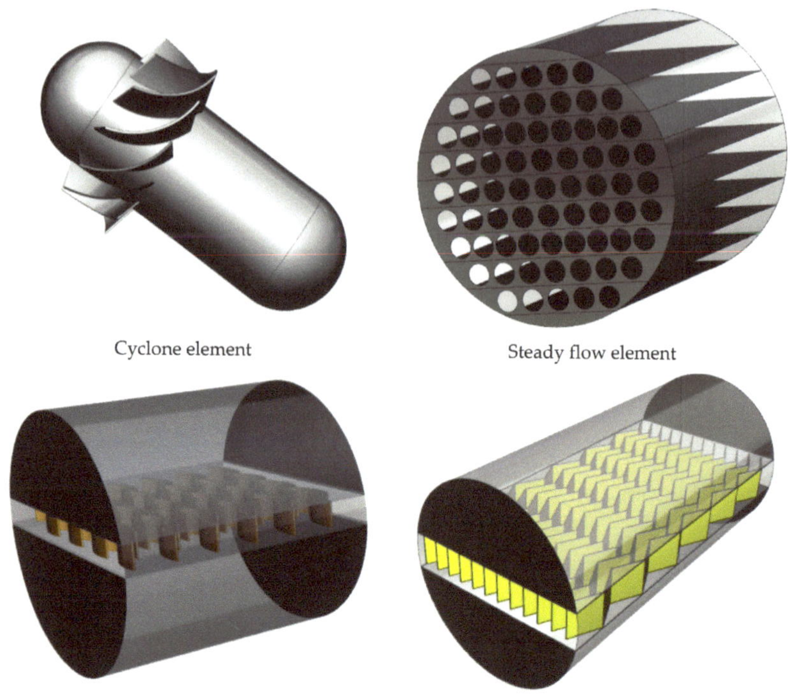


**Figure 1.** Schematic of the separate components.

LDA is an application of the optical Doppler effect that relies on the frequency difference between the scattered light of moving particles and the irradiated light to obtain speed information and determine the particle size by analyzing the phase difference of the scattered light reflected or refracted by spherical particles passing through the measuring body of the laser [22,23]. Due to its laser measurement, the lack of interference from the convection field, wide range of velocity measurements and high accuracy, [24,25], it is now widely used in fluid flow rate measurements [26–29]. In this study, LDA and PDPA equipment were used to measure the liquid content in natural gas, and the equipment model is listed in Table 1.

**Table 1.** Basic device parameters.

| Name | Type Specification |
|---|---|
| Integrated argon-ion laser | LA70-5 |
| Beam splitter | FBL-3 fiberlight$^{TM}$ |
| 2D fiber optic emission probe | TM250 |
| Fiber optic receive probe | RV3070 PDPA |
| Three-channel photodetector assembly | PDM1000-3P |

To ensure the safety of the experiment, air was used as the experimental gas. Figure 2 presents a schematic of the experimental platform. The air is compressed by the air compressor (Figure 3) and enters the experimental apparatus from the inlet pipeline. Figure 4 shows the atomization system, which atomizes the liquid water and injects it into the experimental pipeline through the nozzle to mix with the air and form a gas–liquid two-phase flow. The atomized water particle size ($W_{1d}$) and atomized water content ($W_{1c}$) in the gas–liquid two-phase flow at monitoring point 1 ($M_1$) are measured using LDA and

PDPA devices. After performing a series of gas–liquid separation processes through the cyclone, steady flow, leaf grid and folding plate elements, the LDA and PDPA devices were used at monitoring point 2 ($M_2$) to measure the atomized water particle size ($W_{2d}$) and content ($W_{2c}$) in the gas–liquid two-phase flow through the combined separator. The separation efficiency was calculated using Equation (1). Figure 5 presents an image of the field experimental site.

$$\eta = \frac{W_{1c} - W_{2c}}{W_{1c}} \quad (1)$$

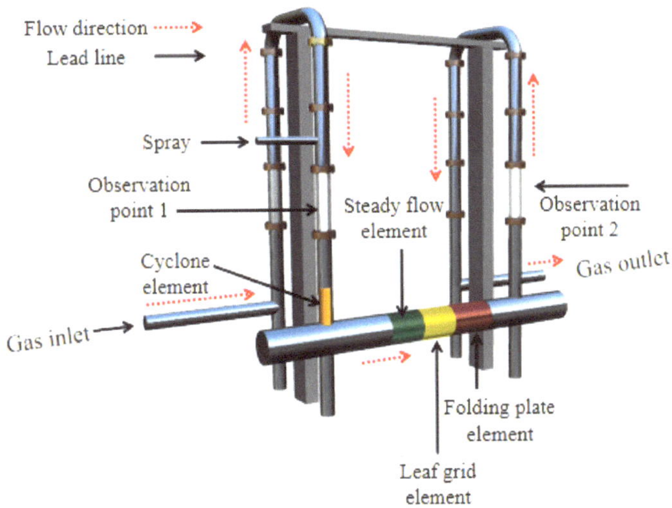


**Figure 2.** Experimental schematic.

**Figure 3.** Air compressor.

**Figure 4.** Atomization generator.

**Figure 5.** Experimental site.

During the experiment, the pipeline pressure was maintained at 0.055 MPa, and the pipeline temperature was 20 °C. Four sets of comparative cases were designed to explore the influence of each element on the liquid separation effect under different flow conditions, as shown in Table 2. Each case was measured in six flow conditions (10, 20, 30, 40, 50 and 60 $Nm^3/h$). During the preliminary experiments, we found that if the spray flow is large, the liquid will form a femoral liquid, and the separation efficiency of various cases in this state is larger, reducing the significance of the study and deviating from the actual working conditions. After repeated debugging, the optimal spray flow rate was determined as 10.44 L/h. At this flow rate, the liquid forms a mist.

**Table 2.** Experimental case table.

| Case | Cyclone Element | Steady Flow Element | Leaf Grid Element | Folding Plate Element |
|---|---|---|---|---|
| 1 | √ | | | |
| 2 | √ | | √ | |
| 3 | √ | √ | √ | |
| 4 | √ | √ | √ | √ |

## 3. Experimental Results and Analysis

### *3.1. Study on the Separation Effect of Cyclone Elements*

To reduce the number of experimental errors, four replicates of the experiment per operating condition were conducted, and the average of the four experiments was taken as the result. Tables 3–6 list the measurement results of the liquid particle size (effective particle size D10) and the liquid content measurements at $M_1$ and $M_2$ under different flow conditions in Case 1. Figure 6 presents the $M_1$ and $M_2$ liquid particle size and content comparative analysis chart. As can be seen in the figure, the impact of the air flow on the atomized water increased as the air flow increased, affecting the diffusion of the atomized water. An increase in air flow led to an increase in the droplet particle size at $M_1$. For example, the particle size at $M_1$ was 21.15 μm at the flow rate of 10 $Nm^3/h$, and the particle size at $M_1$ was 24.85 μm at 60 $Nm^3/h$ (particle size increase of 3.7 μm).

**Table 3.** Liquid particle size in Case 1 at $M_1$ (μm).

| Number of Experiments | 10 $Nm^3/h$ | 20 $Nm^3/h$ | 30 $Nm^3/h$ | 40 $Nm^3/h$ | 50 $Nm^3/h$ | 60 $Nm^3/h$ |
|---|---|---|---|---|---|---|
| 1 | 21.03 | 21.51 | 21.63 | 22.67 | 23.00 | 24.90 |
| 2 | 21.15 | 21.36 | 21.96 | 22.82 | 23.34 | 24.78 |
| 3 | 21.30 | 21.48 | 21.9 | 22.4 | 23.15 | 24.84 |
| 4 | 21.13 | 21.25 | 22.02 | 22.58 | 23.42 | 24.89 |
| Average value | 21.15 | 21.40 | 21.88 | 22.62 | 23.23 | 24.85 |

**Table 4.** Liquid content in Case 1 at $M_1$ ($g/m^3$).

| Number of Experiments | 10 $Nm^3/h$ | 20 $Nm^3/h$ | 30 $Nm^3/h$ | 40 $Nm^3/h$ | 50 $Nm^3/h$ | 60 $Nm^3/h$ |
|---|---|---|---|---|---|---|
| 1 | 957.94 | 406.47 | 255.08 | 215.31 | 155.55 | 133.55 |
| 2 | 960.54 | 435.11 | 249.36 | 225.74 | 164.27 | 128.68 |
| 3 | 955.60 | 427.70 | 265.77 | 205.05 | 152.68 | 125.44 |
| 4 | 948.91 | 419.74 | 249.82 | 218.85 | 159.71 | 123.41 |
| Average value | 955.75 | 422.26 | 255.01 | 216.24 | 158.05 | 127.77 |

**Table 5.** Liquid particle size in Case 1 at $M_2$ (μm).

| Number of Experiments | 10 $Nm^3/h$ | 20 $Nm^3/h$ | 30 $Nm^3/h$ | 40 $Nm^3/h$ | 50 $Nm^3/h$ | 60 $Nm^3/h$ |
|---|---|---|---|---|---|---|
| 1 | 14.00 | 14.47 | 14.41 | 14.55 | 14.34 | 13.92 |
| 2 | 13.95 | 14.57 | 14.51 | 14.41 | 14.36 | 13.92 |
| 3 | 14.14 | 14.46 | 14.44 | 14.51 | 14.37 | 13.87 |
| 4 | 14.01 | 14.44 | 14.47 | 14.42 | 14.36 | 13.99 |
| Average value | 14.03 | 14.49 | 14.46 | 14.47 | 14.36 | 13.93 |

**Table 6.** Liquid content in Case 1 at $M_2$ ($g/m^3$).

| Number of Experiments | 10 $Nm^3/h$ | 20 $Nm^3/h$ | 30 $Nm^3/h$ | 40 $Nm^3/h$ | 50 $Nm^3/h$ | 60 $Nm^3/h$ |
|---|---|---|---|---|---|---|
| 1 | 187.89 | 96.65 | 48.53 | 39.94 | 19.84 | 6.64 |
| 2 | 191.19 | 98.24 | 48.88 | 40.72 | 19.91 | 7.29 |
| 3 | 193.84 | 96.78 | 48.73 | 39.68 | 19.48 | 6.75 |
| 4 | 189.55 | 96.28 | 48.72 | 39.88 | 19.50 | 7.22 |
| Average value | 190.62 | 96.99 | 48.72 | 40.01 | 19.68 | 6.98 |

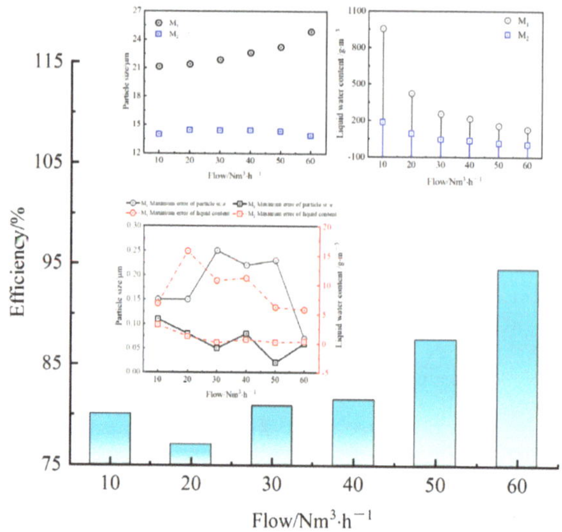


**Figure 6.** Comparison of liquid particle size and content in Case 1 at $M_1$.

The atomized water content was significantly reduced as the air flow increased. The main reason for this phenomenon is because the spray flow rate is constant at 10.44 L/h, and the total flow through $M_1$ rises per unit time when the air flow increases, resulting in the decrease in the relative content of the liquid. After passing through the cyclone element, the liquid particle size decreased from 21.15, 21.40, 22.62, 23.23, 23.23, 24.85 and 24.85 μm to 14.03, 14.49, 14.46, 14.47, 14.36 and 13.93 μm, respectively; the liquid water content decreased from 955.75, 422.26, 255.01, 216.24, 158.05 and 127.55 g/m$^3$ to 190.62, 96.99, 48.72, 40.01, 19.68 and 6.98 g/m$^3$, respectively, indicating that the liquid particle size decreased significantly. Equation (1) was used to calculate the separation efficiency in 10, 20, 30 and 40 Nm$^3$/h flow conditions, which was approximately 80%. Moreover, when the flow rate was higher than 50 Nm$^3$/h, the separation efficiency significantly increased, up to approximately 95% at 60 Nm$^3$/h. The main reason for the increase in the separation efficiency at large flow conditions is that when the flow rate increases, the flow rate of the cyclone element and the intensity of the centrifugal movement increase, resulting in the separation of more droplets. From the experimental results of Case 1, the cyclone element has a good separation effect at a flow rate higher than 50 Nm$^3$/h, but has a lower separation efficiency under small flow conditions.

Additionally, it can be seen from Figure 6 that there are fluctuations in the four measurement results of particle size and liquid content in each working condition. The main reason for the fluctuation is the random error of the experimental measurement, because in the unsteady flow process of the flow field, the flow in the flow field fluctuates. This state is a function of time and the experimental structure is complex. The fluid flow state exhibits turbulent motion and the fluid itself has pulsation values of a random nature. However, from the analysis of the error value, it can be concluded that the measurement error of the experiment was less than 3% (the difference between the average and maximum values of the experimental result was higher than the average value), and the experiment can be considered repeatable.

### 3.2. *Studying the Separation Effect of Cyclone Element and Leaf Grid Element Combination*

Tables 7–10 show the measurement results of the liquid particle size and content at $M_1$ and $M_2$ under different flow conditions in Case 2. Based on the comparison of Tables 3 and 7 and Tables 4 and 8, due to the differences between cases in the downstream of $M_1$, after changing the combination method, the liquid particle size and content at $M_1$

had minimal impact, and some gaps in the data were caused by uncontrollable factors, such as pipeline vibration during the experiment. After the addition of the leaf grid element, the liquid particle size of each flow condition at $M_2$ was not apparent, but the liquid content in the gas–liquid two-phase flow was reduced further than that in Case 1 (Tables 9 and 10). Figure 7 shows the comparison of the separation efficiency of Cases 1 and 2. The measurement error of the four replicated experiments was within 3%, which is considered reproducible. After increasing the leaf grid element, when the gas flow rate was 10, 20, 30, 40, 50 and 60 $Nm^3/h$, the separation efficiency increased from the previous values of 80.06%, 77.03%, 80.89%, 81.50%, 87.55% and 94.54% to 85.24%, 82.04%, 85.32%, 87.14%, 93.11% and 97.85%, respectively. Therefore, with the addition of the leaf grid element, the separation efficiency of the combined separator was significantly improved when the gas flow rate was lower than 60 $Nm^3/h$. Under the action of strong centrifugation, the separation efficiency of the cyclone element was already high when the flow rate was 60 $Nm^3/h$. Therefore, adding a leaf grid element has no apparent effect on the improvement of the separation efficiency under large flow conditions.

**Table 7.** Liquid particle size in Case 2 at $M_1$ (μm).

| Number of Experiments | 10 $Nm^3/h$ | 20 $Nm^3/h$ | 30 $Nm^3/h$ | 40 $Nm^3/h$ | 50 $Nm^3/h$ | 60 $Nm^3/h$ |
|---|---|---|---|---|---|---|
| 1 | 21.98 | 21.37 | 21.9 | 22.13 | 23.27 | 24.66 |
| 2 | 21.40 | 21.46 | 22.01 | 21.8 | 23.38 | 24.67 |
| 3 | 20.95 | 21.56 | 22.16 | 21.75 | 22.92 | 24.67 |
| 4 | 21.36 | 21.47 | 21.95 | 21.92 | 23 | 24.58 |
| Average value | 21.42 | 21.47 | 22.00 | 21.90 | 23.14 | 24.65 |

**Table 8.** Liquid content in Case 2 at $M_1$ ($g/m^3$).

| Number of Experiments | 10 $Nm^3/h$ | 20 $Nm^3/h$ | 30 $Nm^3/h$ | 40 $Nm^3/h$ | 50 $Nm^3/h$ | 60 $Nm^3/h$ |
|---|---|---|---|---|---|---|
| 1 | 928.74 | 406.47 | 260.94 | 213.95 | 167.14 | 128.94 |
| 2 | 955.75 | 435.11 | 277.50 | 209.79 | 160.67 | 123.94 |
| 3 | 952.69 | 427.70 | 262.58 | 214.11 | 165.74 | 133.40 |
| 4 | 951.62 | 419.74 | 245.34 | 214.52 | 166.27 | 130.76 |
| Average value | 947.2 | 422.255 | 261.59 | 213.0925 | 164.955 | 129.26 |

**Table 9.** Liquid particle size in Case 2 at $M_2$ (μm).

| Number of Experiments | 10 $Nm^3/h$ | 20 $Nm^3/h$ | 30 $Nm^3/h$ | 40 $Nm^3/h$ | 50 $Nm^3/h$ | 60 $Nm^3/h$ |
|---|---|---|---|---|---|---|
| 1 | 13.77 | 14.03 | 14.04 | 14.11 | 14.23 | 13.30 |
| 2 | 13.82 | 13.96 | 14.06 | 14.19 | 14.05 | 13.38 |
| 3 | 13.84 | 14.00 | 14.07 | 14.10 | 14.15 | 13.30 |
| 4 | 13.80 | 14.02 | 14.00 | 14.13 | 14.11 | 13.36 |
| Average value | 13.81 | 14.00 | 14.04 | 14.13 | 14.14 | 13.34 |

**Table 10.** Liquid content in Case 2 at $M_2$ ($g/m^3$).

| Number of Experiments | 10 $Nm^3/h$ | 20 $Nm^3/h$ | 30 $Nm^3/h$ | 40 $Nm^3/h$ | 50 $Nm^3/h$ | 60 $Nm^3/h$ |
|---|---|---|---|---|---|---|
| 1 | 139.77 | 75.26 | 39.14 | 27.39 | 11.72 | 2.86 |
| 2 | 140.21 | 76.86 | 38.42 | 27.68 | 11.61 | 2.74 |
| 3 | 140.41 | 76.63 | 38.16 | 27.2 | 11.51 | 2.87 |
| 4 | 138.84 | 74.57 | 37.93 | 27.38 | 10.62 | 2.65 |
| Average value | 139.81 | 75.83 | 38.41 | 27.4125 | 11.365 | 2.78 |

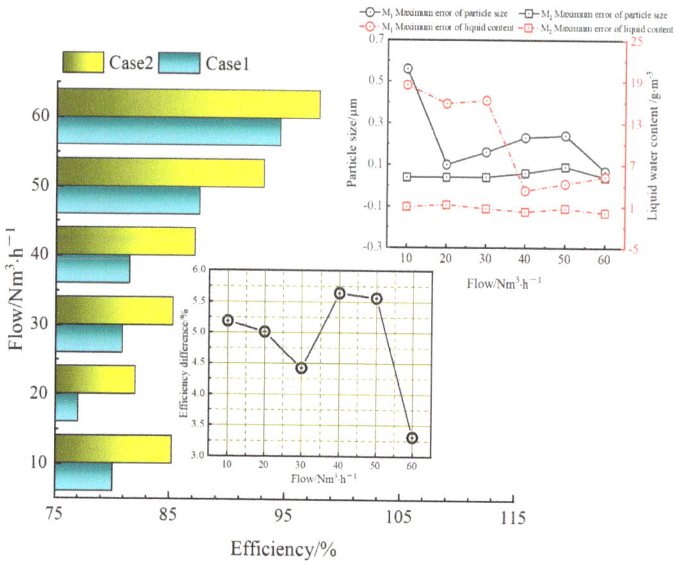


**Figure 7.** Comparison of separation effect between Cases 1 and 2.

### *3.3. Studying the Separation Effect of Cyclone, Leaf Grid and Steady Flow Element Combination*

The measurement results of liquid particle size and content at $M_1$ and $M_2$ under different flow conditions in Case 3 are listed in Tables 11–14. In Table 13, after the gas–liquid two-phase flow through the cyclone, leaf grid and steady flow elements, the liquid particle size was approximately 13 μm, which was slightly reduced compared to those of Cases 1 and 2. Based on the data in the table, Figure 8 presents the comparison of the separation efficiencies of Cases 2 and 3, and the experiment can be considered repeatable if the data fluctuation of the four repeated experiments is within 3%. With the addition of the steady flow element, the separation efficiency of each experimental condition was higher than 90% and the separation efficiency increased by nearly 12% compared to that of Case 2 under a 10 $Nm^3/h$ flow. The reason for this effect can be explained by the two-phase fluid flow characteristics inside the pipeline. The flow presents a strong nonlinearity when the fluid flows through the cyclone element, whereas the steady flow element is an orifice plate structure. When the turbulent fluid hits the steady flow element, some droplets hang on its surface due to the sudden reduction in the overcurrent section, resulting in a sharp contraction of the fluid volume flowing through the steady flow element. The precipitated part of the liquid remains inside the steady flow element, thus improving the separation efficiency. In a large flow working condition, the cyclone element separates most of the liquid; thus, the separation efficiency of the steady flow element for the large flow working condition does not provide significant improvement. However, the addition of a steady flow element greatly increases the separation efficiency of the combined separator under small flow conditions and enhances the applicability of the combined separator.

**Table 11.** Liquid particle size in Case 3 at $M_1$ (μm).

| Number of Experiments | 10 $Nm^3/h$ | 20 $Nm^3/h$ | 30 $Nm^3/h$ | 40 $Nm^3/h$ | 50 $Nm^3/h$ | 60 $Nm^3/h$ |
|---|---|---|---|---|---|---|
| 1 | 21.47 | 21.69 | 21.98 | 21.41 | 23.16 | 24.59 |
| 2 | 21.72 | 21.36 | 21.92 | 21.47 | 22.93 | 24.81 |
| 3 | 21.99 | 21.58 | 21.97 | 22.12 | 23.28 | 24.59 |
| 4 | 21.72 | 21.42 | 21.83 | 21.84 | 22.94 | 24.63 |
| Average value | 21.73 | 21.51 | 21.93 | 21.71 | 23.08 | 24.66 |

**Table 12.** Liquid content in Case 3 at $M_1$ $(g/m^3)$.

| Number of Experiments | 10 $Nm^3/h$ | 20 $Nm^3/h$ | 30 $Nm^3/h$ | 40 $Nm^3/h$ | 50 $Nm^3/h$ | 60 $Nm^3/h$ |
|---|---|---|---|---|---|---|
| 1 | 959.73 | 401.55 | 259.71 | 212.58 | 156.94 | 133.02 |
| 2 | 992.53 | 438.63 | 259.53 | 213.76 | 161.24 | 127.89 |
| 3 | 999.12 | 414.71 | 248.1 | 213.84 | 163.34 | 132.68 |
| 4 | 949.48 | 414.35 | 247.58 | 213.07 | 159.43 | 127.24 |
| Average value | 975.21 | 417.31 | 253.73 | 213.31 | 160.24 | 130.21 |

**Table 13.** Liquid particle size in Case 3 at $M_2$ (μm).

| Number of Experiments | 10 $Nm^3/h$ | 20 $Nm^3/h$ | 30 $Nm^3/h$ | 40 $Nm^3/h$ | 50 $Nm^3/h$ | 60 $Nm^3/h$ |
|---|---|---|---|---|---|---|
| 1 | 13.06 | 13.21 | 13.33 | 13.69 | 13.70 | 13.47 |
| 2 | 13.15 | 13.24 | 13.25 | 13.66 | 13.66 | 13.48 |
| 3 | 13.12 | 13.16 | 13.39 | 13.66 | 13.61 | 13.48 |
| 4 | 13.19 | 13.20 | 13.43 | 13.63 | 13.70 | 13.40 |
| Average value | 13.13 | 13.20 | 13.35 | 13.66 | 13.67 | 13.46 |

**Table 14.** Liquid content in Case 3 at $M_2$ $(g/m^3)$.

| Number of Experiments | 10 $Nm^3/h$ | 20 $Nm^3/h$ | 30 $Nm^3/h$ | 40 $Nm^3/h$ | 50 $Nm^3/h$ | 60 $Nm^3/h$ |
|---|---|---|---|---|---|---|
| 1 | 28.02 | 35.35 | 21.56 | 13.78 | 7.42 | 1.90 |
| 2 | 27.27 | 35.17 | 21.30 | 13.89 | 7.55 | 1.93 |
| 3 | 27.25 | 35.56 | 21.12 | 14.08 | 7.49 | 2.00 |
| 4 | 27.62 | 35.43 | 21.68 | 13.68 | 7.44 | 1.89 |
| Average value | 27.54 | 35.38 | 21.42 | 13.86 | 7.48 | 1.93 |

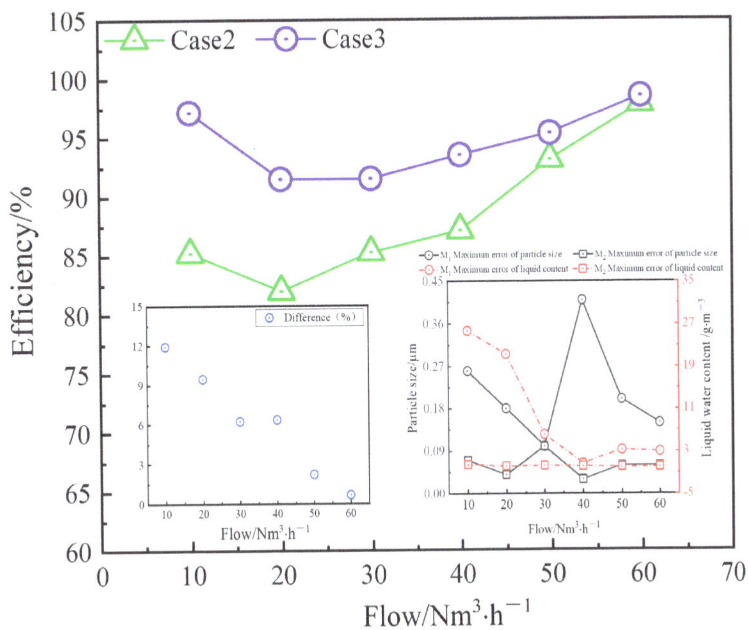


**Figure 8.** Comparison of separation efficiency between Cases 3 and 2.

### 3.4. Studying the Combined Separation Effect of Cyclone, Steady Flow, Leaf Grid and Folding Plate Elements

Tables 15–18 list the measurement results of the liquid particle size and content at $M_1$ and $M_2$ under different flow conditions in Case 4. Based on the data presented in the table, when the gas–liquid two-phase flow passed through the cyclone, steady flow, leaf grid and folding plate elements, the liquid particle size did not change significantly and the particle size under each flow condition was approximately 13 μm. Figure 9 shows the comparison of the efficiency of the four cases. The separation efficiency of Case 4 is higher than that of the other cases, reaching over 95% under each flow condition and as high as 99% at flow rates of 10 and 60 $Nm^3/h$. Due to the addition of the folding plate element, a separation process is added in Case 4, but this only improves the separation efficiency by 0.55% compared to that of Case 3 at the flow condition of 60 $Nm^3/h$; thus, its effect is not apparent. In large flow working conditions, the centrifugal separation generated by the cyclone element increased the separation efficiency to 94.54%. Combined with the cyclone, steady flow and leaf grid elements, the liquid separation effect reached over 98%. Thus, with the addition of the folding plate element, the separation efficiency is somewhat increased. However, when the flow rate is less than 60 $Nm^3/h$, due to the low degree of centrifugation of the cyclone element, the moisture cannot be fully removed. Downstream of the cyclone element, the dehydration effect of the steady flow and leaf grid elements is slightly reduced due to the small turbulence intensity of the gas–liquid two-phase flow. Therefore, this new separation process must be added to further improve the separation efficiency. The results show that under a flow condition of approximately 10–50 $Nm^3/h$, the separation efficiency increased by more than 2% due to the addition of a folding plate element, which provides optimal conditions for efficient separation.

**Table 15.** Liquid particle size in Case 4 at $M_1$ (μm).

| Number of Experiments | 10 $Nm^3/h$ | 20 $Nm^3/h$ | 30 $Nm^3/h$ | 40 $Nm^3/h$ | 50 $Nm^3/h$ | 60 $Nm^3/h$ |
|---|---|---|---|---|---|---|
| 1 | 21.12 | 21.24 | 22.19 | 21.32 | 23.23 | 24.80 |
| 2 | 21.56 | 20.94 | 22.02 | 21.8 | 23.08 | 24.66 |
| 3 | 21.72 | 21.15 | 21.72 | 21.62 | 23.38 | 25.00 |
| 4 | 21.86 | 21.3 | 21.8 | 21.87 | 23.15 | 24.95 |
| Average value | 21.57 | 21.16 | 21.93 | 21.65 | 23.21 | 24.85 |

**Table 16.** Liquid content in Case 4 at $M_1$ ($g/m^3$).

| Number of Experiments | 10 $Nm^3/h$ | 20 $Nm^3/h$ | 30 $Nm^3/h$ | 40 $Nm^3/h$ | 50 $Nm^3/h$ | 60 $Nm^3/h$ |
|---|---|---|---|---|---|---|
| 1 | 910.25 | 432.61 | 245.39 | 217.66 | 163.86 | 131.1 |
| 2 | 988.53 | 422.17 | 272.76 | 202.71 | 168.93 | 131.23 |
| 3 | 932.43 | 405.27 | 253.41 | 206.68 | 163.37 | 133.3 |
| 4 | 990.02 | 433.8 | 250.74 | 209.74 | 160.74 | 132.36 |
| Average value | 955.31 | 423.46 | 255.58 | 209.20 | 164.22 | 132.00 |

**Table 17.** Liquid particle size in Case 4 at $M_2$ (μm).

| Number of Experiments | 10 $Nm^3/h$ | 20 $Nm^3/h$ | 30 $Nm^3/h$ | 40 $Nm^3/h$ | 50 $Nm^3/h$ | 60 $Nm^3/h$ |
|---|---|---|---|---|---|---|
| 1 | 12.73 | 13.34 | 13.8 | 13.91 | 13.99 | 13.75 |
| 2 | 12.83 | 13.37 | 13.81 | 14.02 | 13.98 | 13.77 |
| 3 | 12.68 | 13.31 | 13.75 | 13.98 | 13.95 | 13.68 |
| 4 | 12.67 | 13.37 | 13.37 | 13.92 | 13.87 | 13.84 |
| Average value | 12.73 | 13.35 | 13.68 | 13.96 | 13.95 | 13.76 |

**Table 18.** Liquid content in Case 4 at $M_2$ ($g/m^3$).

| Number of Experiments | 10 $Nm^3/h$ | 20 $Nm^3/h$ | 30 $Nm^3/h$ | 40 $Nm^3/h$ | 50 $Nm^3/h$ | 60 $Nm^3/h$ |
|---|---|---|---|---|---|---|
| 1 | 8.75 | 15.35 | 12.23 | 8.62 | 3.81 | 1.19 |
| 2 | 7.38 | 15.41 | 12.56 | 8.27 | 3.83 | 1.30 |
| 3 | 5.86 | 15.14 | 12.18 | 8.22 | 4.34 | 1.25 |
| 4 | 5.17 | 15.15 | 12.55 | 8.40 | 4.11 | 1.18 |
| Average value | 6.79 | 15.26 | 12.38 | 8.38 | 4.02 | 1.23 |

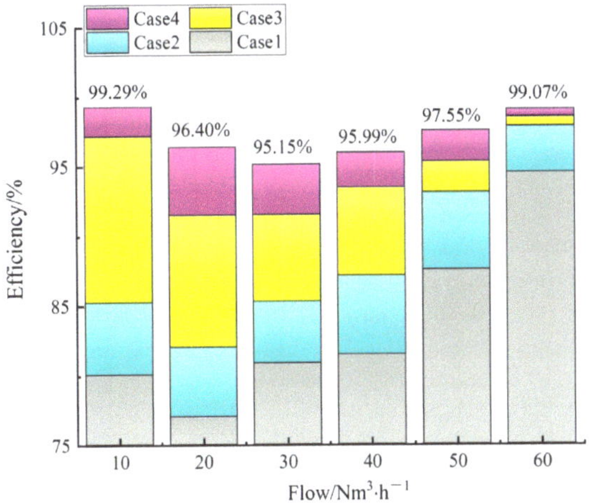


**Figure 9.** Comparison of the separation efficiency in the four cases.

Table 19 shows a comparison between the combined separator and traditional separators. While the separation efficiency of the combined separator is similar to that of the other separators, the combined separator has a wider range of applicability (large and small flows have higher separation efficiencies) and occupies a smaller space. Figure 10 shows the error plot of the four replications in Case 4, and the error result is still within 3%, which verifies that the reproducibility of the experiment is high.

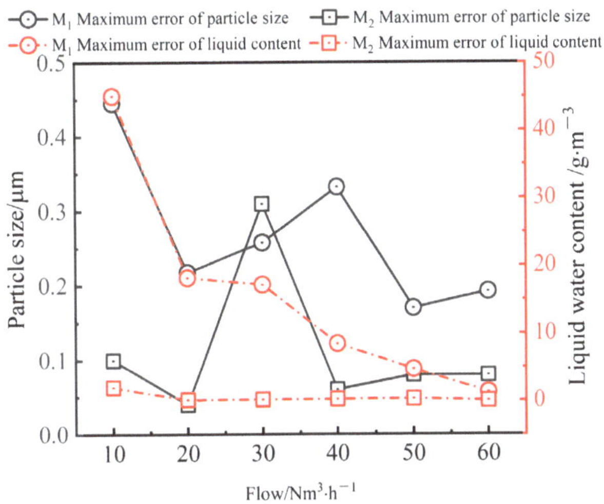


**Figure 10.** Error diagram for four repetitions.

**Table 19.** Comparative analysis of the applicability of all separators.

| Name | Suitable Conditions | Separation Efficiency |
| --- | --- | --- |
| Gravity separator | Gas–liquid separation of droplet size from 60 to 100 μm | Low efficiency, generally used for primary separation |
| Multi-tubular cyclone separator | Heavy loads | 40–50 μm droplets: >98%<br>5–10 μm droplets: >90% |
| Inertial gas–liquid separator | Large particle sizes | Low when particle size is <25 μm |
| Combined separator | Wide application range (suitable for large and small flow rates) | >95% |

## 4. Conclusions

In this study, a new combined separator was designed. The separation assembly includes the cyclone, steady flow, leaf grid and folding plate elements, and the experimental study of different combinations at approximately 10–60 $Nm^3/h$ flow conditions was conducted using LDA and PDPA equipment. The main conclusions of this study are as follows.

(1) When the combined separator only relied on the cyclone element for gas–liquid separation, the separation efficiency was approximately 80% at a flow rate of 10–50 $Nm^3/h$ and approximately 95% at a flow rate of 60 $Nm^3/h$. Thus, this method is only suitable for large flow conditions and its applicability is low.

(2) The gas–liquid separation efficiency under the flow conditions of 10, 20, 30, 40 and 50 $Nm^3/h$ increased by 5.18%, 5.01%, 4.43%, 5.64% and 5.56%, respectively, when the combined mode comprised the cyclone and leaf grid elements. Therefore, the addition of a leaf grid component significantly improves the efficiency of the combined separator under small flow conditions.

(3) When the combined separator comprised the cyclone, steady flow and leaf grid elements, the separation efficiency increased by approximately 12% at a flow rate of 10 $Nm^3/h$. Moreover, the separation efficiency under the working conditions of 20, 30, 40, 50 and 60 $Nm^3/h$ also significantly increased. The separation efficiency of the combined separator was higher than 90% at flow rates of 10–60 $Nm^3/h$, which enhances the applicability of the combined separator.

(4) When the combined separator included the cyclone, steady flow, leaf grid and folding plate elements, the separation efficiency was higher than 95% when the flow rate was in the 10–60 $Nm^3/h$ range, and the separation efficiency exceeded 99% at flow rates of 10 and 60 $Nm^3/h$, indicating that the separator has an efficient separation effect.

**Author Contributions:** Conceptualization, Methodology and Software, L.J.; Data Curation, Writing and Original Draft Preparation, Q.Z.; Visualization and Investigation, H.D.; Supervision, L.Z.; Software and Validation, W.D. All authors have read and agreed to the published version of the manuscript.

**Funding:** This research was funded by the National Key Research and Development Program "Research and Application Demonstration of Complementary Combined Power Generation Technology between Distributed Photovoltaic and Cascade Small Hydropower" (Grant No. 2018YFB0905200); Science and technology research project of Education Department of Jiangxi Province (Grant No. GJJ211941).

**Data Availability Statement:** Not applicable.

**Acknowledgments:** The authors express their gratitude to Deng for providing language help and for his assistance during the experiments.

**Conflicts of Interest:** The authors do not have any conflict of interest to declare.

## References


1. Economides, M.J.; Wood, D.A. The state of natural gas. *J. Nat. Gas Sci. Eng.* **2009**, *1*, 1–13. [CrossRef]
2. Qiu, Y.; Chen, H.; Li, W.; Wu, F.; Li, Z. Optimization of the tracer particle addition method for piv flowmeters. *Processes* **2021**, *9*, 1614. [CrossRef]
3. Ji, L.; Wang, X.; Li, Z.; Li, R. Experimental study on the performance of nw gas-liquid separator of natural gas. *J. Eng. Therm. Energy Power* **2020**, *35*, 116–121.
4. Santos, K.M.C.; Menezes, T.R.; Oliveira, M.R.; Silva, T.S.L.; Santos, K.S.; Barros, V.A.; Melo, D.C.; Ramos, A.L.; Santana, C.C.; Franceschi, E.; et al. Natural gas dehydration by adsorption using MOFs and silicas: A review. *Sep. Purif. Technol.* **2021**, *276*, 119409. [CrossRef]
5. Bahraminia, S.; Anbia, M.; Koohsaryan, E. Dehydration of natural gas and biogas streams using solid desiccants: A review. *Front. Chem. Sci. Eng.* **2021**, *15*, 1050–1074. [CrossRef]
6. Ma, W.; Zhang, Y.; Li, G.; Luo, J.; Chen, Z.; Zhang, H.; Zhao, X. Development status and trend of natural gas dehydration technique at home and abroad. *Pipeline Tech. Equip.* **2011**, *6*, 49–51.
7. Netusil, M.; Ditl, P. Comparison of three methods for natural gas dehydration. *J. Nat. Gas Chem.* **2011**, *20*, 471–476. [CrossRef]
8. Gandhidasan, P.; Al-Farayedhi, A.A.; Al-Mubarak, A.A. Dehydration of natural gas using solid desiccants. *Energy* **2001**, *26*, 855–868. [CrossRef]
9. Kazemi, P.; Hamidi, R. Sensitivity analysis of a natural gas triethylene glycol dehydration plant in Persian Gulf Region. *Pet. Coal* **2011**, *53*, 71–77.
10. Records, L.R.; Seely, D.H. Low temperature dehydration of natural gas. *J. Pet. Technol.* **1951**, *3*, 61–66. [CrossRef]
11. Yu, G.; Dai, C.; Wu, L.; Lei, Z. Natural gas dehydration with ionic liquids. *Energy Fuels* **2017**, *31*, 1429–1439. [CrossRef]
12. Chao, H. Flow Field Simulation and Structure Optimization of a Small Supersonic Gas-Liquid Separator. Master's Thesis, Xi'an Shiyou University, Xi'an, China, 2020.
13. Feng, Y. The Study of the Structure Design and Numerical Simulation of Spiral Flowchannel Cyclone Separator. Master's Thesis, Xi'an Shiyou University, Xi'an, China, 2020.
14. Matsubayashi, T.; Katono, K.; Hayashi, K.; Tomiyama, A. Effects of swirler shape on swirling annular flow in a gas–liquid separator. *Nucl. Eng. Des.* **2012**, *249*, 63–70. [CrossRef]
15. Yu, M.; Zhang, Y.; Chen, G.; Chen, X. Simulation of gas-liquid separation characteristics of waxy natural gas in axial flow vane cyclone separator. *Petro Chem. Equip.* **2019**, *48*, 1–8.
16. Han, C.; Chen, F.; Yang, X.; Zhang, J. Influence of blade parameters on the performance of guide vane cyclone separation tube. *J. Mach. Des.* **2015**, *32*, 72–77. [CrossRef]
17. Sun, Y.; Song, J. Optimization and improvement of gas-liquid separator. *Guangdong Chem. Ind.* **2022**, *49*, 178–180.
18. Li, Z.; Li, W.; Wang, Q.; Xiang, R.; Cheng, J.; Han, W.; Yan, Z. Effects of medium fluid cavitation on fluctuation characteristics of magnetic fluid seal interface in agricultural centrifugal pump. *Int. J. Agric. Biol. Eng.* **2021**, *14*, 85–92. [CrossRef]
19. Li, W.; Li, Z.; Qin, Z.; Yan, S.; Wang, Z.; Peng, S. Influence of the solution pH on the design of a hydro-mechanical magneto-hydraulic sealing device. *Eng. Fail. Anal.* **2022**, *135*, 106091. [CrossRef]
20. Li, W.; Li, Z.; Deng, W.; Ji, L.; Qiu, Y.; Chen, H. Particle image velocimetry flowmeter for natural gas applications. *Flow Meas. Instrum.* **2021**, *82*, 102072.
21. Deng, W.; Pan, S.; Li, Z.; Huang, M. Experimental Study on flow characteristics of natural gas pipelines based on PIV. *J. Eng. Therm. Energy Power* **2020**, *35*, 171–177.
22. Eder, A.; Durst, B.; Jordan, M. Laser-Doppler Velocimetry—Principle and Application to Turbulence Measurements. In *Optical Measurements*; Springer: Berlin/Heidelberg, Germany, 2001; pp. 117–138.
23. Le Duff, A.; Plantier, G.; Valiere, J.C.; Bosch, T. Velocity measurement in a fluid using LDV: Low-cost sensor and signal processing design. In Proceedings of the Sensors, Orlando, FL, USA, 12–14 June 2002.
24. Guo, X.; Zhang, B.; Li, L.; Liu, B.; Fu, T. Experimental investigation of flow structure and energy separation of Ranque–Hilsch vortex tube with LDV measurement. *Int. J. Refrig.* **2019**, *101*, 106–116. [CrossRef]
25. Li, J.; Xu, L.; Peng, Y.; Zhao, X. Experimental and numerical simulation study on the flow characteristics of the draft tube in Francis turbine. *Machines* **2022**, *10*, 230.
26. Tsuji, Y.; Morikawa, Y.; Shiomi, H. LDV measurements of an air-solid two-phase flow in a vertical pipe. *J. Fluid Mech.* **1984**, *139*, 417–434. [CrossRef]
27. Norberg, C. LDV-measurements in the near wake of a circular cylinder. ASME Paper No. FEDSM98-521. In Proceedings of the Advances in the Understanding of Bluff Body Wakes and Vortex-Induced Vibration, Washington, DC, USA, 21–25 June 1998; pp. 41–45.
28. Pedersen, N.; Larsen, P.S.; Jacobsen, C.B. Flow in a centrifugal pump impeller at design and off-design conditions—part I: Particle image velocimetry (PIV) and laser Doppler velocimetry (LDV) measurements. *J. Fluids Eng.* **2003**, *125*, 61–72. [CrossRef]
29. Jang, C.M.; Furukawa, M.; Inoue, M. Analysis of vortical flow field in a propeller fan by LDV measurements and LES—part I: Three-dimensional vortical flow structures. *J. Fluids Eng.* **2001**, *123*, 748–754. [CrossRef]

processes

MDPI

Article

# Dehydration and Rehydration Kinetics Modeling in the Phytochemical, Aroma, and Antioxidant Capacity of Tree Tomato Fruit Dried with Microwaves and Freeze Driers: A Comparative Study

**Marc Antoine Ndisanze * and Ilkay Koca**

Institute of Graduate School, Ondokuz Mayis University, Samsun 55139, Turkey; itosun@omu.edu.tr
* Correspondence: ndisamarc@ines.ac.rw; Tel.: +250-788784005

**Abstract:** In the present study, we investigated and compared the effect of microwaves and freeze-drying methods on the dehydration and rehydration kinetics in the phenolic, anthocyanin, aroma profiles, and antioxidant properties of tree tomato fruit (*Solanum betaceum*). The tree tomatoes were dried using microwaves at 350 W, 500 W, and 650 W, and then freeze-dried. The obtained drying curves were processed to find the most suitable mathematical modeling among the different moisture ratio expressions. Total phenolics, total anthocyanins, total flavonoids total carotenoids, vitamin C, Ferric Reducing Antioxidant Power (FRAP), and 2,2-diphenyl-1-picrylhydrazyl (DPPH) were tested. Using High performance Liquid Chromatography (HPLC), phenolic and anthocyanin compound profiles were identified. The aroma profile was analyzed using gas chromatography-MS. The Midilli model, among others, precisely describes the dehydration methodology of all used drying methods with the coefficient of determination $R^2$ = 0.99. On the other hand, the Weibull model precisely describes the rehydration process of the used drying methods ($R^2$ = 0.99). Physical changes (color, shrinkage) were also studied. The freeze-dried tree tomatoes had a high number of phenolic compounds with 3.94 ± 0.26 mg GAE/g and total carotenoid compounds with 0.48 ± 0.04 µg/g. Epicathechin was the most abundant compound among the tested phenolics, followed by Cathechin. The Pelargonidin-3-glucoside was the most abundant anthocyanin whereas in freeze-dried tree tomatoes, 1.22 ± 0.01 mg/g. Fifty-four aroma compounds were detected and quantified. Among others, Eucalyptol was one of the most abundant aroma compounds analyzed in dried tree tomato fruit. Freeze-dried tree tomatoes retained most of the antioxidant and flavor compounds analyzed.

**Keywords:** tree tomatoes; dehydration; mathematical modeling; antioxidant; aroma

**Citation:** Ndisanze, M.A.; Koca, I. Dehydration and Rehydration Kinetics Modeling in the Phytochemical, Aroma, and Antioxidant Capacity of Tree Tomato Fruit Dried with Microwaves and Freeze Driers: A Comparative Study. *Processes* **2022**, *10*, 1437. https://doi.org/10.3390/pr10081437

Academic Editors: Jan Havlík and Anet Režek Jambrak

Received: 20 June 2022
Accepted: 20 July 2022
Published: 22 July 2022



## 1. Introduction

Fruit is one of the richest sources of active bio-compounds among other foods and their distribution on the planet is neither equitable nor uniform. [1]. Fruit consumption is highly recommended, since through its consumption, the human body gains important vitamins, minerals, fiber, tocopherols, and polyphenols [2]. It is well known that eating habits can have an influence on the prevention of metabolic diseases such as hypertension, diabetes, and obesity. Various studies described many polyphenols compounds found in foods, such as fruit, vegetables, tea, and seeds, as having antioxidants properties [3,4]. These polyphenol compounds are recognized to prevent damage caused by oxidative stress, such as cell and DNA damage [5–7]. Epidemiological studies revealed that there is a correlation between high consumption of phenolic compounds in the diet and reduced risks of cardiovascular diseases. Probably the main antioxidant activity that has been associated with phenols is their ability to scavenge free radicals [8].

Therefore, various techniques are being used to deliver fruit to the consumer with maximum quality. As mostly well known, fruit contains approximately 80–99% free water which makes them highly perishable. The drying process is one of the preservative methods used to extend the shelf life of fruit and better retain their nutrients by the means of reducing free water in food. The drying method is chosen based on the particularities of the product characteristics and socio-economic considerations as well as the energy need as fossil fuel prices are increasing [9]. Unfortunately, some studies pointed out the losses of some active biological compounds such as vitamin C and carotenoids [10,11], total polyphenols contents, and total flavonoids content [12] during drying. The long drying times at relatively high temperatures during the falling rate periods mostly lead to thermal degradation of some heat sensitive fruit compounds; consequently, microwave drying uses a shortened drying time and improves the final quality of the dried products [11]. Furthermore, for conserving some components of fruit, thermolabile, some techniques, such as drying using microwaves and freeze-drying, have been developed with the aim of offering products with extended storage stability and adequate convenience for the consumer while reducing the nutrients losses.

Freeze-drying, also known as lyophilization, is among the best drying techniques used for producing better quality of dried and solid foods. It is a preferred method for drying foods containing high heat sensitivity and oxidation-prone compounds, since it operates at very low temperatures and under high vacuum [13]. The application of freeze drying to various fruit and vegetables, such as lemon, apple, guava, strawberry, blackberry, pumpkin, tomato, asparagus, coffee, tea, garlic, ginger, maple syrup, etc., has already been reported in the literature [14–17]. Compared to freeze drying, the microwave drying system caused some reductions on one hand, and the increase of some bioactive compounds on the other hand, as reported also by [18–22].

Distinct mathematical models describing the drying process have been proposed to optimize the process and the set up the effective dryers. Moreover, the prediction of drying rates for distinct dryers and moisture diffusion parameters of fruit products are important components of microwave drying simulation models and are essential for an efficient moisture transfer analysis [23].

At present, there is no study that investigated the effect of a drying method on the bioactive compounds of tree tomatoes, which are a good source of polyphenols, vitamin C, flavonoids, carotenoids, anthocyanin, and also exhibit associated antioxidant capacity as per DPPH and FRAP [4,24,25]. In the present study, the comparison between the physical properties and bioactive compounds of tree tomato fruit dried with freeze dryers and microwaves was conducted, the used methodologies and the obtained results are described in the next sections.

## 2. Materials and Methods

### 2.1. Material

Fresh tree tomatoes were cultivated and collected from Rulindo District, Rwanda in the long rainy season (always called season B) April 2019. All samples were stored in a refrigerator at 4 °C and 90% relative humidity prior to be brought to Ondokuz Mayis University (Samsun, Tukey) in laboratories for food engineering for further experiments. Tree tomatoes with the same size and color were selected to ensure uniformity of the physical-chemical characteristics of the samples.

### 2.2. Drying Process

The samples were washed, peeled, and sliced (5 mm thickness) prior to the drying process. The drying process was made in microwaves (Arçelik MW 674 S 20 lt, Istanbul-Türkiye) at the power of 350 W, 500 W, and 650 W for the first method whereas the moisture ratio was being checked with every one minute difference. The second method was the freeze drying method using a freeze dryer (LABCONCO; FREEZONE 12PLUS, Kansas, MO, USA).

2.2.1. Physical Properties of Dried Tree Tomatoes

Color Measurement

The Color measurement was made with the Konica Minolta CR-400 color measurement and DP-400 data processing device (Japan) with CIE (Commission International L'Eclairage, English: International Commission of Illumination, ICI) Color Scale ($L$, $a$, $b$). Briefly, the colorimeter was calibrated by placing the tip of measuring head flat against the surface of the white and black calibration place. After standardization, "$L$" (lightness), "$a$" (redness/greenness), and "$b$" (yellowness/blueness) values were measured on the surface of fresh and dried tree tomatoes slices. For analyzing the color change after drying, total color difference ($\Delta E$), Chroma and hue angle (H°) were calculated as the equation below describing [26,27]:

$$Chroma = \sqrt{(\mathrm{a})^2 + (\mathrm{b})^2} \tag{1}$$

$$Hue\ angle = arctan\frac{b}{a} \tag{2}$$

$$the\ \Delta E = \sqrt{(L - Lo)^2 + (a - ao)^2 + (b - bo)^2} \tag{3}$$

where $L_0$, $a_0$, and $b_0$ are the color parameters of fresh samples; $L$, $a$, $b$ are the color parameters of dried tree tomatoes slice samples.

Determination of Morphological Features

Morphological features were analyzed using the Scan Electron Microscopy SEM (Jeol 7001F FEG gun, Tokyo-Japan) device. Briefly, the dehydration of tree tomatoes was performed in an acetone series, the critical point dried using carbon dioxide and then mounted directly on stubs using double-side adhesive tape, and sputter-coated with gold. Observations were made with a Scan Electron Microscopy SEM [28].

Shrinkage Analysis

In this analysis, tree tomato slices were placed in the area divided into 1 $mm^2$ squares as a reference under the glass pane, and photographs were taken 10 cm above each sample slice by the phone's camera (Samsung Galaxy J5 prime, Seoul-Korea). Area change measurements were made from the photographs taken with the help of the AutoCAD package program. The chance in area was measured, compared to the initial sample, and expressed in percentage [29].

### 2.3. *Mathematical Modeling of Drying Kinetics*

The obtained drying curves were processed to find the most suitable model among the various moisture ratio expressions are given in Table 1. The moisture ratio was expressed as per Formula (4) [30,31]:

$$MR = \frac{(Mt - Mo)}{Me - Mo} \tag{4}$$

where $MR$ is the moisture ratio, $M_t$ is the moisture content at any point in time (g water/g dry matter), and $Mo$ is the initial moisture content (g water/g dry matter),

Model parameters were determined using non-linear regression analysis. The indicator used to assess the compliance of the tested models with experimental data are the coefficient of determination ($R^2$), Sum of Square Errors (SSE), root mean squared error (RMSE) for measuring the accuracy, and chi-square ($\chi^2$), which is for measuring the difference between the observed and expected frequencies of the outcomes of a set of events or variables. To complete this analysis, the MATLAB R2013a (1.22.6.30) software was used. Table 1 presents the mathematical models used. The root mean square error, RMSE, is a widely used measure of the difference between the values predicted by the model and the actual observed values. RMSE is a good measure of accuracy and serves to combine residuals into a single measure of predictive power. It can be calculated using the following [32]:

$$\chi^2 = \frac{\sum_{i=1}^{N}(MRo - MRe)^2}{N - n} \quad (5)$$

$$\text{RMSE} = \left[\frac{1}{N}\sum_{i=1}^{N}(MRo - MRe)^2\right]^{1/2} \quad (6)$$

$$\text{SSE} = \sum_{i=1}^{N}(MRo - MRe)^2 \quad (7)$$

where $N$ is the number of observations, $n$ is the constant number, $MRo, i$. predicted moisture content values, $MRe, i$ are the experimental moisture content values.

**Table 1.** Mathematical models applied to the drying curves.

| Models Name | Models Formula | References |
|---|---|---|
| Page | MR = exp(−kt$^n$) | [33] |
| Two-term exponential | MR = a exp(−kt) + (1 − a) exp(−kat) | [34] |
| Logarithmic | MR = a exp(−kt) + b | [35] |
| Wang and Singh | MR = 1 + at + bt$^2$ | [36] |
| Approximation of Diffusion | MR = a exp(−kt) + (1 − a) exp(−kbt) | [37] |
| Midilli | MR = a exp(−kt$^n$) + bt | [38] |

MR is the moisture ratio; t is the time; and α, b, c, k, and n are the constants of models.

## 2.4. Rehydration Process

To determine the rate of rehydration, a certain number of dried products was taken, 50 times the weight of the sample (1 g of dried fruit in 50 g of distilled water), placed in distilled water at 25 °C (room temperature) and kept in pure water for 12 h. The amount of water absorbed by the samples was recorded at 1-h intervals [39].

## 2.5. Mathematical Modeling of Rehydration Kinetics

As with previous studies, this study used various empirical models to evaluate the mechanism of rehydration. The rehydration rate equation is as follows:

$$\text{Rehydration rate} = \frac{Mr}{\text{Md}} \quad (8)$$

where $Mr$: Wet product weight (g); Md: The dried product weight (g).

The rehydration patterns are described in Table 2. In these equations; M(t) is the sample moisture at rehydration time t (g $H_2O$/g dry matter), Mo is the initial moisture content of the dried sample (g $H_2O$/g dry matter), Me is the equilibrium moisture (g $H_2O$/g dry matter), t is the rehydration time, a [min. (g dry matter/g $H_2O$] Peleg rate constant, and b [g dry matter/g $H_2O$] Peleg capacitance constant. The experimental results of rehydration of tree tomatoes dried with microwaves at different powers (350 W, 500 W, and 650 W) were corrected using empirical models. The mathematical simulation was coded using the 2013a version of the MATLAB software. Four criteria were used to assess the suitability of the models used: the sum of squares of the standard error (SSE), coefficient of determination ($R^2$), root mean square error (RMSE), and chi-square ($\chi^2$).

$$\text{SSE} = \sum_{i=1}^{N}(Mei - Mci)^2 \quad (9)$$

$$\text{RMSE} = \left[\frac{1}{N}\sum_{i=1}^{N}(Mei - Mci)^2\right]^{1/2} \quad (10)$$

$$\chi^2 = \frac{\sum_{i=1}^{N}(Mei - Mci)^2}{N - z} \quad (11)$$

where *Mci* is the calculated moisture content (g water/g dry matter.), *Mei* is the experimental moisture content (g water/g dry matter.), *z*; constant number and *N* data number [40].

**Table 2.** Mathematical models applied to the rehydration curves.

| Model Name | Model Formula | References |
|---|---|---|
| Peleg | $M(t) = M0 + \frac{t}{a+bt}$ | [41] |
| Weibull | $M(t) = Me + (M0 - Me)\exp\left[-\left(\frac{t}{b}\right)^a\right]$ | [42] |
| First order kinetic | $M(t) = Me + (M0 - Me)\exp(-at)$ | [43] |
| Exponential Model | $M(t) = Me + (M0 - Me)\exp(-at^k)$ | [44] |
| Proposed Model | $M(t) = a\ \exp\left(-\frac{b}{(1+t)^k}\right)$ | [45] |
| Exponential Related Equation | $M(t) = Me(1 - \exp(-at))$ | [46] |

M(t) is the moisture ration in the function of time, *a*, *b* and *k* are the constants of models.

### 2.6. *Natural Antioxidants and Determination of Antioxidant Activity*

#### 2.6.1. Ascorbic acid Determination

Vitamin C was determined according to the method described by Benassi at al. [47] and AOAC [48] with slight modifications. Tree tomato samples were extracted using oxalic acid solution (0.4 g oxalic acid in 1 L distilled water). 150 µL of the fruit extracts was added to 1350 µL of 0.01% 2,6-dichlorophenolindophenol. The blank was prepared by mixing 150 µL oxalic acid with 1350 µL 2,6-dichlorophenolindophenol. The absorbance was read at 520 nm against the blank. The reduction ratio was calculated with Equation (12) and the values of vitamin C were determined as mg/g using a calibration curve of ascorbic acid at the concentrations ranged from 0.01 mg/g to 0.1 mg/g.

$$\text{Reduction } (\%) = \frac{\text{Absorbance of the control} - \text{Absorbance of the extract}}{\text{Absorbance of the control}} \quad (12)$$

#### 2.6.2. Determination of Total Phenolic Compounds

The contents of total phenols (TPC) were determined according to the Folin-Ciocalteu methods [49,50]. Briefly, fruit extracts were diluted to a suitable concentration for analysis. Half a milliliter of extract, 1 mL of 1 M Folin-Ciocalteu reagent and 1 mL of 20% (*w*/*v*) $Na_2CO_3$ were mixed. After 2 h of incubation in the dark at room temperature, the mixture was centrifuged for 10 min (8000 rpm). The supernatant was read at 765 nm using spectrophotometer (Thermospectronic-Helios Gamma, Paisley-UK). Different concentrations of Gallic acid (10–90 µg/mL) were determined to be a calibration curve and the results were presented as mg Gallic acid equivalents per gram, on dry weight basis (GAE)/g dw.

#### 2.6.3. Determination of Total Flavonoids

Total flavonoid compounds were determined using a slightly modified method described by Zhishen et al. [51]. Briefly, 1 mL of the diluted fruit extract solution was added to 0.3 mL of 5% $NaNO_2$ and left to stand for 5 min, then 0.5 mL of 5% $AlCl_3$ was added. The mixture was kept for 6 min before adding 0.5 mL of 1 M NaOH. After 10 min, the absorbance was read at 510 nm using a spectrophotometer (Thermospectronic-Helios Gamma, Paisley-UK). The total content of flavonoids was estimated from a calibration curve using epicatechin as a standard. Results were presented as mg epicatechin equivalents (ECE) mg/g dry weight basis.

#### 2.6.4. Determination of Total Anthocyanin Compounds

The amount of total anthocyanin compounds was determined using the pH difference method and calculated as per cyanidin-3-glucoside equivalence [52]. Briefly, after extraction, the samples were fragmented and diluted with pure water; then, they were mixed with pH 4.5 and pH 1.0 buffer solutions, filtered, and absorbance was read at 510 nm and 700 nm on a spectrophotometer (Thermospectronic-Helios Gamma, Paisley-UK). Results were presented in mg cyanidin-3-glucoside equivalents (CGE) per mg/g dry matter weight.

#### 2.6.5. Determination of Total Carotenoids

Total carotenoid compounds were extracted and analyzed according to the method described by Hernández et al. [53] with minor modifications. Homogenized dried tree tomato samples were extracted with 25 mL of cold acetone and filtered under vacuum until the color disappeared. The extract was gradually added to 50 mL of ethyl ether in a decanting funnel. The organic phase was treated several times with anhydrous $Na_2SO_4$ (20 g/L) to remove residual water. Then the optical density was measured at 450 nm using a spectrophotometer (Thermospectronic-Helios Gamma, Paisley-UK). The results were expressed as mg of β-carotene/100 g, dry matter basis.

### *2.7. Determination of Antioxidant Capacity*

Determination of antioxidant capacity of the tree tomato fruit was performed by 2,2-diphenyl-1-picrylhydrazyl (DPPH), radical scavenging, and ferric reducing activity (FRAP) assays. The FRAP assay value was estimated by a calibration curve of the standard solution of $FeSO_4$ then expressed as mmol $FeSO_4$ equivalents per g (mmol ISE/g). FRAP was calculated from a standard curve using iron sulfate at the concentrations ranged from 0.12 to 0.59 mmol/g [54]. The DPPH assay was performed according to the method adopted by Zannou et al. [55] and Kalisz et al. [56]. Briefly, 50 µL of the appropriately diluted extract was mixed with 1 mL of DPPH solution (0.06 mM in 80% methanol). The mixture was kept in the dark for 1 h at room temperature (approximately 25 °C), then the absorbance was read at 517 nm. The DPPH standard solution was used as a control, and the scavenging ratio was calculated with Equation (12). The values of DPPH radical scavenging were determined as mmol trolox/g.

### *2.8. Individual Anthocyanin and Phenolic Compound Profiles Analysis*

Anthocyanin and phenolic compound profiles analyses were performed using high performance liquid chromatography (HPLC) coupled with a mass spectrometric detector (LC-MS/MS, Shimadzu LC-MS 8040) with an Inertsil ODS-4 column (3 µm, 4.6 mm × 50 mm) (GL Sciences Kat No: 5020-04042), flow rate 1 mL/min, oven temperature 30 °C, injection volume: 20 µL, with mobile phase A: 94% 2 mM sodium acetate and 6% acetic acid. Mobile phase B: 100% acetonitrile. The MS/MS system operated at a capillary temperature of 300 °C, an evaporator temperature of 350 °C, and a gas pressure in the cladding of 30 arb. Discharge current 4 µA. 20 µL of each sample was filtered through 0.45 µm nylon filters and injected into a reverse phase C18 column (ODS-hypersil 5 µm, 4.6 mm × 250 mm). Identified compounds were quantified using a mixture at concentrations of 0, 50, 75, 100, 150, and 200 ppm of epicatechin, catechin, resveratrol, p-coumaric acid, salicylic acid, gallic acid, sinapic acid, quercetin-3 standards-glucoside, cyanine chloride, cyanidin-3-glucoside, and pelargonidin-3-glucoside as standards [57].

### *2.9. Analysis of Aroma Compounds Profile*

The analysis of volatile components was determined using a gas chromatography (GC) system (Agilent Brand 7890B, 7010B MS, USA). The solid phase microextraction (SPME) method of extraction was used. Briefly, 3.0 g of the sample was placed in a 20 mL vial and kept at 50 °C for 15 min, then a solid phase microextraction apparatus (SPME) for 50/30 µm divinylbenzene/carboxene was used. Polydimethylsiloxane (DVB/CAR/PDMS) coated fiber and volatile components were absorbed within 30 min. DB-Wax (60 m × 0.25 mm, id × 0.25 µm, J&W Scientific-Folsom, CA, USA) was then introduced into the capillary column by desorption for 5 min. The injection temperature was 250 °C, the column temperature was raised to 90 °C. at a rate of 3 °C per minute after holding for 4 min at 40 °C, and then to 130 °C by increments of 4 °C per minute. The temperature was brought to 240 °C by increasing by 5 °C and maintained at this temperature for 8 min. Helium (He) was used as the carrier gas. The electron energy is 70 eV and the mass range is 30–600 $m/z$. The split ratio was 1:10 [58].

### 2.10. Statistical Analysis

Experimental results of dehydration and rehydration of dehydrated tree tomatoes were compared with empirical models. The fitting was coded using MATLAB R2013a software (1.22.6.30). Four criteria were used to assess the quality of fitting and evaluate the quality of fitting of each model: linear regression coefficient ($R^2$), root mean square error (RMSE), sum of squared errors (SSE), and chi-square ($\chi^2$). One-way statistical analysis was performed by ANOVA with Duncan's test using SPSS (version 21). The significance of the results was assessed at $p \leq 0.05$.

## 3. Results and Discussion

### 3.1. Dried Tree Tomatoes Specification

#### 3.1.1. Color

In the present research, L (lightness), a (redness/greenness), and b (yellowness/blueness), ΔE, Hue Angle and Chroma values were analyzed to characterize the color and the results are presented in Table 3. Delta-E (ΔE) representing the distance between two colors; whereas in this study the comparison was between the fresh tree tomato fruit to the dried ones. Considering the obtained ΔE values, there is a big difference between the fresh tree tomato and the dried ones as when the ΔE value is greater than 5, it indicates a large color difference [59].

**Table 3.** Color characteristic and morphological characteristics of dried tree tomatoes.

| Methods | L | a | b | ΔE | Hue Angle | Chroma |
|---|---|---|---|---|---|---|
| 350 W | 20.58 ± 1.31 b | 6.86 ± 1.69 a | 4.28 ± 0.65 b | 21.56 ± 2.25 a | 31.51 ± 6.73 d | 8.12 ± 2.04 c |
| 500 W | 23.37 ± 4.49 ab | 7.34 ± 1.69 a | 5.86 ± 0.27 ab | 20.66 ± 1.51 ab | 36.92 ± 1.65 c | 9.51 ± 3.00 b |
| 650 W | 21.66 ± 0.34 b | 6.63 ± 0.22 a | 7.73 ± 0.65 a | 19.16 ± 0.43 b | 49.33 ± 1.98 a | 10.19 ± 0.59 ab |
| FD | 27.71 ± 0.47 a | 8.31 ± 0.21 a | 8.29 ± 0.26 a | 13.49 ± 0.27 c | 44.93 ± 1.61 b | 11.74 ± 0.08 a |

There is no statistical difference between the averages shown with the same letter in the same row ($p > 0.05$).

According to the results of the present study, the color parameters lightness (L value), redness-greenness (a value) and yellowness-blueness (b value) changed when we compared the dried tree tomatoes with the fresh tree tomato fruit samples. Accordingly, the highest lightness L was observed in the tree tomatoes dried with the freeze dryer with 27.71 ± 0.47. The change in lightness was not significant ($p > 0.05$) between the samples dried with microwaves at different powers. The literature pointed out the degradation of color pigment due to non-enzymatic browning as a result of Maillard and caramelization of sugar components of the fruit [60–62]. The freeze-drying method affected less the tree tomatoes color as indicated by the lowest ΔE value (13.49 ± 0.27). Previous studies pointed out the same thoughts and support the present study stating the change in color of food material because of enzymatic and non-enzymatic reactions taking place during drying process [13,63].

#### 3.1.2. Shrinkage and Surface Morphological Features of Dries Tree Tomatoes

Food drying usually results in product deterioration not only from a sensorial point of view but also from a physicochemical and nutritional one. The surface morphology of tree tomato fruit was examined using Scan Electron microscopy (SEM). The present study revealed that microwave drying affected more than the freeze-drying technique the morphological properties of tree tomato fruit samples, as presented in Figures 1 and 2. In microwave drying technique, as the drying power was increasing, the shrinkage values also increased. This may be due to the reduction of intercellular space. The water molecules evaporated, which may make the fruit tissue shrink. The freeze-drying technique does not affect much the size of the tree tomato fruit as shown in Figures 1 and 2, which may be due to the fact that the tissues are not brutally cracked by the water friction while being forced out by the evaporation process.

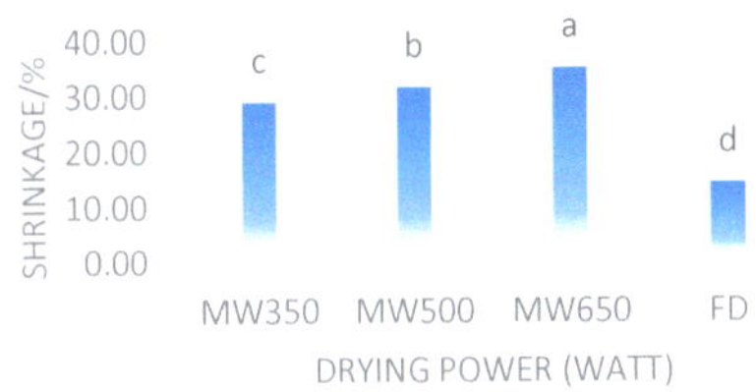


**Figure 1.** The shrinkage of tree tomato fruit after drying. MW: Microwaves, FD: Freeze-dried.

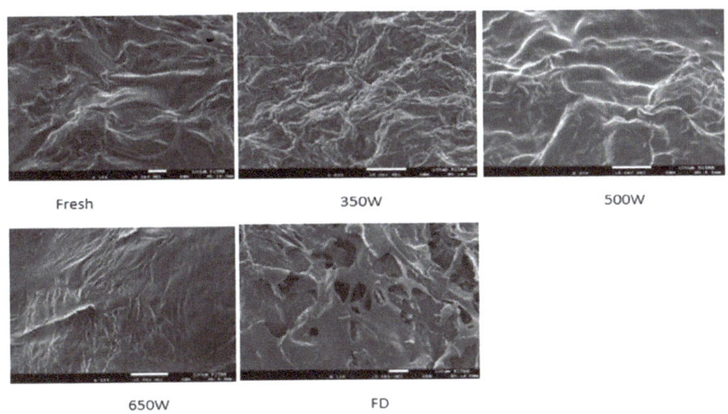


**Figure 2.** Surface morphological structure of fresh and dried tree tomato fruit taken by SEM.

In the present study, the highest shrinkage values were observed in the tree tomato fruit dried with microwaves at 650 W followed by those dried with microwaves at 500 W with 35.19% and 31.67%, respectively. Freeze-dried fruit showed a low shrinkage level with 14.34% compared to other tested methods. The freeze-drying process left the pronounced holes as shown in Figure 2 from SEM as the water molecules first condensed in ice and then evaporated as solid. This study is supported by previous studies [19,61,64,65].

### 3.2. *Mathematical Modeling of Dehydration and Rehydration Kinetics*

Drying kinetics as expressed by the moisture ratio were modeled according to the models of Page, Two-term Exponential, logarithmic, Wang and Singh, Midilli, and Approximation of Diffusion. According to the results of the present study, the values of coefficient of determination $R^2$ ranged from 0.98–0.99 within all used models, which means the suitability of the used methods to the tested models. The values of the constants in these models were regressed against these variables using multiple regression analysis. Changes in the reduced moisture ratio (MR) as a function of the time spend by the process of microwave drying of tree tomato fruit are summarized in the figures.

As indicated in Figure 3, as the drying power increased, the drying time decreased. The shortest time was observed in tree tomato fruit dried with microwaves at 650 Watts in 35 min while the longest was within the samples dried with microwaves at 350 Watts in 73 min. The all-possible combinations of these variables were tested and were included in the mathematical analysis described in Table 4.

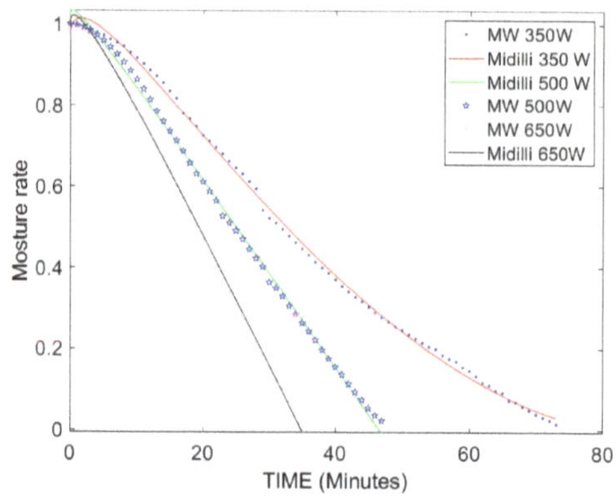


**Figure 3.** Drying curves of the Microwave drying process of tree tomato fruit.

**Table 4.** Dehydration kinetics models.

| Method | Model Name | SSE | $R^2$ | RMSE | $X^2$ | Model Constants |
|---|---|---|---|---|---|---|
| MW350 | Wang and Singh | 0.10 | 0.99 | 0.04 | 0.00 | a = −0.015746 b = 2.388 × $10^{-05}$ |
| | Approximation of Diffusion | 0.05 | 0.99 | 0.03 | 0.00 | a = 2.188 b = −2.6626 k = 0.0094144 |
| | Page | 0.01 | 0.99 | 0.01 | 0.00 | k = 0.0017301 y = 1.7232 |
| | Logarithmic | 0.05 | 0.99 | 0.03 | 0.00 | a = 2.1816 b = −1.0988 k = 0.0094554 |
| | Midilli | 0.01 | 0.99 | 0.01 | 0.00 | a = 0.97272 b = −0.053816 k = −0.091877 n = 0.6355 |
| | Two-term exponential | 0.05 | 0.99 | 0.03 | 0.00 | a = 2.1307 k = 0.042673 |
| MW500 | Wang and Singh | 0.03 | 0.99 | 0.03 | 0.00 | a = −0.01716 b = −9.6673 × $10^{-05}$ |
| | Approximation of Diffusion | 0.02 | 0.99 | 0.02 | 0.00 | a = 8.1429 b = −1.824 k = 0.0029664 |
| | Page | 0.02 | 0.99 | 0.02 | 0.00 | k = 0.0016625 y = 1.8959 |
| | Logarithmic | 0.01 | 0.99 | 0.02 | 0.00 | a = 73.5186 b = −72.4614 k = 0.00030639 |
| | Midilli | 0.01 | 0.99 | 0.01 | 0.00 | a = 0.99761 b = −0.023706 k = −0.052838 n = 0.17678 |
| | Two-term exponential | 0.08 | 0.98 | 0.04 | 0.00 | a = 2.1606 k = 0.058609 |
| MW650 | Wang and Singh | 0.02 | 0.99 | 0.03 | 0.00 | a = −0.020247 b = −0.00025716 |
| | Approximation of Diffusion | 0.02 | 0.99 | 0.03 | 0.00 | a = 9.9362 b = 1.7107 k = 0.0032205 |
| | Page | 0.02 | 0.99 | 0.02 | 0.00 | k = 0.001962 y = 2.0056 |
| | Logarithmic | 0.02 | 0.99 | 0.02 | 0.00 | a = 94.5642 b = −93.4931 k = 0.00031942 |
| | Midilli | 0.01 | 0.99 | 0.01 | 0.00 | a = 0.99494 b = −0.03611 k = −0.057308 n = 0.39814 |
| | Two-term exponential | 0.08 | 0.98 | 0.05 | 0.00 | a = 2.1813 k = 0.076869 |

α, b, c, k, n, and y are the constants of models.

In the modeling of drying kinetics, it is assumed that the model which shows the lower SSE, RMSE, and $X^2$ and highest $R^2$ is the best-suited one for both dehydration and rehydration kinetics. This study suggested mostly the Midilli model for dehydration and Weibull for rehydration which gave the best simulation of the curves of dried tree tomatoes based on the percentage of demonstrated variances. The rehydration kinetics curve is best described by the Peleg Model and is summarized in Figure 4 and Table 5. This study revealed that the time required for rehydration depends on the power used for drying. The longest time was observed in tree tomato fruit dried in microwaves at 650 in 108 min followed by the tree tomatoes dried using microwaves at 500 W in 105 min. The present study is supported by recent studies [66–70].

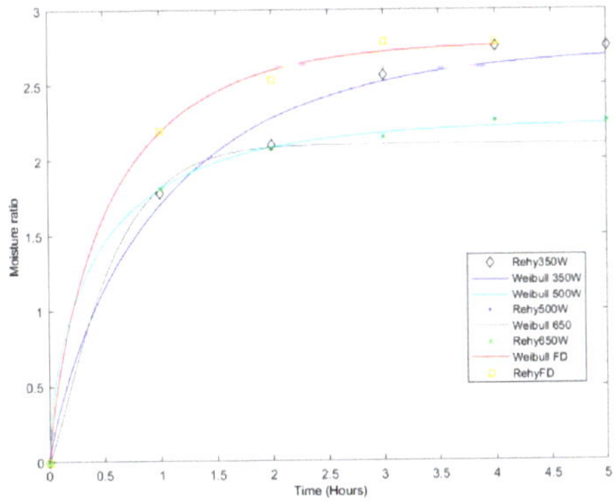


**Figure 4.** Rehydration curve of microwaves dried and freeze-dried tree tomato fruit.

**Table 5.** Rehydration kinetics mathematical modeling.

| Methods | Models | SSE | $R^2$ | RMSE | $X^2$ | Model Constants |
|---|---|---|---|---|---|---|
| MW 350 | Peleg | 0.06 | 0.99 | 0.11 | 0.01 | a = 5.9582 b = 0.37274 |
| | First order kinetic | 0.05 | 0.99 | 0.09 | 0.01 | a = 0.057467 |
| | Exponential related equation | 0.05 | 0.99 | 0.09 | 0.01 | a = 0.057466 |
| | Exponential model | 0.04 | 0.99 | 0.09 | 0.01 | a = 0.041472, k = 1.1014 |
| | Weibull | 0.04 | 0.99 | 0.09 | 0.01 | a = 1.1019, b = 17.9891 |
| | proposed model | 0.05 | 0.99 | 0.11 | 0.01 | a = 2.438, b = 19.0225, k = 1.204 |
| MW 500 | Peleg | 0.03 | 0.98 | 0.07 | 0.00 | a = 10.1066, b = 0.62977 |
| | First order kinetic | 0.01 | 0.99 | 0.04 | 0.00 | a = 0.052314 |
| | Exponential related equation | 0.01 | 0.99 | 0.04 | 0.00 | a = 0.052315 |
| | Exponential model | 0.01 | 0.99 | 0.04 | 0.00 | a = 0.028411, k = 1.1927 |
| | Weibull | 0.01 | 0.99 | 0.04 | 0.00 | a = 1.1934, b = 19.7981 |
| | proposed model | 0.00 | 1.00 | 0.02 | 0.00 | a = 1.3466, b = 168.7216, k = 1.9786 |
| MW 650 | Peleg | 0.01 | 1.00 | 0.03 | 0.00 | a = 2.4679, b = 0.44546 |
| | First order kinetic | 0.02 | 1.00 | 0.05 | 0.00 | a = 0.092312 |
| | Exponential related equation | 0.02 | 1.00 | 0.05 | 0.00 | a = 0.092313 |
| | Exponential model | 0.01 | 1.00 | 0.03 | 0.00 | a = 0.21323, k = 0.71353 |
| | Weibull | 0.01 | 1.00 | 0.03 | 0.00 | a = 0.71379, b = 8.7242 |
| | proposed model | 0.01 | 1.00 | 0.03 | 0.00 | a = 2.1926, b = 7.598, k = 1.1675 |
| FD | Peleg | 0.01 | 1.00 | 0.04 | 0.00 | a = 0.13338, b = 0.32209 |
| | First order kinetic | 0.01 | 1.00 | 0.06 | 0.00 | a = 1.491 |
| | Exponential related equation | 0.01 | 1.00 | 0.06 | 0.00 | a = 1.491 |
| | Exponential model | 0.01 | 1.00 | 0.05 | 0.00 | a = 1.5407, k = 0.80783 |
| | Weibull | 0.01 | 1.00 | 0.05 | 0.00 | a = 0.80837, b = 0.58589 |
| | proposed model | 0.02 | 1.00 | 0.10 | 0.01 | a = 2.7692, b = 4.3711, k = 4.1696 |

FD: freeze drying, MW: Microwaves, a, b and k are model's constants.

### *3.3. Phytochemical Profiles and Antioxidant Activity of Dried Tree Tomatoes*

In this study, two types of drying methods, freeze drying and microwaves (at 350 W, 500 W, 650 W), were used and the effects of the drying methods on the antioxidant properties was assessed. The freeze drying preserved more TPC and total carotenoids (TCC) than other tested drying methods with 3.94 ± 0.26 mg GAE/g and 0.48 ± 0.04 µg/g, respectively (Table 6). This amount is significantly higher ($p < 0.05$) than that of other tested drying methods. This study also revealed that the higher microwaves' drying power preserved more TPC and TFC than the lower power. The tree tomatoes dried in microwaves at 650 W had significantly ($p < 0.05$) higher TPC and TFC than other tested drying powers (Table 6). Those results are supported by the study of Hayat et al. [71], which the increase of TPC with the increase of microwaves power. The freeze-drying method well-preserved the anthocyanin compounds and vitamin C significantly ($p < 0.05$) more than the microwaves drying methods. Our findings were in concordance with previous studies on pomegranate peel and persimmon, which reported that freeze drying retained higher TPC, TFC, and antioxidant capacity than other drying methods, such as vacuum oven and hot air [72,73]. Those findings are also supported by the study performed by Ng. et al. [74], Saifullah et al. [75], and Papoutsis et al. [76]. The antioxidant capacity of the dried tree tomato fruit as indicated by the DPPH and FRAP were statistical significantly ($p < 0.05$) higher in the tree tomatoes dried by microwaves. DPPH was higher in the tree tomato fruit dried with microwaves at 650 W with 57.48 ± 0.90 mmol trolox/g and the lowest was in the samples dried in freeze dryer with 46.14 ± 1.39 mmol trolox/g.

**Table 6.** Antioxidant properties of dried tree tomato fruit.

| Methods | TPC (mg GAE/g) | TCC (μg/g) | TFC (mg/g) | Anth (mg/g) | VitC (mg/g) | FRAP (mmol ISE/g) | DPPH (mmol Trolox/g) |
|---|---|---|---|---|---|---|---|
| FD | 3.94 ± 0.26 a | 0.48 ± 0.04 a | 0.95 ± 0.05 b | 0.85 ± 0.02 a | 1.671 ± 0.02 ab | 71.69 ± 1.13 b | 46.14 ± 1.39 cd |
| MW350 W | 2.37 ± 0.34 c | 0.25 ± 0.09 b | 1.56 ± 0.10 b | 0.12 ± 0.00 c | 1.58 ± 0.02 c | 71.11 ± 0.44 b | 51.50 ± 1.74 b |
| MW500 W | 2.10 ± 0.31 d | 0.17 ± 0.07 cd | 1.26 ± 0.09 d | 0.11 ± 0.01 c | 1.15 ± 0.078 d | 81.34 ± 3.35 a | 48.09 ± 1.11 c |
| MW650 W | 2.90 ± 0.42 b | 0.15 ± 0.04 d | 1.82 ± 0.40 a | 0.25 ± 0.04 b | 1.78 ± 0.04 a | 35.17 ± 1.11 c | 57.48 ± 0.90 a |

There is no statistical difference between the averages shown with the same letter in the same row ($p > 0.05$).

This study also analyzed the phenolic profile of dried tree tomatoes in which Gallic acid, catechin, epicatechin, p-coumaric acid; sinapic acid, quercetin-3-glucoside, salicylic acid, and resveratrol were analyzed. Epicatechin was the most abundant phenolic compound but mostly in samples dried with microwaves at 500 W which was significantly ($p < 0.05$) greater than other tested drying methods (Table 7). The value of epicatechin was 2.790 ± 0.038 mg/g in the sample dried with microwaves at 500 W followed by those dried in microwaves at 650 W with 2.615 ± 0.039 mg/g. The present study revealed the lowest concentration of epicatechin in the samples dried in the freeze dryer. In contrast to this study, Çoklar and Akbulut [77] performed a study on the black grapes and revealed a decrease of 25.3% of grapes' epicatechin content after freeze drying, while sun and oven drying affected more the grapes' epicatechin. Moreover, freeze drying preserved significantly ($p < 0.05$) sinapic acid, Resveratrol, and quercetin-3-glucoside more than other tested drying methods (Table 7) while Gallic acid, p-coumaric acid, salicylic acid, and Catechin were better preserved within the samples dried in microwaves at 650 W (Table 7). This indicate the correlation between higher TPC and DPPH radical scavenging activity of the sample dried in microwaves at 650 W.

**Table 7.** Phenolic profile of dried tree tomatoes.

| Meth | Gallic Acid (mg/g) | p-Coumaric Acid (mg/g) | Sinapic Acid (mg/g) | Salicylic Acid (mg/g) | Resveratrol (mg/g) | Q.-3-Glu (mg/g) | Catechin (mg/g) | Epicatechin (mg/g) |
|---|---|---|---|---|---|---|---|---|
| 350 W | 0.602 ± 0.008 a | 0.003 ± 0.000 b | 0.057 ± 0.002 cd | 0.063 ± 0.004 ab | 0.009 ± 0.000 e | 0.057 ± 0.002 bc | 1.071 ± 0.039 b | 2.596 ± 0.041 b |
| 500 W | 0.444 ± 0.004 b | 0.002 ± 0.000 b | 0.081 ± 0.002 c | 0.030 ± 0.004 b | 0.010 ± 0.000 cd | 0.081 ± 0.003 b | 0.808 ± 0.026 c | 2.790 ± 0.038 a |
| 650 W | 0.649 ± 0.004 a | 0.013 ± 0.001 a | 0.038 ± 0.007 b | 0.074 ± 0.009 a | 0.009 ± 0.000 e | 0.038 ± 0.001 c | 1.448 ± 0.006 a | 2.615 ± 0.039 ab |
| FD | 0.105 ± 0.003 c | 0.001 ± 0.000 b | 0.122 ± 0.006 a | 0.058 ± 0.002 bc | 0.012 ± 0.002 a | 0.122 ± 0.008 a | 0.158 ± 0.013 d | 0.118 ± 0.00 c |

There is no statistical difference between the averages shown with the same letter in the same row ($p > 0.05$).

The anthocyanin profile also was analyzed in the dried tree tomato fruit samples where Cyanin chloride, Cyanidin-3-glucoside, Pelargonidin-3-glucoside were detected. The freeze drying method showed significantly ($p < 0.05$) to better preserve the all tested anthocyanin compounds (Table 8). There no statistical significance in the concentration of cyanine chlorite found in all microwaves dried samples. Pelargonidine-3-glucosite was the most abundant anthocyanin compound followed by Cyanidin-3-glucoside in all tested dried tree tomato fruit samples (Table 8). The highest amount of Pelargonidine-3-glucosite was tested in samples dried in freeze dryer with 1223.82 ± 5.96 mg/kg followed by samples dried in microwaves at 650 W power. Freeze-dried samples also contained significantly ($p < 0.05$) highest Cyanidin-3-glucoside and Cyanine chlorite with 99.28 ± 1.51 mg/kg and

19.34 ± 0.8 mg/kg, respectively. This study is in agreement with the study by Wojdyło et al. [60], who revealed a decrease of anthocyanin compound, especially Pelargonidine-3-glucosite, in the strawberry fruit dried with microwaves compared to those dried in freeze dryer.

**Table 8.** The anthocyanin profile of dried tree tomatoes.

| Drying Methods | Cyanine Chlorite (mg/kg) | Cyanidin-3-Glucoside (mg/kg) | Pelargonidine-3-Glucosite (mg/kg) |
|---|---|---|---|
| 350 W | 2.62 ± 0.01 b | 8.79 ± 0.22 c | 71.51 ± 1.26 c |
| 500 W | 2.34 ± 0.00 b | 11.02 ± 0.03 b | 124.29 ± 0.38 c |
| 650 W | 2.31 ± 0.01 b | 16.11 ± 0.86 b | 251.28 ± 3.40 b |
| FD | 19.34 ± 0.84 a | 99.28 ± 1.51 a | 1223.82 ± 5.96 a |

There is no statistical difference between the averages shown with the same letter in the same row ($p > 0.05$).

### 3.4. The Aroma Profile of the Dried Tree Tomatoes

Fifty-four aroma compounds were analyzed in the dried tree tomato fruit sample with different concentration according to drying method and power. The samples contained distinct types of aroma compound, including alcohol, organic acids, esters, terpenes, aldehydes, and ketones, as well as other organic substances (Table 9). Eucalyptol is one of the most abundant aroma compounds analyzed in dried tree tomato fruit samples whereas it was mostly abundant in samples dried in microwaves at 350 W. This study revealed that some aromatic compounds which may be not heat labile were found to be absent in the samples dried in microwaves but present in the samples dried in the freeze dryer and vice versa. Those include some alcohols, such as 3-Decyn-2-ol, Dodecanol, 3,7-dimethyl-1,7-octanediol, and Aldehydes and ketones, such as cyclohexenone and benzofuranone (Table 9). During drying, new aroma compounds were developed perhaps due to some chemical reaction. Those may include some aldehyde and ketones, which are the product of caramelization of sugars. We can site here some aroma compounds, such as furfural, which were more pronounced when the drying power was increasing as follows: 389.60 ± 11.95 μg/kg, 419.98 ± 52.98 μg/kg, 433.51 ± 14.96 μg/kg and 26.81 ± 5.45 μg/kg for samples dried in microwaves at 350 W, 500 W, 650 W, and FD. respectively. For other aromatic compounds imparted with the drying power, we can cite some aldehyde and ketones such as Decanal, Butanal-2-methyl, Isovaleraldehyde (3-methylbutanal), 5-Ethoxydihydro-2(3H)-furanone, 5-Methyl-2(5H)-furanone, H-pyrrol-2-yl-Ethanone, and Pyran-4-one (Table 9) which are mostly food flavorings. D-Limonene presence contributed to give a pleasant lemon-like odor [78,79] to the dried tree tomato fruit samples. Apart from the citrus aroma, the dried tree tomato fruit also possessed other chemical compositions, such as p-cimene, myrcene, α-pinene, Carene-1S-3R-6R-gamma-terpinene, and phellandrene which gives a herbacious, pine, and woody aroma and flavor [80,81]. The presence of linalool is the aromatic compound which makes the freeze-dried tree tomato fruit have a very nice and distinct fragrance as supported by the previous study of Letizia et al. [82].

**Table 9.** Aroma compounds isolated from dried tree tomatoes (μg/kg).

| Alcohol | | | | | |
|---|---|---|---|---|---|
| Compounds | | 350 W | 500 W | 650 W | FD |
| 1. | 3-Decyn-2-ol | nd | nd | nd | 33.62 ± 1.54 a |
| 2. | Dodecanol | nd | nd | nd | 6.12 ± 1.87 a |
| 3. | Eucalyptol | 213.84 ± 80.94 a | 47.48 ± 20.74 c | 54.35 ± 0.27 b | 63.58 ± 2.46 b |
| 4. | p-Mentha1,5-dien-8-ol | 11.46 ± 0.24 b | 12.26 ±1.87 b | 9.82 ± 0.40 c | 25.82 ±7.22 a |
| 5. | Dehydro-1-8-cineole | nd | 1.71 ± 1.41 b | nd | 7.43 ± 2.67 a |
| 6. | 3,7-dimethyl-1,7-octanediol | nd | nd | nd | 29.86 ± 0.00 |
| 7. | Furanmethanol | 7.03 ± 0.23 b | 8.34 ± 0.22 a | 7.46 ± 0.67 b | nd |

**Table 9.** *Cont.*

| | | | | | |
|---|---|---|---|---|---|
| **Alcohol** | | | | | |
| **Acids.** | | | | | |
| 8. | Pyridine carboxylic acid | 5.62 ± 0.57 a | nd | 4.17 ± 0.41 b | nd |
| 9. | Cyclopentylacetic acid | nd | 6.28 ± 0.71 a | 5.15 ± 1.18 b | nd |
| **Esters.** | | | | | |
| 10. | Amyl acetate | nd | 2.21 ± 0.46 a | 2.99 ± 2.99 a | nd |
| 11. | Benzoic acid, methyl ester | nd | 1.79 ± 1.79 c | 5.02 ± 0.48 b | 18.19 ± 0.42 a |
| 12. | Butanedioic acid, methyl-, ester | 17.84 ± 17.84 a | 4.56 ± 0.36 b | nd | nd |
| 13. | Butanoic acid, 4-ethyl ester | 30.47 ± 5.22 c | 47.11 ± 7.70 b | 51.43 ± 40.58 a | 7.18 ± 0.88 d |
| 14. | Ethyl Acetate | nd | nd | nd | 12.75 ± 0.44 |
| 15. | Hexanoic acid, methyl ester | 49.28 ± 0.21 b | 37.26 ± 23.84 c | 35.87 ± 25.06 c | 135.28 ± 1.00 a |
| 16. | Propanoic acid, 2-methyl-, 1-methylethyl ester | 5.09 ± 0.51 b | 6.26 ± 0.47 a | 4.06 ± 2.46 b | nd |
| 17. | Octanoic-acid-ethyl-ester | 5.16 ± 1.94 b | 2.30 ± 1.31 c | 5.33 ± 4.16 b | 12.72 ± 6.27 a |
| 18. | Linalyl acetate | nd | nd | nd | 10.16 ± 0.85 |
| **Terpene** | | | | | |
| 19. | Carene-1S-3R-6R-gamma-Terpinene | 4.25 ± 2.92 bc | 8.37 ± 5.84 a | nd | 5.97 ± 0.79 b |
| 20. | D-Limonene | 11.77 ± 2.16 c | 14.25 ± 10.02 b | 11.39 ± 6.66 c | 17.25 ± 2.49 a |
| 21. | Octane-2-methyl-N-N-Dimethyl piperazine | 5.24 ± 1.71 b | 1.39 ± 0.74 c | 4.82 ± 3.31 bc | 7.40 ± 7.40 a |
| 22. | p-cymene | 38.16 ± 0.93 a | 30.87 ± 5.51 b | 35.79 ± 7.09 ab | 27.99 ± 1.88 c |
| 23. | Phellandrene | nd | 2.61 ± 0.55 a | nd | nd |
| 24. | α-Pinene | | | | |
| 25. | β-Pinene/beta myrcene | nd | 4.48 ± 2.95 b | 5.08 ± 3.60 a | 3.70 ± 0.08 c |
| 26. | 2-Methyl-1,3,5-hexatriene | nd | nd | 4.66 ± 2.05 | nd |
| 27. | Linalool | nd | nd | nd | 9.90 ± 6.30 a |
| **Aldehydes and ketones** | | | | | |
| 28. | Acetaldehyde | 8.77 ± 0.76 bc | 9.12 ± 0.01 b | 8.02 ± 0.08 bc | 21.73 ± 16.77 a |
| 29. | TATP Butanal | 8.22 ± 0.70 c | 14.29 ± 1.59 b | 17.00 ± 0.91 a | nd |
| 30. | Pentanal | nd | 13.82 ± 6.36 | 12.20 ± 1.90 | nd |
| 31. | Heptanal | nd | 1.45 ± 0.86 a | nd | nd |
| 32. | Octanal | nd | 10.68 ± 6.77 a | nd | 1.84 ± 1.84 b |
| 33. | Nonanone | nd | 2.97 ± 1.92 b | nd | 19.61 ± 19.61 a |
| 34. | Sulcatone (6-methyl-5-hepten-2-one) | nd | 14.78 ± 11.61 a | nd | 4.64 ± 0.01 b |
| 35. | Nonanal | 11.25 ± 2.35 c | 29.90 ± 4.62 b | 12.92 ± 0.75 c | 48.35 ± 6.69 a |
| 36. | Decanal | 7.57 ± 0.84 b | 9.26 ± 0.10 a | 9.04 ± 0.44 a | nd |
| 37. | Furfural | 389.60 ± 11.95 c | 419.98 ± 52.98 ab | 433.51 ± 14.96 a | 26.81 ± 5.45 d |
| 38. | Ethanone | 13.39 ± 0.45 c | 17.95 ± 1.03 a | 17.67 ± 0.84 a | 15.73 ± 4.11 b |
| 39. | Butanal-2-methyl | 21.78 ± 0.75 c | 42.24 ± 9.11 b | 56.58 ± 11.76 a | nd |
| 40. | Isovaleraldehyde (3-methylbutanal) | 53.24 ± 5.62 b | 55.04 ± 36.46 b | 98.43 ± 16.99 a | nd |
| 41. | 5-Ethoxydihydro-2(3H)-furanone | 52.24 ± 0.87 c | 65.51 ± 7.21 b | 72.41 ± 3.30 a | nd |
| 42. | 5-Methyl-2(5H)-furanone | 6.28 ± 0.21 ab | 7.45 ± 0.70 a | 6.36 ± 0.48 ab | nd |
| 43. | 2-furan carboxaldehyde 5-methy | 44.08 ± 1.04 b | 53.70 ± 4.94 a | 53.55 ± 7.42 a | 6.17 ± 2.10 c |
| 44. | H-pyrrol-2-yl-Ethanone | 31.07 ± 8.79 c | 37.25 ± 2.02 a | 33.57 ± 11.51 b | nd |
| 45. | Pyran-4-one | 19.31 ± 8.84 a | 2.89 ± 2.89 c | 16.24 ± 6.10 ab | nd |
| 46. | Hydroxymethylfurfural (HMF) | 75.74 ± 6.63 a | 22.63 ± 1.31 c | 45.69 ± 11.16 b | nd |
| 47. | 2,5-Furandicarboxaldehyde | 12.07 ± 3.14 a | 6.40 ± 6.40 b | 11.72 ± 4.11 a | 1.91 ± 1.91 c |
| 48. | Cyclohexenone | nd | nd | nd | 9.50 ± 5.52 a |
| 49. | Benzofuranone | nd | nd | nd | 6.60 ± 4.53 a |
| **Others** | | | | | |
| 50. | Carbondioxide | 42.00 ± 0.21 b | 41.12 ± 1.56 b | 44.25 ± 5.15 a | 14.20 ± 7.52 c |
| 51. | Borane methyl sulfide complex | 5.35 ± 0.30 c | 12.63 ± 0.54 a | 6.61 ± 0.69 b | nd |
| 52. | Tetradecane | 30.97 ± 10.35 b | 16.83 ± 0.06 d | 23.63 ± 1.55 c | 34.76 ± 12.03 a |
| 53. | Benzenepropanamine | nd | nd | 4.64 ± 0.56 b | 17.96 ± 6.20 a |
| 54. | 2,2,4,6,6-pentamethylheptane | nd | nd | nd | 45.10 ± 3.06 a |

There is no statistical difference between the averages shown with the same letter in the same row ($p > 0.05$). nd: not detected.

## 4. Conclusions

The general objective of this study was to compare the phytochemical, antioxidant properties and aroma profiles, dehydration and rehydration kinetics, and antioxidant properties of tree tomato fruit dried using microwaves at different powers and a freeze dryer. The results obtained from this study revealed that the increase of drying power in the microwaves' drying method reduces the time of drying for the tree tomato fruit. The used drying methods well fit with the tested models for both dehydration and rehydration kinetics with the coefficient of variation $R^2$ approximately 0.99. The freeze drying method effects less the physical characteristics such as the color and the shape of the tree tomatoes more than microwaves drying method at different power. The phytochemical contents of phenolic compounds, flavonoids, vitamin C, Carotenoids and anthocyanin compound was changing according to the drying methods. Phenolic and flavonoids compounds increased with the increase of the drying power in microwaves' drying methods. Anthocyanin compounds were better preserved in the freeze drying method more than in other used methods. Epicatechin was the most abundant phenolic among the tested phenolic compounds. Pelargonidine-3-glucosite was the most abundant anthocyanin compound present in all dried tree tomato fruit samples and was greater in freeze-dried samples. Fifty-four aroma compounds were isolated in the dried tree tomato sample and were different from one another according to the used drying methods. The determined aromatic compounds are playing distinct and/or important roles in giving the distinct flavor and aroma to the dried tree tomato fruit.

**Author Contributions:** Conceptualization, M.A.N. and I.K.; data curation, M.A.N.; formal analysis, M.A.N.; investigation, M.A.N.; methodology, M.A.N. and I.K.; project administration, I.K.; Supervision, I.K.; writing—original draft, M.A.N.; writing—review & editing, I.K. All authors have read and agreed to the published version of the manuscript.

**Funding:** This research was funded by [Ondokuz Mayis University project office] grant number [PYO. MUH.1904.20.008]. The APC was funded by the corresponding author.

**Institutional Review Board Statement:** Not applicable.

**Informed Consent Statement:** Not applicable.

**Data Availability Statement:** Not applicable.

**Acknowledgments:** The authors acknowledge the help of Ondokuz Mayis University whose laboratories were used. The authors acknowledge also the help of Ondokuz Mayis University's BAP project office for supporting this research (Project No: PYO. MUH.1904.20.008).

**Conflicts of Interest:** The authors declare no conflict of interest.

## References

1. Jackson, D.; Looney, N.E.; Morley-Bunker, M. The distribution of fruits. In *Temperate and Subtropical Fruit Production*, 3rd ed.; CABI: Worcester, MA, USA, 2011.
2. Canova, L.; Bobbio, A.; Manganelli, A.M. Predicting fruit consumption: A multi-group application of the Theory of Planned Behavior. *Appetite* **2019**, *145*, 104490. [CrossRef] [PubMed]
3. Espin, S.; Gonzalez-Manzano, S.; Taco, V.; Poveda, C.; Ayuda-Durán, B.; Gonzalez-Paramas, A.M.; Santos-Buelga, C. Phenolic composition and antioxidant capacity of yellow and purple-red Ecuadorian cultivars of tree tomato (*Solanum betaceum* Cav.). *Food Chem.* **2016**, *194*, 1073–1080. [CrossRef]
4. Diep, T.; Pook, C.; Yoo, M. Phenolic and Anthocyanin Compounds and Antioxidant Activity of Tamarillo (*Solanum betaceum* Cav.). *Antioxidants* **2020**, *9*, 169. [CrossRef]
5. Zhu, X.; Ge, Y.; Wu, T.; Zhao, K.; Chen, Y.; Wu, B.; Zhu, F.; Zhu, B.; Cui, L. Co-infection with respiratory pathogens among COVID-2019 cases. *Virus Res.* **2020**, *285*, 198005. [CrossRef] [PubMed]
6. Chen, C.W.; Ho, C.T. Antioxidant properties of polyphenols extracted from green and black teas. *J. Food Lipids* **1995**, *2*, 35–46. [CrossRef]
7. Majidinia, M.; Bishayee, A.; Yousefi, B. Polyphenols: Major regulators of key components of DNA damage response in cancer. *DNA Repair* **2019**, *82*, 102679. [CrossRef]
8. Hu, J.; Wang, Z.; Tan, B.K.; Christian, M. Dietary polyphenols turn fat "brown": A narrative review of the possible mechanisms. *Trends Food Sci. Technol.* **2020**, *97*, 221–232. [CrossRef]
9. Changrue, V.; Raghavan, V.G.; Orsat, V.; Vijaya Raghavan, G. Microwave drying of fruits and vegetables. *Stewart Postharvest Rev.* **2006**, *2*, 1–7.
10. Onwude, D.; Bahrami, F.; Shrivastava, C.; Berry, T.; Cronje, P.; North, J.; Kirsten, N. Physics driven digital twins to quantify the impact of pre and postharvest variability on the end quality evolution of orange fruit. *engrXiv* **2022**. [CrossRef]

11. Shonte, T.T.; Duodu, K.; de Kock, H.L. Effect of drying methods on chemical composition and antioxidant activity of underutilized stinging nettle leaves. *Heliyon* **2020**, *6*, e03938. [CrossRef]
12. Thamburaj, S.; Rajagopal, V.; Palanivel, R.; Pugazhendhi, S. Effect of different drying treatments on total polyphenolics content and in-vitro biological properties of Ficus benghalensis fruit: A comparative study. *Biocatal. Agric. Biotechnol.* **2022**, *39*, 102249. [CrossRef]
13. Dziki, D.; Polak, R.; Rudy, S.; Krzykowski, A.; Gawlik-Dziki, U.; Różyło, R.; Miś, A.; Combrzyński, M. Simulation of the process kinetics and analysis of physicochemical properties in the freeze drying of kale. *Int. Agrophysics* **2018**, *32*, 49–56. [CrossRef]
14. Chu, Y.; Wei, S.; Ding, Z.; Mei, J.; Xie, J. Application of Ultrasound and Curing Agent during Osmotic De-hydration to Improve the Quality Properties of Freeze-Dried Yellow Peach (*Amygdalus persica*) Slices. *Agriculture* **2021**, *11*, 1069. [CrossRef]
15. Bhatta, S.; Janezic, T.S.; Ratti, C. Freeze-Drying of Plant-Based Foods. *Foods* **2020**, *9*, 87. [CrossRef]
16. Fante, L.; Noreña, C.P.Z. Quality of hot air dried and freeze-dried of garlic (*Allium sativum* L.). *J. Food Sci. Technol.* **2013**, *52*, 211–220. [CrossRef]
17. Ciurzyńska, A.; Lenart, A.; Gręda, K.J. Effect of pre-treatment conditions on content and activity of water and colour of freeze-dried pumpkin. *LWT-Food Sci. Technol.* **2014**, *59*, 1075–1081. [CrossRef]
18. Farina, V.; Cinquanta, L.; Vella, F.; Niro, S.; Panfili, G.; Metallo, A.; Cuccurullo, G.; Corona, O. Evolution of Carotenoids, Sensory Profiles and Volatile Compounds in Microwave-Dried Fruits of Three Different Loquat Cultivars (*Eriobotrya japonica* Lindl.). *Mater. Veg.* **2020**, *75*, 200–207. [CrossRef]
19. Lv, W.; Li, D.; Lv, H.; Jin, X.; Han, Q.; Su, D.; Wang, Y. Recent development of microwave fluidization technology for drying of fresh fruits and vegetables. *Trends Food Sci. Technol.* **2019**, *86*, 59–67. [CrossRef]
20. Sriwichai, T.; Sookwong, P.; Siddiqui, M.W.; Sommano, S.R. Aromatic profiling of Zanthoxylum myri-acanthum (makwhaen) essential oils from dried fruits using different initial drying techniques. *Ind. Crops Prod.* **2019**, *133*, 284–291. [CrossRef]
21. Santana, I.; Castelo-Branco, V.N.; Guimarães, B.M.; Silva, L.D.O.; Peixoto, V.O.D.S.; Cabral, L.M.C.; Freitas, S.P.; Torres, A.G. Hass avocado (*Persea americana* Mill.) oil enriched in phenolic compounds and tocopherols by expeller-pressing the unpeeled microwave dried fruit. *Food Chem.* **2019**, *286*, 354–361. [CrossRef]
22. Coklar, H.; Akbulut, M.; Kilinc, S.; Yildirim, A.; Alhassan, I. Effect of Freeze, Oven and Microwave Pretreated Oven Drying on Color, Browning Index, Phenolic Compounds and Antioxidant Activity of Hawthorn (*Crataegus orientalis*) Fruit. *Not. Bot. Horti Agrobot. Cluj-Napoca* **2018**, *46*, 449–456. [CrossRef]
23. Darvishi, H.; Asl, A.R.; Asghari, A.; Azadbakht, M.; Najafi, G.; Khodaei, J. Study of the drying kinetics of pepper. *J. Saudi Soc. Agric. Sci.* **2014**, *13*, 130–138. [CrossRef]
24. Viera, W.; Shinohara, T.; Samaniego, I.; Sanada, A.; Terada, N.; Ron, L.; Koshio, K. Phytochemical Composition and Antioxidant Activity of Passiflora spp. Germplasm Grown in Ecuador. *Plants* **2022**, *11*, 328. [CrossRef] [PubMed]
25. Llerena, W.; Samaniego, I.; Navarro, M.; Ortíz, J.; Angós, I.; Carrillo, W. Effect of modified atmosphere packaging (MAP) in the antioxidant capacity of arazá (*Eugenia stipitata* McVaugh), naranjilla (*Solanum quitoense* Lam.), and tree tomato (*Solanum betaceum* Cav.) fruits from Ecuador. *J. Food Process. Preserv.* **2020**, *44*, e14757. [CrossRef]
26. Pathare, P.; Opara, U.L.; Al-Said, F.A.-J. Colour Measurement and Analysis in Fresh and Processed Foods: A Review. *Food Bioprocess Technol.* **2012**, *6*, 36–60. [CrossRef]
27. Bai, J.-W.; Sun, D.-W.; Xiao, H.-W.; Mujumdar, A.; Gao, Z.-J. Novel high-humidity hot air impingement blanching (HHAIB) pretreatment enhances drying kinetics and color attributes of seedless grapes. *Innov. Food Sci. Emerg. Technol.* **2013**, *20*, 230–237. [CrossRef]
28. Juan, R.; Pastor, J.; Fernandez, I. SEM and light microscope observations on fruit and seeds in Scrophulariaceae from southwest Spain and their systematic significance. *Ann. Bot.* **2000**, *86*, 323–338. [CrossRef]
29. Liu, Z.; He, C.; Guo, C.; Chen, F.; Bhandari, B.; Zhang, M. Dehydration-triggered shape transformation of 4D printed edible gel structure affected by material property and heating mechanism. *Food Hydrocoll.* **2021**, *115*, 106608. [CrossRef]
30. Diamante, L.M.; Munro, P.A. Mathematical modelling of hot air drying of sweet potato slices. *Int. J. Food Sci. Technol.* **2007**, *26*, 99–109. [CrossRef]
31. Yaldiz, O.; Ertekin, C.; Uzun, H. Mathematical modeling of thin layer solar drying of sultana grapes. *Energy* **2001**, *26*, 457–465. [CrossRef]
32. Menges, H.O.; Ertekin, C. Mathematical modeling of thin layer drying of Golden apples. *J. Food Eng.* **2006**, *77*, 119–125. [CrossRef]
33. Page, G.E. *Factors Influencing the Maximum Rates of Air Drying Shelled Corn in Thin Layers*; Purdue University: West Lafayette, IN, USA, 1949.
34. Sharaf-Eldeen, Y.I.; Blaisdell, J.L.; Hamdy, M.Y. A Model for Ear Corn Drying. *Trans. ASAE* **1980**, *23*, 1261–1271. [CrossRef]
35. Yagcioglu, A.D.A.C.F. Drying characteristic of laurel leaves under different conditions. In Proceedings of the 7th International Congress on Agricultural Mechanization and Energy, Adana, Turkey, 26–27 May 1999; Faculty of Agriculture, Cukurova University: Adana, Turkey, 1999; pp. 565–569.
36. Singh, T.J.; Wang, J.H. The modulator-dependent protein kinase. A multifunctional protein kinase activatable by the Ca2+-dependent modulator protein of the cyclic nucleotide system. *J. Biol. Chem.* **1978**, *253*, 3387–3390.
37. Sharaf-Eldeen, Y.I.; HaMWy, M.Y.; Blaisdell, J.L. Falling rate drying of fully exposed biological materials: A review of mathematical models. *ASAE* **1979**, *79*, 6522–6543.

38. Midilli, A.D.N.A.N.; Kucuk, H.A.Y.D.A.R.; Yapar, Z.İ.Y.A. A new model for single-layer drying. *Dry. Technol.* **2002**, *20*, 1503–1513. [CrossRef]
39. Aral, S.; Beşe, A.V. Convective drying of hawthorn fruit (*Crataegus* spp.): Effect of experimental parameters on drying kinetics, color, shrinkage, and rehydration capacity. *Food Chem.* **2016**, *210*, 577–584. [CrossRef]
40. Vega-Gálvez, A.; Di Scala, K.; Rodríguez, K.; Lemus-Mondaca, R.; Miranda, M.; López, J.; Perez-Won, M. Effect of air-drying temperature on physico-chemical properties, antioxidant capacity, colour and total phenolic content of red pepper (*Capsicum annuum*, L. var. Hungarian). *Food Chem.* **2009**, *117*, 647–653. [CrossRef]
41. Peleg, M. An Empirical Model for the Description of Moisture Sorption Curves. *J. Food Sci.* **1988**, *53*, 1216–1217. [CrossRef]
42. Goula, A.M.; Adamopoulos, K.G. Modeling the Rehydration Process of Dried Tomato. *Dry. Technol.* **2009**, *27*, 1078–1088. [CrossRef]
43. Apar, D.K.; Demirhan, E.; ÖZBEK, B.; Dadali, G. Rehydration kinetics of microwave-dried okras as affected by drying conditions. *J. Food Process. Preserv.* **2009**, *33*, 618–634. [CrossRef]
44. Saguy, I.S.; Marabi, A.; Wallach, R. New approach to model rehydration of dry food particulates utilizing principles of liquid transport in porous media. *Trends Food Sci. Technol.* **2005**, *16*, 495–506. [CrossRef]
45. Vega-Galvez, A.; Notte-Cuello, E.; Lemus-Mondaca, R.; Zura, L.; Miranda, M. Mathematical modelling of mass transfer during rehydration process of Aloe vera (*Aloe barbadensis* Miller). *Food Bioprod. Process.* **2009**, *87*, 254–260. [CrossRef]
46. Noshad, M.; Mohebbi, M.; Shahidi, F.; Mortazavi, S.A. Kinetic modeling of rehydration in air-dried quinces pretreated with osmotic dehydration and ultrasonic. *J. Food Process. Preserv.* **2011**, *36*, 383–392. [CrossRef]
47. Benassi, M.D.T.; Antunes, A.J. A comparison of metaphosphoric and oxalic acids as extractants solutions for the determination of vitamin C in selected vegetables. *Arq. Biol. Tecnol.* **1988**, *31*, 507–513.
48. AOAC, H.W. *International A: Official Methods of Analysis of the AOAC International*; The Association: Arlington County, VA, USA, 2000.
49. Singleton, V.L.; Orthofer, R.; Lamuela-Raventós, R.M. Analysis of total phenols and other oxidation substrates and antioxidants by means of folin-ciocalteu reagent. In *Methods in Enzymology*; Academic Press: London, UK, 1999; Volume 299, pp. 152–178.
50. Grobelna, A.; Kalisz, S.; Kieliszek, M. The Effect of the Addition of Blue Honeysuckle Berry Juice to Apple Juice on the Selected Quality Characteristics, Anthocyanin Stability, and Antioxidant Properties. *Biomolecules* **2019**, *9*, 744. [CrossRef]
51. Zhishen, J.; Mengcheng, T.; Jianming, W. The determination of flavonoid contents in mulberry and their scavenging effects on superoxide radicals. *Food Chem.* **1999**, *64*, 555–559. [CrossRef]
52. Lee, J.; Durst, R.W.; Wrolstad, R.E. Determination of total monomeric anthocyanin pigment content of fruit juices, beverages, natural colorants, and wines by the pH differential method: Collaborative study. *J. AOAC Int.* **2005**, *88*, 1269–1278. [CrossRef]
53. Hernández-Carrión, M.; Hernando, I.; Quiles, A. High hydrostatic pressure treatment as an alternative to pasteurization to maintain bioactive compound content and texture in red sweet pepper. *Innov. Food Sci. Emerg. Technol.* **2014**, *26*, 76–85. [CrossRef]
54. Benzie, I.F.; Strain, J.J. The ferric reducing ability of plasma (FRAP) as a measure of "antioxidant power": The FRAP assay. *Anal. Biochem.* **1996**, *239*, 70–76. [CrossRef]
55. Zannou, O.; Koca, I.; Aldawoud, T.; Galanakis, C.M. Recovery and stabilization of anthocyanins and phenolic antioxidants of roselle (*Hibiscus sabdariffa* L.) with hydrophilic deep eutectic solvents. *Molecules* **2020**, *25*, 3715. [CrossRef]
56. Kalisz, S.; Oszmiański, J.; Kolniak-Ostek, J.; Grobelna, A.; Kieliszek, M.; Cendrowski, A. Effect of a variety of polyphenols compounds and antioxidant properties of rhubarb (*Rheum rhabarbarum*). *LWT* **2019**, *118*, 108775. [CrossRef]
57. Zannou, O.; Pashazadeh, H.; Galanakis, C.M.; Alamri, A.S.; Koca, I. Carboxylic acid-based deep eutectic solvents combined with innovative extraction techniques for greener extraction of phenolic compounds from sumac (*Rhus coriaria* L.). *J. Appl. Res. Med. Aromat. Plants* **2022**, *30*, 100380. [CrossRef]
58. Li, H.; Qin, D.; Wu, Z.; Sun, B.; Sun, X.; Huang, M.; Zheng, F. Characterization of key aroma compounds in Chinese Guojing sesame-flavor Baijiu by means of molecular sensory science. *Food Chem.* **2019**, *284*, 100–107. [CrossRef] [PubMed]
59. Cserhalmi, Z.; Sass-Kiss, Á.; Tóth-Markus, M.; Lechner, N. Study of pulsed electric field treated citrus juices. *Innov. Food Sci. Emerg. Technol.* **2006**, *7*, 49–54. [CrossRef]
60. Wojdyło, A.; Figiel, A.; Oszmianski, J. Effect of drying methods with the application of vacuum micro-waves on the bioactive compounds, color, and antioxidant activity of strawberry fruits. *J. Agric. Food Chem.* **2009**, *57*, 1337–1343. [CrossRef]
61. Figiel, A.; Michalska, A. Overall quality of fruits and vegetables products affected by the drying processes with the assistance of vacuum-microwaves. *Int. J. Mol. Sci.* **2016**, *18*, 71. [CrossRef]
62. Calín-Sánchez, Á.; Lipan, L.; Cano-Lamadrid, M.; Kharaghani, A.; Masztalerz, K.; Carbonell-Barrachina, A.; Figiel, A. Comparison of Traditional and Novel Drying Techniques and Its Effect on Quality of Fruits, Vegetables and Aromatic Herbs. *Foods* **2020**, *9*, 1261. [CrossRef]
63. Kaur, R.; Kaur, K. Effect of processing on color, rheology and bioactive compounds of different sweet pepper purees. *Plant Foods Hum. Nutr.* **2020**, *75*, 369–375. [CrossRef]
64. Jeong, H.K.; Lee, D.; Kim, H.P.; Baek, S.H. Structure analysis and antioxidant activities of an amylopectin-type polysaccharide isolated from dried fruits of Terminalia chebula. *Carbohydr. Polym.* **2019**, *211*, 100–108. [CrossRef]
65. Abbasi, H.; Layeghiniya, N.; Mohammadi, S.; Karimi, S. Effect of fruit thickness on microwave drying characteristics of *Myrtus communis* L. *Iran. J. Chem. Chem. Eng.* **2022**. [CrossRef]

66. Simal, S.; Femenia, A.; Garau, M.C.; Rosselló, C. Use of exponential, Page's and diffusional models to simulate the drying kinetics of kiwi fruit. *J. Food Eng.* **2005**, *66*, 323–328. [CrossRef]
67. Toğrul, İ.T.; Pehlivan, D. Modelling of thin layer drying kinetics of some fruits under open-air sun drying process. *J. Food Eng.* **2004**, *65*, 413–425. [CrossRef]
68. Castro, A.; Mayorga, E.; Moreno, F. Mathematical modelling of convective drying of fruits: A review. *J. Food Eng.* **2018**, *223*, 152–167. [CrossRef]
69. Bassey, E.J.; Cheng, J.H.; Sun, D.W. Improving drying kinetics, physicochemical properties and bioactive compounds of red dragon fruit (*Hylocereus species*) by novel infrared drying. *Food Chem.* **2022**, *375*, 131886. [CrossRef] [PubMed]
70. Roy, M.; Bulbul, A.I.; Hossain, M.A.; Shourove, J.H.; Ahmed, S.; Sarkar, A.; Biswas, R. Study on the drying kinetics and quality parameters of osmotic pre-treated dried Satkara (*Citrus macroptera*) fruits. *J. Food Meas. Charact.* **2021**, *16*, 471–485. [CrossRef]
71. Hayat, K.; Abbas, S.; Hussain, S.; Shahzad, S.A.; Tahir, M.U. Effect of microwave and conventional oven heating on phenolic constituents, fatty acids, minerals and antioxidant potential of fennel seed. *Ind. Crops Prod.* **2019**, *140*, 111610. [CrossRef]
72. Mphahlele, R.R.; Fawole, O.A.; Makunga, N.P.; Opara, U.L. Effect of drying on the bioactive compounds, antioxidant, antibacterial and antityrosinase activities of pomegranate peel. *BMC Complement. Altern. Med.* **2016**, *16*, 143. [CrossRef]
73. Kamiloglu, S.; Toydemir, G.; Boyacioglu, D.; Beekwilder, J.; Hall, R.D.; Capanoglu, E. A Review on the Effect of Drying on Antioxidant Potential of Fruits and Vegetables. *Crit. Rev. Food Sci. Nutr.* **2015**, *56*, S110–S129. [CrossRef]
74. Ng, Z.X.; Yong, P.H.; Lim, S.Y. Customized drying treatments increased the extraction of phytochemicals and antioxidant activity from economically viable medicinal plants. *Ind. Crops Prod.* **2020**, *155*, 112815. [CrossRef]
75. Saifullah, M.; McCullum, R.; McCluskey, A.; Vuong, Q. Effects of different drying methods on extractable phenolic compounds and antioxidant properties from lemon myrtle dried leaves. *Heliyon* **2019**, *5*, e03044. [CrossRef]
76. Papoutsis, K.; Pristijono, P.; Golding, J.B.; Stathopoulos, C.E.; Bowyer, M.C.; Scarlett, C.J.; Vuong, Q.V. Effect of vacuum-drying, hot air-drying and freeze-drying on polyphenols and antioxidant capacity of lemon (*Citrus limon*) pomace aqueous extracts. *Int. J. Food Sci. Technol.* **2017**, *52*, 880–887. [CrossRef]
77. Oklar, H.; Akbulut, M. Effect of sun, oven and freeze-drying on anthocyanins, phenolic compounds and antioxidant activity of black grape (*Ekşikara*) (*Vitis vinifera* L.). *S. Afr. J. Enol. Vitic.* **2017**, *38*, 264–272.
78. Mahdavi, S.A.; Sadeghi, R.; Faridi, A.; Hedayati, S.; Shaddel, R.; Dima, C.; Jafari, S.M. Nanodelivery systems for d-limonene; techniques and applications. *Food Chem.* **2022**, *384*, 132479. [CrossRef] [PubMed]
79. Lotfabadi, S.V.; Mortazavi, S.A.; Yeganehzad, S. Study on the release and sensory perception of encapsulated d -limonene flavor in crystal rock candy using the time–intensity analysis and HS-GC/MS spectrometry. *Food Sci. Nutr.* **2020**, *8*, 933–941. [CrossRef]
80. Li, K.; Zhou, R.; Jia, W.W.; Li, Z.; Li, J.; Zhang, P.; Xiao, T. Zanthoxylum bungeanum essential oil induces apoptosis of HaCaT human keratinocytes. *J. Ethnopharmacol.* **2016**, *186*, 351–361. [CrossRef]
81. Tatsadjieu, L.N.; Ngang, J.E.; Ngassoum, M.B.; Etoa, F.X. Antibacterial and antifungal activity of Xylo-pia aethiopica, Monodora myristica, Zanthoxylum xanthoxyloıdes and Zanthoxylum leprieurii from Cameroon. *Fitoterapia* **2003**, *74*, 469–472. [CrossRef]
82. Letizia, C.; Cocchiara, J.; Lalko, J.; Api, A. Fragrance material review on linalool. *Food Chem. Toxicol.* **2003**, *41*, 943–964. [CrossRef]

processes

MDPI

Article

# Controlled Germination of Faba Beans: Drying, Thermodynamic Properties and Physical-Chemical Composition

Lumara Tatiely Santos Amadeu [1], Alexandre José de Melo Queiroz [1,*], Rossana Maria Feitosa de Figueirêdo [1], João Paulo de Lima Ferreira [1], Wilton Pereira da Silva [1], Josivanda Palmeira Gomes [1], Yaroslávia Ferreira Paiva [2], Caciana Cavalcanti Costa [3], Henrique Valentim Moura [1], Dyego da Costa Santos [4], Ana Raquel Carmo de Lima [5] and Hanndson Araujo Silva [2]

1 Department of Agricultural Engineering, Federal University of Campina Grande, Campina Grande 58429-900, Brazil; lumaratatielyea@gmail.com (L.T.S.A.); rossanamff@gmail.com (R.M.F.d.F.); joaop_l@hotmail.com (J.P.d.L.F.); wiltonps@uol.com.br (W.P.d.S.); josivanda@gmail.com (J.P.G.); valentim_henrique@hotmail.com (H.V.M.)
2 Science and Technology Center, Federal University of Campina Grande, Campina Grande 58429-900, Brazil; yaroslaviapaiva@gmail.com (Y.F.P.); hanndson@gmail.com (H.A.S.)
3 Center for Sustainable Development of the Semiarid, Federal University of Campina Grande, Sumé 58540-000, Brazil; caciana.cavalcanti@professor.ufcg.edu.br
4 Department of Technology in Agroindustry, Federal Institute of Education, Science and Technology of Rio Grande do Norte, Pau dos Ferros 59900-000, Brazil; dyego.csantos@gmail.com
5 Department of Technology in Agroindustry, Federal Institute of Education, Science and Technology of de Alagoas, Batalha 57420-000, Brazil; ana.carmo@ifal.edu.br
* Correspondence: alexandre.melo@professor.ufcg.edu.br

**Citation:** Amadeu, L.T.S.; Queiroz, A.J.d.M.; Figueirêdo, R.M.F.d.; Ferreira, J.P.d.L.; Silva, W.P.d.; Gomes, J.P.; Paiva, Y.F.; Costa, C.C.; Moura, H.V.; Santos, D.d.C.; et al. Controlled Germination of Faba Beans: Drying, Thermodynamic Properties and Physical-Chemical Composition. *Processes* **2022**, *10*, 1460. https://doi.org/10.3390/pr10081460

Academic Editor: Blaž Likozar

Received: 8 July 2022
Accepted: 22 July 2022
Published: 26 July 2022

**Publisher's Note:** MDPI stays neutral with regard to jurisdictional claims in published maps and institutional affiliations.



**Abstract:** The objective of this work was to determine the drying kinetics and the thermodynamic properties of the drying process of germinated seeds from faba beans of the Olho-de-Vó Preta (OVP), Raio-de-Sol (RS) and Branca (B) varieties. Additionally, the physicochemical properties of the germinated seeds and subsequent dried flours were determined. A thin layer of seeds were dried using a convective dryer at temperatures of 50, 60, 70 and 80 °C. Mathematical models were applied to the drying experimental data. The samples were further characterized for water content, water activity, ash, pH, alcohol-soluble acidity, total and reducing sugars, proteins, and starch. Page and Midilli models revealed the best predictions of the drying kinetics for all evaluated conditions. The effective diffusion coefficient increased with increasing temperature and presented magnitude in the order of $10^{-9}$ $m^2/s$. The activation energy presented results in the range of 19 and 27 kJ/mol, falling within the range reported for agricultural products. The entropy and enthalpy values were higher in the OVP, followed by RS, higher than in the B variety. The increase in drying temperature resulted in a reduction of enthalpy and entropy and an increase in Gibbs free energy, indicating that the drying process is endothermic and requires external energy. Samples have acidic pH and acidity decreased with drying; the RS and B varieties had higher sugar contents; the B variety had the highest protein contents, and these were obtained from the *in natura* germinated samples; in the B variety the highest starch content was obtained. All flours showed good characteristics, presenting themselves as an alternative for diversifying the supply of beans.

**Keywords:** *Phaseolus lunatus* L.; malting; drying kinetics; mathematical modeling; flour; nutrition

## 1. Introduction

The faba bean (*Phaseolus lunatus* L.) is also known as the lima bean [1], which is a legume grown mainly in the Northeast region of Brazil. Its production corresponded to 11,381 tons in 2019, in 36,252 hectares, with the states of Ceará (4.614 t), Paraiba (2.910 t) and Rio Grande do Norte (1.274 t) being the three largest producers in Brazil [2]. Despite being a culture considered for subsistence, it presents significant commercial exploitation due to the high commercial value of the grains and is considered the second most consumed bean species in Brazil [3,4]. The faba bean seeds are high in protein (16 to 21%) and carbohydrates (55 to 61%), low in fat (1 to 2.3%), having fiber levels from 3.2 to 6.8%, elevated levels of minerals such as K, Zn, Ca and Fe and low levels of Na and P [5]. The beans contain protease inhibitors, which help fight the development of cancer cells [6].

With the growing trend in the search for healthy and nutritious products, techniques that increase the nutritional value of foods at a low cost are of permanent interest. The improvement of the nutritional value of seeds can be done through the controlled use of the germination process, which conveniently alters the quality of the original raw material, increasing commercial potential and application possibilities. After germination, the seeds have high water content, a factor that compromises quality and confers high perishability to the product, requiring prior treatment that prevents rapid deterioration, enabling further processing or commercialization.

Among the technologies that enable the extension of the useful life of grains and seeds, convective drying is the most widely used, offering attractive costs and operational simplicity. Convective drying is one of the most used technologies for the preservation of agricultural products, combining benefits of increased shelf life, weight, and volume reduction, consequently reducing costs with packaging, transport and storage. Furthermore, drying can preserve the nutritional quality of faba bean, while providing easy availability to an increased added value product [7]. However, the inadequate use of convective drying can harm the physical, sensory, and nutritional quality of the product. Thus, the drying process should be planned accordingly to avoid damage, serving the control and administration of various stages of agribusiness [8,9].

The behavior of a product subjected to a dehydration process is evaluated using drying kinetics, which involves different conditions such as temperature and air velocity [10]. The drying process can be represented by mathematical modeling, helping to improve and design equipment, in addition to providing information for the optimization of the process for each raw material [11,12]. Beyond the drying kinetics, the determination of thermodynamic properties plays a significant role in the process, as it helps to calculate the energy needed to remove the moisture from the sample. These properties help with evaluating the physical phenomena that occurs on the surface of the products and contribute to the scale-up of the process [13]. Studies on drying kinetics deserve interest due to the different biological structures involved in the heat and mass transfer of each product [14].

Even though several studies related to bean culture can be found, different varieties still lack work. Additionally works related to the convective drying of germinated seeds are scarce, including faba bean varieties. Therefore, the objective of our work was to determine the drying kinetics of germinated faba bean seeds, from the varieties Olho-de-Vó, Raio-de-Sol and Branca at temperatures of 50, 60, 70 and 80 °C, to determine the thermodynamic properties of the process and to characterize physico-chemical properties of the germinated seeds and subsequent dried flours.

## 2. Materials and Methods

### 2.1. Material

Three varieties (Orelha-de-Vó Preta, Raio-de-Sol and Branca) of faba bean seeds (*Phaseolus lunatus* L.) produced in the region of the city of Campina Grande were used (Geographical coordinates: 7°13′50″ S and 35°52′52″ O), Paraiba, Brazil. The seeds were received and hand-picked, choosing the intact ones and with uniform size.

### 2.2. Processing and Germination of Seeds

About 100 g of seeds of each faba bean variety were sanitized by immersion in a 7% ($w/v$) sodium hypochlorite solution at a 1:10 ratio (seeds/solution) at room temperature (28 ± 1 °C) for 5 min. Soon after sanitization, the seeds were washed with distilled water, drained, and placed in trays at room temperature to eliminate surface water. The seeds were germinated following the recommendations of the Rules for Seed Analysis [15], using germitest paper. Fifty-six seeds were distributed per leaf, kept in BOD-type chambers at 25 °C, for 72 h (Olho-de-vó preta) and 96 h (Raio-de-Sol and Branca). Germination times were chosen based on previous studies [16,17] and preliminary germination tests (data not shown). The seeds were irrigated every 24 h, applying about 25 mL of distilled water.

### 2.3. Drying Procedure

The germinated faba bean seeds were crushed in a domestic blender and spread on stainless steel screened trays, in a thin layer, with an approximate height of 0.64 cm. The samples were subjected to drying in a convective dryer at temperatures of 50, 60, 70 and 80 °C and drying air speed of 1.0 m/s. The dryings were performed in triplicate, with the samples being weighed on an analytical balance with a precision of 0.0001 g, at regular times of 5, 10, 20, 30 and 60 min, until a constant mass was obtained. The water content at the end of the drying kinetics was determined gravimetrically by drying in an oven at 105 °C for 24 h [18].

### 2.4. Data Modeling

The drying data were used after converting the moisture loss data into the dimensionless moisture content ratio (MR) parameter, according to Equation (1).

$$MR = \frac{M_t - M_e}{M_i - M_e} \quad (1)$$

where, MR is the moisture content ratio (dimensionless); M is the moisture content (% dry basis); $M_e$ is the equilibrium moisture content (% dry basis); $M_0$ is the initial moisture content (% dry basis). Then, the mathematical models presented in Table 1 were fitted to the drying kinetics data, represented by the ratio of moisture content as a function of drying time, using the computer program Statistica version 7.0 (StatSoft® Inc., Tulsa, OK, USA) using regression nonlinear and the Quasi-Newton method.

**Table 1.** Mathematical models used to estimate the drying kinetics curves of the faba beans.

| Model Name | Equation | | References |
|---|---|---|---|
| Newton | $MR = \exp(-kt)$ | (2) | [19] |
| Page | $MR = \exp(-kt^n)$ | (3) | [20] |
| Henderson and Pabis | $MR = a\ \exp(-kt)$ | (4) | [21] |
| Modified Henderson and Pabis | $MR = a\ \exp(-kt) + b\exp(-k_0 t) + c\ \exp(-k_1 t)$ | (5) | [22] |
| Thompson | $MR = epx(-a - (a^2 + 4bt)^{0,5})/2b$ | (6) | [23] |
| Logarithmic | $MR = a\ \exp(-kt) + c$ | (7) | [24] |
| Two terms | $MR = (-k_0 t) + b\ \exp(-k_1 t)$ | (8) | [25] |
| Midilli | $MR = a\ \exp(-kt^n) + bt$ | (9) | [26] |
| Approximation of Diffusion | $MR = a\ \exp(-kt) + (1-a)\exp(-kbt)$ | (10) | [27] |
| Two-term exponential | $MR = a\ \exp(-kt) + (1-a)\exp(-kat)$ | (11) | [27] |
| Verma | $MR = a\ \exp(-kt) + (1-a)\exp(-k_1 at)$ | (12) | [28] |

MR—moisture content ratio (dimensionless); a, b, c, k, k0, k1, n—model parameters; t—drying time (min).

The mathematical models were evaluated for the quality of the adjustments, taking as parameters the magnitude of the coefficient of determination ($R^2$), the mean square deviation (MSD) (Equation (13)) and the chi-square ($\chi^2$) (Equation (14)).

$$MSD = \left[\frac{1}{N}\sum_{i=1}^{N}\left(MR_{pred,i} - MR_{exp,i}\right)^2\right]^{\frac{1}{2}} \quad (13)$$

$$\chi^2 = \frac{1}{N-n}\sum_{i=1}^{N}\left(MR_{pred,i} - MR_{exp,i}\right)^2 \quad (14)$$

where, $MR_{pred}$ is the moisture content ratio predicted by the model; $MR_{exp}$ is the experimental moisture content ratio; N is the number of observations; n is the number of model constants.

## 2.5. Determination of the Effective Diffusion Coefficient and Activation Energy

The effective diffusion coefficients were determined by fitting the mathematical model of liquid diffusion, with approximation of four terms (Equation (15)), to the experimental data of the drying kinetics, considering the uniform initial moisture distribution, constant diffusivity and external resistance and negligible volume contraction. This model is the analytical solution of Fick's second law considering the geometric form of the material as approximated to a flat plate [29].

$$MR = \frac{M_t - M_e}{M_i - M_e} = \frac{8}{\pi^2} \sum_{n=0}^{\infty} \frac{1}{(2n+1)^2} \exp\left[-(2n+1)^2 \pi^2 D_{ef} \frac{t}{4L^2}\right] \tag{15}$$

where, $D_{ef}$ is the effective diffusion coefficient ($m^2/s$); n is the number of terms in the equation; L is the characteristic dimension (half sample thickness) (m); t is the time(s).

The influence of temperature on the effective diffusion coefficients was evaluated using an Arrhenius-type equation (Equation (16)).

$$D_{ef} = D_0 \left(-\frac{E_e}{RT}\right) \tag{16}$$

where, $D_0$ is the pre-exponential factor ($m^2/s$); $E_a$ is the activation energy (kJ/mol); R is the universal gas constant (0.008314 kJ/mol K); T is the absolute temperature (K).

## 2.6. Determination of the Thermodynamic Properties

The thermodynamic properties of enthalpy (Equation (17)), entropy (Equation (18)) and Gibbs free energy (Equation (19)) in the drying process of germinated faba bean seeds were quantified using the equations described by Silva et al. [30].

$$\Delta H = E_a - RT \tag{17}$$

$$\Delta S = R\left[\ln D_0 - \ln\left(\frac{k_b}{h_p}\right) - \ln(T)\right] \tag{18}$$

$$\Delta G = \Delta H - T\Delta S \tag{19}$$

where, $\Delta H$ is the specific enthalpy (J/mol); $\Delta S$ is the specific entropy (J/mol K); $\Delta G$ is the Gibbs free energy (J/mol); Kb is the Boltzmann constant ($1.38 \times 10^{-23}$ J/K); hp is the Planck's constant ($6.626 \times 10^{-34}$ J/s); T is the absolute temperature (K).

## 2.7. Physicochemical Characterization

The germinated seeds and the flours from the dried germinated seeds were characterized, in triplicate, for each physicochemical property. According to the analytical procedures of Adolfo Lutz Institute [18], the moisture content was determined by the gravimetric method in an oven at 105 °C/24 h; alcohol-soluble acidity, by titration with 0.1 M NaOH; pH determined in digital potentiometer; ash content, by incineration in a muffle at 550 °C. Water activity was determined at 25 °C by direct reading using an Aqualab meter (3TE model, Decagon Devices, São José dos Campos, Brazil). The reducing sugars were determined by the method of dinitrosalicylic acid [31]; total sugars by the anthrone method [32]; starch by the methodology of the Adolf Lutz Institute [18], which is based on the acid hydrolysis of starch and on the titration of the Fehling solution; and the protein content, quantified according to the Kjeldahl method, where the total nitrogen content is determined, with the total protein being determined by multiplying the total nitrogen content by a factor of 6.25.

### 2.8. Statistical Analysis

The results of the physicochemical analyses were expressed as the mean ± standard. The data analysis was performed in a completely randomized design and the differences between treatment means were determined using one-way analysis of variance (ANOVA) and applying the Tukey test at 5% probability, using Assistat software version 7.7 Beta (Federal University of Campina Grande, Campina Grande, Paraíba, Brazil) [33].

## 3. Results

### 3.1. Mathematical Modeling of Drying Kinetics

Table 2 shows the coefficients of determination ($R^2$), the mean square deviations (MSD) and the chi-squares ($\chi^2$) obtained for each mathematical model adjusted to the drying kinetics of the germinated seeds of faba beans, varieties Orelha-de-Vó Preta (OVP), Raio-de-Sol (RS) and Branca (B), at temperatures of 50, 60, 70 and 80 °C. It is observed that for the three varieties under study, all mathematical models presented $R^2 > 0.9900$, MSD $\leq 0.040$ and $\chi^2 \leq 13.2843 \times 10^{-4}$. For a model to satisfactorily represent a drying process, it is essential that the coefficient of determination ($R^2$) is greater than 0.99, and that the mean square deviations (MSD) and chi-squares ($\chi^2$) have the lowest possible values [34]. It is observed that all models, as they present $R^2$ above 0.990 and values close to zero for MSD and $\chi^2$, can be used to represent the behavior under drying from 50 to 80 °C of the three varieties of germinated seeds of faba beans. However, of the 11 models assessed, the Page and Midilli models stood out for presenting the highest $R^2$ ($\geq$0.9997) and the lowest MSD ($\leq$0.0064) and $\chi^2$ ($\leq 0.4647 \times 10^{-4}$), demonstrating excellent fits to the experimental data.

**Table 2.** Coefficients of determination ($R^2$), mean square deviations (MSD) and chi-squares ($\chi^2$) of the mathematical models adjusted to the drying kinetics data of the germinated seeds of faba beans (varieties OVP, RS, B), at temperatures of 50, 60, 70 and 80 °C.

| Models | T (°C) | OVP | | | RS | | | B | | |
|---|---|---|---|---|---|---|---|---|---|---|
| | | $R^2$ | MSD | $\chi^2$ | $R^2$ | MSD | $\chi^2$ | $R^2$ | MSD | $\chi^2$ |
| Newton | 50 | 0.9991 | 0.0119 | 1.4696 | 0.9984 | 0.0165 | 2.8440 | 0.9957 | 0.0262 | 7.1549 |
| | 60 | 0.9975 | 0.0193 | 3.8834 | 0.9982 | 0.0162 | 2.7445 | 0.9953 | 0.0270 | 7.5771 |
| | 70 | 0.9980 | 0.0176 | 3.2418 | 0.9975 | 0.0190 | 3.7827 | 0.9961 | 0.0241 | 6.0580 |
| | 80 | 0.9970 | 0.0204 | 4.3462 | 0.9968 | 0.0212 | 4.6884 | 0.9940 | 0.0302 | 9.5150 |
| Page | 50 | 0.9999 | 0.0040 | 0.1757 | 0.9999 | 0.0037 | 0.1518 | 0.9999 | 0.0038 | 0.1585 |
| | 60 | 0.9999 | 0.0048 | 0.2529 | 0.9999 | 0.0041 | 0.1823 | 0.9998 | 0.0054 | 0.3176 |
| | 70 | 0.9998 | 0.0058 | 0.3726 | 0.9998 | 0.0055 | 0.3307 | 0.9997 | 0.0064 | 0.4415 |
| | 80 | 0.9997 | 0.0060 | 0.3912 | 0.9997 | 0.0064 | 0.4505 | 0.9998 | 0.0062 | 0.4183 |
| Henderson and Pabis | 50 | 0.9994 | 0.0096 | 0.9975 | 0.9989 | 0.0124 | 1.6673 | 0.9970 | 0.0203 | 4.4385 |
| | 60 | 0.9981 | 0.0156 | 2.6597 | 0.9987 | 0.0130 | 1.8245 | 0.9965 | 0.0219 | 5.2130 |
| | 70 | 0.9984 | 0.0157 | 2.7129 | 0.9980 | 0.0171 | 3.1914 | 0.9969 | 0.0216 | 5.0900 |
| | 80 | 0.9974 | 0.0188 | 3.8938 | 0.9974 | 0.0192 | 4.0333 | 0.9952 | 0.0271 | 7.9877 |
| Modified Henderson and Pabis | 50 | 0.9994 | 0.0096 | 1.2073 | 0.9989 | 0.0124 | 2.0007 | 0.9970 | 0.0203 | 5.2840 |
| | 60 | 0.9981 | 0.0156 | 3.2856 | 0.9987 | 0.0130 | 2.2086 | 0.9965 | 0.0219 | 6.2556 |
| | 70 | 0.9984 | 0.0157 | 3.3512 | 0.9980 | 0.0171 | 3.9006 | 0.9969 | 0.0216 | 6.2211 |
| | 80 | 0.9974 | 0.0188 | 4.8673 | 0.9974 | 0.0192 | 4.9823 | 0.9952 | 0.0271 | 9.7628 |
| Thompson | 50 | 0.9974 | 0.0207 | 4.6441 | 0.9981 | 0.0174 | 3.2902 | 0.9953 | 0.0280 | 8.4414 |
| | 60 | 0.9970 | 0.0221 | 5.3326 | 0.9974 | 0.0206 | 4.6071 | 0.9928 | 0.0350 | 13.2843 |
| | 70 | 0.9979 | 0.0180 | 3.5467 | 0.9973 | 0.0196 | 4.1802 | 0.9959 | 0.0247 | 6.6588 |
| | 80 | 0.9927 | 0.0315 | 10.9041 | 0.9957 | 0.0247 | 6.6753 | 0.9924 | 0.0339 | 12.5472 |

**Table 2.** *Cont.*

| Models | T (°C) | OVP | | | RS | | | B | | |
|---|---|---|---|---|---|---|---|---|---|---|
| | | $R^2$ | MSD | $\chi^2$ | $R^2$ | MSD | $\chi^2$ | $R^2$ | MSD | $\chi^2$ |
| Logarithmic | 50 | 0.9994 | 0.0096 | 1.0440 | 0.9989 | 0.0124 | 1.7365 | 0.9971 | 0.0200 | 4.5178 |
| | 60 | 0.9982 | 0.0155 | 2.7747 | 0.9987 | 0.0129 | 1.9020 | 0.9967 | 0.0216 | 5.2967 |
| | 70 | 0.9984 | 0.0157 | 2.8464 | 0.9980 | 0.0171 | 3.3336 | 0.9969 | 0.0214 | 5.2362 |
| | 80 | 0.9974 | 0.0188 | 4.0958 | 0.9974 | 0.0192 | 4.2343 | 0.9953 | 0.0268 | 8.2042 |
| Two terms | 50 | 0.9994 | 0.0096 | 1.0925 | 0.9989 | 0.0124 | 1.8189 | 0.9970 | 0.0203 | 4.8277 |
| | 60 | 0.9981 | 0.0156 | 2.9397 | 0.9987 | 0.0130 | 1.9983 | 0.9965 | 0.0219 | 5.6874 |
| | 70 | 0.9984 | 0.0157 | 2.9984 | 0.9980 | 0.0171 | 3.5105 | 0.9969 | 0.0216 | 5.5990 |
| | 80 | 0.9974 | 0.0188 | 4.3265 | 0.9974 | 0.0192 | 4.4578 | 0.9952 | 0.0271 | 8.7865 |
| Midilli | 50 | 0.9999 | 0.0037 | 0.1628 | 0.9999 | 0.0033 | 0.1315 | 0.9999 | 0.0035 | 0.1467 |
| | 60 | 0.9999 | 0.0046 | 0.2511 | 0.9999 | 0.0039 | 0.1771 | 0.9998 | 0.0050 | 0.2922 |
| | 70 | 0.9998 | 0.0055 | 0.3663 | 0.9998 | 0.0053 | 0.3409 | 0.9997 | 0.0062 | 0.4647 |
| | 80 | 0.9998 | 0.0057 | 0.3928 | 0.9997 | 0.0061 | 0.4523 | 0.9998 | 0.0060 | 0.4252 |
| Approximation of Diffusion | 50 | 0.9999 | 0.0037 | 0.1547 | 0.9999 | 0.0033 | 0.1233 | 0.9999 | 0.0038 | 0.1615 |
| | 60 | 0.9999 | 0.0044 | 0.2193 | 0.9999 | 0.0038 | 0.1599 | 0.9998 | 0.0059 | 0.3987 |
| | 70 | 0.9998 | 0.0055 | 0.3429 | 0.9983 | 0.0157 | 2.8032 | 0.9976 | 0.0191 | 4.1524 |
| | 80 | 0.9970 | 0.0204 | 4.8037 | 0.9977 | 0.0180 | 3.7214 | 0.9960 | 0.0247 | 6.9530 |
| Two-term exponential | 50 | 0.9991 | 0.0122 | 1.6292 | 0.9983 | 0.0169 | 3.0821 | 0.9956 | 0.0265 | 7.5949 |
| | 60 | 0.9974 | 0.0195 | 4.1739 | 0.9981 | 0.0165 | 2.9612 | 0.9952 | 0.0273 | 8.0462 |
| | 70 | 0.9979 | 0.0180 | 3.5331 | 0.9975 | 0.0190 | 3.9547 | 0.9960 | 0.0244 | 6.4809 |
| | 80 | 0.9998 | 0.0059 | 0.3800 | 0.9997 | 0.0061 | 0.4106 | 0.9939 | 0.0305 | 10.1270 |
| Verna | 50 | 0.9999 | 0.0037 | 0.1547 | 0.9999 | 0.0033 | 0.1233 | 0.9957 | 0.0262 | 7.7511 |
| | 60 | 0.9975 | 0.0193 | 4.2718 | 0.9982 | 0.0162 | 2.9940 | 0.9953 | 0.0270 | 8.2360 |
| | 70 | 0.9980 | 0.0176 | 3.5660 | 0.9975 | 0.0190 | 4.1430 | 0.9961 | 0.0241 | 6.6349 |
| | 80 | 0.9979 | 0.0170 | 3.3346 | 0.9968 | 0.0212 | 5.1572 | 0.9940 | 0.0302 | 10.4212 |

OVP—Orelha-de-Vó Preta; RS—Raio-de-Sol; B—Branca.

Among the models that presented the best fits, although Midilli had the smallest mean squared deviations and reduced chi-squares, the Page model, being simpler and using only two parameters, simplifying its application in mathematical simulations, is the recommended model to represent drying in a thin layer of the germinated faba bean seeds. The Midilli and Page models were also reported as satisfactory for estimating the drying kinetic curves by Lisboa et al. [35] for mulatto beans (*Phaseolus vulgaris* L.), at temperatures of 40, 50, 60 and 70 °C, with drying air speed of 1.0 m/s; and by Rahmanian-Koushkaki et al. [36] for corn seeds, at temperatures of 40, 50 and 60 °C, in which among the models tested, the Page was the most suitable. Figure 1 shows the experimental points and curves estimated by the Page model for the ratio of water content as a function of drying time for germinated seeds of faba beans of the OVP, RS and B varieties, at temperatures of 50, 60, 70 and 80 °C.

The best use of the energy spent on drying is observed to be the period between 50 to 100 min, with moisture content ratios close to zero from 100 min onwards, even at a temperature of 50 °C. According to Zielinska and Michalska et al. [37] this reflects the high-moisture content available under weak molecular binding and as the process progressed, drying rates decreased, possibly due to increased internal resistance to heat and mass transfer. This behavior was also reported by Chielle et al. [38] on papaya seeds (*Carica papaya* L.), Hasan et al. [39] on rice seeds. Increasing the temperature increases the difference between the vapor pressure of the drying air and the samples, therefore, higher temperatures result in greater and faster water removal, as observed in watermelon seeds [40], canola [41] and common beans [42].

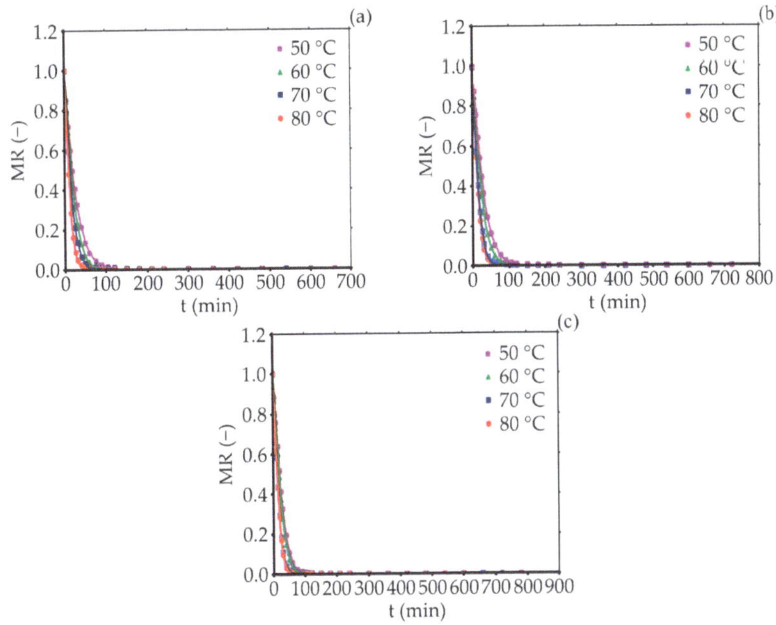


**Figure 1.** Experimental and estimated values by the Page model for the ratio of water content as a function of drying time of germinated faba bean seeds at temperatures of 50, 60, 70 and 80 °C: (**a**) OVP; (**b**) RS; and (**c**) B.

The average time required to complete the drying process of the 'Orelha-de-Vó' variety (OVP) ranged from 480 to 660 min for temperatures between 50 and 80 °C, for the 'Raio-de-Sol' variety (RS) the process lasted between 540 and 720 min and, finally, in the 'Branca' variety (B), between 600 and 720 min. Similar drying times were verified by Ferreira et al. [43] when analyzing the drying kinetics of germinated pumpkin seeds, which reported drying times between 470 min (70 °C) and 720 min (50 °C). Lisboa et al. [35], in the mathematical description of the drying curves of mulatto beans (*Phaseolus vulgaris* L.), observed an average time of 1300, 1000, 880 and 640 min to complete the drying process at temperatures of 40, 50, 60 and 70 ° C, respectively.

In Figure 2, we have the representation of the experimental moisture content ratio values and the values predicted by the Page model. The good prediction, represented by the curve, is verified by its superposition with the experimental points determined in the drying kinetics, corroborating the satisfactory results of $R^2$, MSD and $\chi^2$.

Table 3 shows the parameters of the Page model adjusted to the drying kinetics data of germinated seeds of faba beans of different varieties at temperatures of 50, 60, 70 and 80 °C OVP and intermediate values in the RS. For the drying parameter "k", the lowest values, while comparing the same temperatures, were found for the Branca variety, and the highest in the OVP, but close to RS. According to Lisboa et al. [35], this constant represents the effect of external drying conditions, indicating that the drying rate increases with an increase in air temperature. The authors reported similar behavior to those observed here when drying mulatto beans (*Phaseolus vulgaris* L.) at temperatures of 40, 50, 60 and 70 °C, on what "k" and "n" showed as an increase as the temperature was increased.

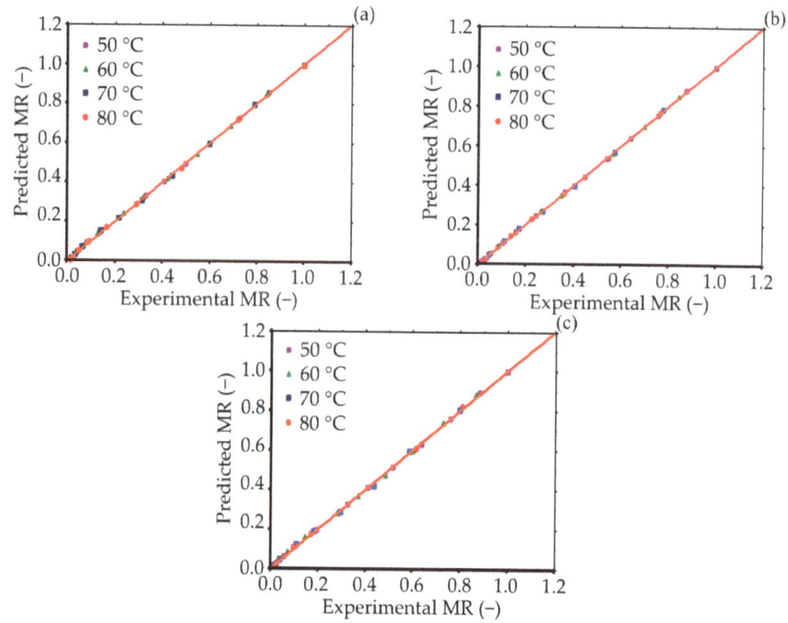


**Figure 2.** Relationship between the values predicted by the Page model and the experimental water content ratio values in the drying of germinated faba bean seeds at temperatures of 50, 60, 70 and 80 °C: (**a**) OVP; (**b**) RS; and (**c**) B.

**Table 3.** Parameters of the Page model adjusted to experimental data of drying kinetics of germinated seeds of faba bean varieties, at temperatures of 50, 60, 70 and 80 °C.

| Variety | T (°C) | k | n |
|---|---|---|---|
| OVP | 50 | 0.0257 | 1.1067 |
| | 60 | 0.0233 | 1.2085 |
| | 70 | 0.0339 | 1.1829 |
| | 80 | 0.0426 | 1.2444 |
| RS | 50 | 0.0194 | 1.1591 |
| | 60 | 0.0242 | 1.1731 |
| | 70 | 0.0342 | 1.2149 |
| | 80 | 0.0351 | 1.2453 |
| B | 50 | 0.0138 | 1.2923 |
| | 60 | 0.0151 | 1.3030 |
| | 70 | 0.0273 | 1.2749 |
| | 80 | 0.0211 | 1.3669 |

OVP—Orelha-de-Vó Preta; RS—Raio-de-Sol; B—Branca.

### 3.2. *Diffusion Coefficient and Activation Energy*

Table 4 shows the average effective diffusion coefficients ($D_{ef}$) obtained from drying germinated faba beans at temperatures of 50, 60, 70 and 80 °C. It is observed that the values of the diffusion coefficients ranged from 1.7890 to 4.6411 × $10^{-9}$ $m^2/s$, for the varieties RS at 50 °C and OVP at 80 °C, respectively, lying within the range mentioned by Madamba et al. [44] for food products from $10^{-11}$ to $10^{-9}$ $m^2/s$. The increase in temperature caused the increase in the diffusion coefficient. This behavior can be explained by the greater agitation of the water molecules, which reduces their attraction forces and their resistance to flow, thus facilitating the diffusion of water to the surface of the sample [45,46]. During the drying of Gandu beans (*Cajanus cajan* (L.) Mills.), Silva et al. [42]

reported effective diffusivity with values between 2.1 × $10^{-10}$ and 6.8 × $10^{-10}$ $m^2/s$, for a range between 40 and 70 °C, demonstrating an increase with the increase in drying air temperature, as observed for the germinated seeds of faba beans.

**Table 4.** Effective diffusion coefficients obtained from drying germinated seeds of faba beans at temperatures of 50, 60, 70 and 80 °C.

| Variety | T (°C) | $D_{ef}$ × $10^{-9}$ ($m^2/s$) | $R^2$ |
|---|---|---|---|
| OVP | 50 | 1.9763 | 0.9887 |
| | 60 | 2.4541 | 0.9849 |
| | 70 | 3.2291 | 0.9874 |
| | 80 | 4.6411 | 0.9851 |
| RS | 50 | 1.7890 | 0.9860 |
| | 60 | 2.2939 | 0.9865 |
| | 70 | 3.5491 | 0.9868 |
| | 80 | 3.9182 | 0.9859 |
| B | 50 | 1.9761 | 0.9809 |
| | 60 | 2.2026 | 0.9810 |
| | 70 | 3.4098 | 0.9841 |
| | 80 | 3.4303 | 0.9805 |

OVP—Orelha-de-Vó Preta; RS—Raio-de-Sol; B—Branca; $R^2$—coefficient of determination.

The linearized moisture diffusion coefficients were plotted as a function of the inverse of the absolute drying temperature (Figure 3), and its dependence on the drying air temperature was satisfactorily represented by an Arrhenius type equation, which presented values of $R^2 \geq 0.8783$.

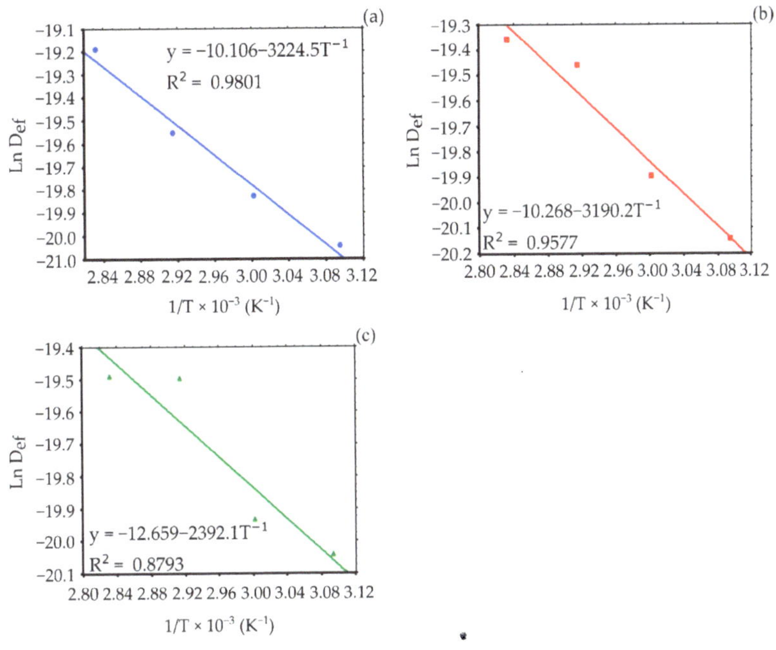


**Figure 3.** Arrhenius representation for the average effective diffusion coefficients obtained from drying germinated seeds of faba bean at temperatures of 50, 60, 70 and 80 °C for the varieties: (**a**) Orelha-de-Vó Preta; (**b**) Raio-de-Sol; and (**c**) Branca.

Table 5 shows the adjustment parameters of the Arrhenius equation for germinated seeds of faba beans. The activation energy to start the drying process of the germinated seeds of faba beans in the evaluated temperature range was similar for the OVP and RS varieties, and higher than the B variety, being all within the range described by Zogzas et al. [47], in which the activation energy for agricultural products can range from 12.7 to 110 kJ/mol. These values were higher than those reported by Ferreira et al. [43] in the drying of germinated pumpkin seeds, which ranged from 2.73 to 8.11 kJ/mol. These results indicate that the drying process of germinated faba bean seeds requires more energy for the diffusion of moisture to start.

**Table 5.** Fitting parameters of the Arrhenius equation for germinated faba bean seeds.

| Variety | $E_a$ (kJ/mol) | $D_0$ ($m^2$/s) | $R^2$ |
|---|---|---|---|
| OVP | 26.8085 | $4.0834 \times 10^{-5}$ | 0.9801 |
| RS | 26.5233 | $3.4727 \times 10^{-5}$ | 0.9577 |
| B | 19.8879 | $3.1788 \times 10^{-6}$ | 0.8783 |

OVP—Orelha-de-Vó Preta; RS—Raio-de-Sol; B—Branca; Ea—activation energy; $D_0$—pre-exponential factor; $R^2$—coefficient of determination.

### 3.3. Thermodynamic Properties

In Table 6, the average thermodynamic properties of dry germinated faba beans at different temperatures are presented. It is observed that the increase in the temperature of the drying air promotes a reduction in enthalpy (ΔH), indicating, according to Morais et al. [48], that at higher temperatures there is less energy demand for the occurrence of dehydration of the samples. Furthermore, according to Shafaei et al. [49], positive enthalpy values indicate an endothermic process, that is, a process in which heat absorption occurs. According to Silva et al. [30], the reductions observed in entropy (ΔS), with increasing temperature are related with the relative order of the system, where at lower temperatures there is less excitation of the water molecules, expressing, therefore, a greater degree of order. Negative entropy values can be attributed to the existence of structural changes in the adsorbent [50]. Gibbs free energy was directly proportional to the increase in temperature and showed positive values in the range evaluated. Positive values indicate an exogenous reaction, in which an external agent providing energy to the environment is needed for the reaction to occur, indicating a consistent result, since desorption is not a spontaneous reaction [49,51]. This thermodynamic property represents the maximum amount of energy released in a process under constant temperature and pressure that is available to be used, representing the balance between enthalpy and entropy [52]. Silva et al. [53] and Lisboa et al. [35] reported reductions in enthalpy and entropy and increases in Gibbs free energy as the drying temperature of soybeans was increased at 20, 30, 40 and 50 °C and of Mulatto beans (*Phaseolus vulgaris* L.) at temperatures of 40, 50, 60 and 70 °C.

### 3.4. Physicochemical Characterization

Table 7 shows the results of the physicochemical characterization of the germinated seeds of fresh faba beans and flours from the dry samples at temperatures of 50, 60, 70 and 80 °C. Drying reduced the water content of the samples to values between 1.55 and 5.18% at 80 and 50 °C, respectively. In all cases, these values are compatible with safe storage, with a reduction of about 96.34% when comparing the *in natura* and dried germinated seeds. Among the flours, the moisture content is below the upper limit recommended by the technical regulation for cassava flour [54] and wheat flour [55], which is 13 and 15%, respectively. Ferreira et al. [43] evaluating germinated pumpkin seeds (*Cucurbita moschata* D.), variety 'Jacarezinho' dried at 50, 60 and 70 °C, obtained water content values ranging from 1.10 to 5.93% (bs) being close to values obtained for the samples of the present work.

**Table 6.** Thermodynamic properties of germinated faba beans dried at temperatures of 50, 60, 70 and 80 °C.

| Variety | T (°C) | ΔH (kJ/mol) | ΔS (kJ/mol K) | ΔG (kJ/mol) |
|---|---|---|---|---|
| OVP | 50 | 24.1218 | −0.3296 | 130.6312 |
| | 60 | 24.0387 | −0.3299 | 133.9284 |
| | 70 | 23.9555 | −0.3301 | 137.2281 |
| | 80 | 23.8724 | −0.3303 | 140.5303 |
| RS | 50 | 23.8367 | −0.3309 | 130.7812 |
| | 60 | 23.7535 | −0.3312 | 134.0919 |
| | 70 | 23.6704 | −0.3314 | 137.4051 |
| | 80 | 23.5872 | −0.3317 | 140.7208 |
| B | 50 | 17.2013 | −0.3508 | 130.5696 |
| | 60 | 17.1181 | −0.3511 | 134.0791 |
| | 70 | 17.0350 | −0.3513 | 137.5911 |
| | 80 | 16.9518 | −0.3516 | 141.1056 |

OVP—Orelha-de-Vó Preta; RS—Raio-de-Sol; B—Branca; ΔH—enthalpy; ΔS—entropy; ΔG—Gibbs free energy.

**Table 7.** Physicochemical parameters of fresh fava bean germinated seeds and flour from germinated samples dried at different temperatures.

| Drying Conditions | Water Content (g/100 g w.b.) | Water Activity (Decimal) | Ashes (g/100 g d.b.) | pH | Alcohol-Soluble Acidity (mL NaOH/100 g d.b.) | Total Sugars (g/100 g d.b.) | Reducing Sugars (g/100 g d.b.) | Crude Protein (g/100 g d.b.) | Starch (g/100 g d.b.) |
|---|---|---|---|---|---|---|---|---|---|
| OVP/FG | 52.73 ± 0.13 c | 0.996 ± 0.00 a | 4.11 ± 0.17 cde | 4.25 ± 0.02 h | 3.90 ± 0.00 c | 4.50 ± 0.34 i | 2.47 ± 0.04 h | 22.53 ± 0.85 ab | 45.41 ± 0.22 h |
| OVP/50 | 4.70 ± 0.07 e | 0.211 ± 0.00 d | 4.26 ± 0.10 cd | 4.44 ± 0.02 g | 0.58 ± 0.00 e | 10.54 ± 0.04 e | 2.26 ± 0.02 i | 24.62 ± 0.65 a | 54.35 ± 0.80 bcd |
| OVP/60 | 3.98 ± 0.05 f | 0.169 ± 0.01 f | 4.41± 0.16 abc | 4.48 ± 0.01 f | 0.38 ± 0.00 g | 10.49 ± 0.05 e | 2.29 ± 0.02 i | 24.07 ± 1.35 a | 52.36 ± 1.74 def |
| OVP/70 | 2.69 ± 0.08 hi | 0.130 ± 0.00 h | 4.69 ± 0.09 ab | 4.49 ± 0.01 f | 0.38 ± 0.00 ij | 10.22 ± 0.07 ef | 2.35 ± 0.03 hi | 22.52 ± 0.50 ab | 55.31 ± 0.67 b |
| OVP/80 | 2.53 ± 0.07 i | 0.119 ± 0.00 i | 4.80 ± 0.05 a | 4.50 ± 0.02 f | 0.38 ± 0.00 jl | 9.85 ± 0.11 f | 2.40 ± 0.01 hi | 23.74 ± 0.87 a | 49.75 ± 0.67 g |
| RS/FG | 56.27 ± 0.21 b | 0.998 ± 0.00 a | 4.12 ± 0.13 cde | 4.28 ± 0.02 h | 4.21 ± 0.00 b | 6.79 ± 0.03 g | 3.20 ± 0.10 fg | 23.20 ± 0.73 a | 55.58 ± 0.30 b |
| RS/50 | 4.22 ± 0.06 f | 0.244 ± 0.00 b | 3.85 ± 0.03 ef | 4.60 ± 0.01 de | 0.38 ± 0.00 g | 12.28 ± 0.06 c | 3.16 ± 0.01 g | 13.83 ± 0.64 d | 53.00 ± 0.75 cde |
| RS/60 | 3.23 ± 0.06 g | 0.199 ± 0.00 e | 3.56 ± 0.15 f | 4.70 ± 0.01 c | 0.38 ± 0.00 h | 12.10 ± 0.11 c | 3.37 ± 0.07 e | 10.71 ± 0.38 e | 52.66 ± 0.60 de |
| RS/70 | 2.32 ± 0.06 ij | 0.146 ± 0.00 g | 3.89 ± 0.11 def | 4.80 ± 0.01 b | 0.38 ± 0.00 jl | 11.72 ± 0.06 d | 3.89 ± 0.02 b | 9.89 ± 0.64 e | 55.06 ± 0.67 bc |
| RS/80 | 1.55 ± 0.04 l | 0.118 ± 0.00 i | 3.91 ± 0.13 def | 4.84 ± 0.01 a | 0.37 ± 0.00 m | 11.66 ± 0.12 h | 4.20 ± 0.05 a | 13.91 ± 0.23 d | 50.17 ± 0.96 fg |
| B/FG | 62.64 ± 0.21 a | 0.996 ± 0.00 a | 4.20 ± 0.12 cde | 4.27 ± 0.03 h | 4.91 ± 0.00 a | 6.26 ± 0.04 h | 3.65 ± 0.07 cd | 25.35 ± 0.63 a | 61.88 ± 0.32 a |
| B/50 | 5.18 ± 0.02 d | 0.222 ± 0.00 c | 4.37 ± 0.02 bc | 4.46 ± 0.02 fg | 0.58 ± 0.00 d | 14.27 ± 0.20 a | 3.75 ± 0.02 bc | 8.40 ± 0.45 e | 51.15 ± 0.69 efg |
| B/60 | 4.06 ± 0.02 f | 0.193 ± 0.00 e | 4.39 ± 0.15 bc | 4.56 ± 0.02 e | 0.58 ± 0.00 f | 13.90 ± 0.08 a | 3.58 ± 0.04 d | 10.04 ± 0.66 e | 53.43 ± 0.00 bcd |
| B/70 | 3.00 ± 0.04 gh | 0.135 ± 0.00 h | 4.08 ± 0.21 cde | 4.58 ± 0.01 de | 0.38 ± 0.00 hi | 13.25 ± 0.04 b | 3.32 ± 0.06 ef | 19.47 ± 0.48 c | 49.98 ± 0.68 g |
| B/80 | 2.14 ± 0.05 j | 0.111 ± 0.00 i | 4.07 ± 0.11 cd e | 4.62 ± 0.00 d | 0.38 ± 0.00 l | 10.41 ± 0.19 e | 3.42 ± 0.02 e | 19.84 ± 0.78 bc | 49.54 ± 0.67 g |

OVP/FG—orelha-de-vó preta fresh germinated; OVP/50—orelha-de-vó preta dry at 50 °C; OVP/60—orelha-de-vó preta dry at 60 °C; OVP/70—orelha-de-vó preta dry at 70 °C; OVP/80—orelha-de-vó preta dry at 80 °C; RS/FG—raio-de-sol fresh germinated; RS/50—raio-de-sol dry at 50 °C; RS/60—raio-de-sol dry at 60 °C; RS/70—raio-de-sol dry at 70 °C; RS/80—raio-de-sol dry at 80 °C; B/FG—branca fresh germinated; B/50—branca dry at 50 °C; B/60—branca dry at 60 °C; B/70—branca dry at 70 °C; B/80—branca dry at 80 °C. The values are means ± standard deviation of the determination in triplicate. Means with the same letter in the same column do not present statistical difference according to Tukey's test at 5% probability.

The values obtained for the water activity (Table 7) follow the behavior of the moisture content, decreasing with the increase in the drying temperature. Low values of $a_w$ contribute to the preservation of the product, as they reduce the availability of water for the proliferation of microorganisms and development of enzymatic reactions, favoring preservation and storage. Water activities lower than 0.8 reduce the development of bacteria and below 0.6 reduce the development of fungi, yeast and mold [56]. Considering these values, it is concluded that the faba bean flours have a $a_w$ in the safety range against these agents at room temperature. Values like those for faba bean flour were found by Santos et al. [57] in red rice grain flours (*Oryza sativa* L.), dried at 40, 50, 60, 70 and 80 °C, obtained water activity values ranging from 0.101 to 0.229. Olagunju et al. [58] evaluated the water activity in ground, fermented and roasted bamboo flour (*Vigna underground* (L.) Green) during storage (lasting for 20 days) and obtained values ranging from 0.09 to 0.95, from 0.34 to 1.02 and from 0.42 to 0.89, respectively.

Ash contents ranged from 3.56 to 4.80%, with statistical equality between the fresh samples and the highest percentage in the 'orelha-de-vó' variety flour, followed by 'Branca'. Duenas et al. [59] evaluating black bean (4.3%) and pea (3.6%) seeds, both germinated for seven days at 20 °C, identified mean values of 4.3% and 3.6%, respectively, close to the ash content of the in natura bean samples. Ash values lower than those in the present study were quantified by Singh et al. [60], who reported 1.41, 1.27 and 1.16% for germinated soybean seed flours at temperatures of 25, 30 and 35 °C and dried at 45 °C in a convective dryer; and by Xu et al. [61], analyzing seed flours germinated for 72 and 96 h of chickpeas, lentils and yellow peas, which obtained mean ash values of 3.19 and 3.33%, 2.59 and 2.52% and 2.71 and 2.73%, respectively.

The samples presented an increase in pH as a function of the increase in drying temperature, remaining in the range of 4.44 to 4.84, similar to the in natura material. Variety RS had the highest values, followed by variety B and OVP. Higher values were identified by Silva et al. [62] on dried and freeze-dried alfalfa sprouts, pH 6.63 and 5.73, respectively; and by Santos et al. [57] who found values ranging from 6.72 to 6.79, for temperatures between 40 and 80 °C, in red rice flour. According to Kadam and Balasubramanian [63], pH below 4.5 lead to reduced growth of microorganisms, as is the case with the OVP sample flour, while samples with pH above 4.5 are prone to microbiological development and proliferation, in this case that flours of the RS and B varieties are included, with values slightly above this limit. An inverse behavior to the pH was observed for the alcohol-soluble acidity, which ranged from 0.37 to 0.58 mL NaOH/100 g (d.b.) in the flours, with lower results present in the RS variety flour (0.38–0.037 mL NaOH/100 g). The reduction in this parameter with drying may be related, according to Araújo et al. [64], to the oxidation of organic acids with increasing drying temperature. Reis et al. [65], studying the stability of the physicochemical properties of acerola flour, also observed a reduction in titratable acidity with increasing drying temperature. Contrary to Santos et al. [66] when evaluating the acidity of black rice grains, they observed an increase among the samples dried at temperatures between 40 and 80 °C.

The flours showed an increase in the content of total sugars in relation to the raw material and, among the flours, the increase in temperature promoted a gradual reduction in these. Lower results for total sugars were identified by Amadeu et al. [67] in kibbled seeds germinated for 48 h, of 5.67 g/100 g (d.b.), and for the flour of seeds germinated and dried at 70 °C, of 3.07 g/100 g (d.b.). Queiroz et al. [68] quantified for lychee seed flour the mean value of total sugars of 16.57 g/100 g (d.b.), therefore higher than those of fresh and dried faba bean samples. In the content of reducing sugars, the OVP variety maintains the relationship observed in total sugars, with values lower than the other varieties. Comparing the RS and B varieties, it is observed, as in the total sugars, an alternation between values, with statistically higher and lower results, indicating similarity between the samples. It is verified, with drying, by an increase in reducing sugars, with an increasing trend as the temperature of the drying air increases, except in the sample of variety B. The increase in the reducing sugar content with drying is commonly explained by the transformation of compounds into products of the Maillard reaction, which involves reactions of reducing sugars with amino groups [69]. Amadeu et al. [67] reported reducing sugar values of 5.97 and 1.59 g/100 g (d.b.) for germinated fresh pumpkin seeds and for seed flour, respectively. Moongngarm and Saetung [70] reported values for reducing sugars of 10.9 g/100 g (d.b.) and totals of 14.6 g/100 g (d.b.) in germinated husk rice powder, exceeding those determined in faba bean flour.

Among the fresh samples, the Branca variety presented higher protein values than the OVP and RS, which have statistically similar values to each other. With drying, OVP maintained similar levels of protein between the fresh sample and the flours. Samples RS and B showed fluctuation in values, but with statistically significant reductions between the fresh sample and the flours. According to Driscoll [71], the reduction in protein content with increasing drying temperature can be explained by protein denaturation, with a decrease in its solubility as a result of higher temperatures. Duenas et al. [59] identified lower values for protein content, when evaluating the composition of germinated bean (*Phaseolus vulgaris* L.) and lentil (*Lens culinaris* L.) seed flours, being 15.7 and 18.9%, respectively. Xu et al. [61] obtained protein content of 26.06, 33.13 and 27.8 for chickpea (*Cicer aretinium* L.), lentil (*Lens culinaris* Merr.) and yellow pea (*Pisum sativum* L.) flours, respectively. According to the Brazilian Table of Food Composition (TACO) [72], the protein content of wheat and corn flour corresponds to 9.8 and 7.2%, respectively, indicating that the faba bean germinated, both in nature and in the form of flour, can adequately replace those traditionally consumed flours as a protein source.

The starch content among the samples germinated in natura was higher in the Branca variety, surpassing that of RS, which was higher than that of the OVP. Lower starch values than those found in natura faba beans were quantified by Xu et al. [61] evaluating, for six days, the starch content of chickpea, lentil, and yellow pea sprouts, all in nature, with values ranging from 38.51 to 43.81 g/100 g. It is observed that in the OVP sample the starch content of the flours was higher than in the fresh sample; while in the samples of the RS and B varieties, in general, the starch contents of the flours were lower than in the fresh samples. Starch values were reported by Cornejo et al. [73] in *Amaranthus quitensis* (black species) and *Amaranthus caudatus* (white species) flours, with mean values of 54.69 g/100 g and 27.07 g/100 g, respectively.

## 4. Conclusions

The Page and Midilli models presented the best adjustment parameters for the drying kinetics of the three faba bean varieties. The effective diffusivity coefficients were in the order of magnitude of values reported in the literature for this property ($10^{-11}$ to $10^{-9}$), increasing with increasing temperature and without significant differences between the varieties; the activation energies showed equivalent results in the Olho-de-Vó Preta and Raio-de-Sol varieties, surpassing the Branca variety by almost 30%, ranging from 19.89 to 26.81 kJ/mol. The entropy and enthalpy values were higher in the Olho-de-Vó Preta variety, followed by Raio-de-Sol, higher than in the Branca variety. The three varieties had approximate Gibbs free energy values; the increase in drying temperature resulted in a reduction in enthalpy and entropy and an increase in Gibbs free energy, corroborating that the drying process of germinated faba bean seeds is endothermic and requires external energy input. All samples showed acidic pH and the acidity was reduced with drying; the Raio-de-Sol and Branca varieties had higher sugar contents and the total sugars were increased with drying; the highest protein contents were determined in the Branca variety and in the in nature germinated samples; in the Branca variety the highest starch content was also verified.

**Author Contributions:** Conceptualization, L.T.S.A., A.J.d.M.Q. and R.M.F.d.F.; data curation, L.T.S.A., Y.F.P., H.V.M. and H.A.S.; formal analysis, W.P.d.S., J.P.G. and C.C.C.; investigation, J.P.d.L.F. and D.d.C.S.; methodology, L.T.S.A., A.J.d.M.Q. and J.P.d.L.F.; software, W.P.d.S. and J.P.G.; supervision, A.J.d.M.Q. and R.M.F.d.F.; validation, C.C.C., D.d.C.S. and A.R.C.d.L.; visualization, Y.F.P., H.V.M. and A.R.C.d.L.; writing—original draft, L.T.S.A.; writing—review and edit-ing, A.J.d.M.Q., R.M.F.d.F. and J.P.d.L.F.; funding acquisition, A.J.d.M.Q. All authors have read and agreed to the published version of the manuscript.

**Funding:** This research was funded by the Conselho Nacional de Desenvolvimento Científico e Tecno-lógico (CNPq): Process number 305972/2019-7 (Brazilian Research Agencie).

**Data Availability Statement:** Data can be digitized from the graphs or requested to the corresponding author.

**Acknowledgments:** The authors are grateful to the Federal University of Campina Grande (Brazil) for the research infrastructure.

**Conflicts of Interest:** The authors declare no conflict of interest.

## References

1. Kokalis-Burelle, N.; McSorley, R.; Wang, K.-H.; Saha, S.K.; McGovern, R.J. Rhizosphere Microorganisms Affected by Soil Solarization and Cover Cropping in *Capsicum Annuum* and *Phaseolus Lunatus* Agroecosystems. *Appl. Soil Ecol.* **2017**, *119*, 64–71. [CrossRef]
2. IBGE. Instituto Brasileiro de Geografia e Estatística. Produção Agrícola Municipal. IBGE, 2019. Available online: http://www.sidra.ibge.gov.br (accessed on 4 February 2019).
3. E Lacerda, R.R.; do Nascimento, E.S.; de Lacerda, J.T.J.G.; da Silva Pinto, L.; Rizzi, C.; Bezerra, M.M.; Pinto, I.R.; Filho, S.M.P.; de Paulo Texeira Pinto, V.; Filho, G.C.; et al. Lectin from Seeds of a Brazilian Lima Bean Variety (*Phaseolus lunatus* L. Var. Cascavel) Presents Antioxidant, Antitumour and Gastroprotective Activities. *Int. J. Biol. Macromol.* **2017**, *95*, 1072–1081. [CrossRef] [PubMed]
4. Da Silva, S.I.A.; Souza, T.; Santos, D.; da Silva Souza, R.F. Avaliação Dos Componentes de Produção Em Variedades Crioulas de Fava Cultivadas No Agreste da Paraíba. *Rev. De Ciências Agrárias* **2019**, *42*, 731–742.
5. Oshodi, A.A.; Aletor, V.A. Functional Properties of Haemagglutinins (Lectins) Extracted from Some Edible Varieties of Lima Beans (*Phaseolus lunatus* Linn). *Int. J. Food Sci. Nutr.* **1993**, *44*, 133–136. [CrossRef]
6. Abarshi, M.M.; Abubakar, A.L.; Garba, A.; Mada, S.B.; Ibrahim, A.B.; Maruthi, M.N. Molecular Detection and Characterisation of Horsegram Yellow Mosaic Virus (HgYMV) Infecting Lima Bean (*Phaseolus lunatus*) in India. *Niger. J. Biotechnol.* **2017**, *33*, 41–48. [CrossRef]
7. Suriya, M.; Baranwal, G.; Bashir, M.; Reddy, C.K.; Haripriya, S. Influence of Blanching and Drying Methods on Molecular Structure and Functional Properties of Elephant Foot Yam (*Amorphophallus paeoniifolius*) Flour. *LWT—Food Sci. Technol.* **2016**, *68*, 235–243. [CrossRef]
8. Hidar, N.; Ouhammou, M.; Mghazli, S.; Idlimam, A.; Hajjaj, A.; Bouchdoug, M.; Jaouad, A.; Mahrouz, M. The Impact of Solar Convective Drying on Kinetics, Bioactive Compounds and Microstructure of Stevia Leaves. *Renew. Energy* **2020**, *161*, 1176–1183. [CrossRef]
9. Santos, K.C.; Guedes, J.S.; Rojas, M.L.; Carvalho, G.R.; Augusto, P.E.D. Enhancing Carrot Convective Drying by Combining Ethanol and Ultrasound as Pre-Treatments: Effect on Product Structure, Quality, Energy Consumption, Drying and Rehydration Kinetics. *Ultrason. Sonochem.* **2021**, *70*, 105304. [CrossRef]
10. Onwude, D.I.; Hashim, N.; Janius, R.B.; Nawi, N.M.; Abdan, K. Modeling the Thin-Layer Drying of Fruits and Vegetables: A Review. *Compr. Rev. Food Sci. Food Saf.* **2016**, *15*, 599–618. [CrossRef]
11. Castro, A.M.; Mayorga, E.Y.; Moreno, F.L. Mathematical Modelling of Convective Drying of Fruits: A Review. *J. Food Eng.* **2018**, *223*, 152–167. [CrossRef]
12. Santos, N.C.; Almeida, R.L.J.; da Silva, G.M.; Monteiro, S.S.; André, A.M.M.C.N. Effect of Ultrasound Pre-Treatment on the Kinetics and Thermodynamic Properties of Guava Slices Drying Process. *Innov. Food Sci. Emerg. Technol.* **2020**, *66*, 102507. [CrossRef]
13. De Oliveira, D.E.C.; Resende, O.; Chaves, T.H.; Souza, K.A.; de Souza Smaniotto, T.A. Propriedades Termodinâmicas Das Sementes de Pinhão-Manso. *Biosci. J.* **2014**, *30*, 147–157.
14. De Farias Leite, D.D.; de Melo Queiroz, A.J.; de Figueirêdo, R.M.F.; Lima, L.S.L. Mathematical Drying Kinetics Modeling of Jackfruit Seeds (*Artocarpus heterophyllus* Lam.). *Rev. Ciência Agronômica* **2019**, *50*, 361–369. [CrossRef]
15. Ministério da Agricultura, Pecuária e Abastecimento. *Regras Para Análise De Sementes*; Ministério da Agricultura: Brasília, Brazil, 2009; 399p.
16. Saleh, H.M.; Hassan, A.A.; Mansour, E.H.; Fahmy, H.A.; El-Bedawey, A.E.-F.A. Melatonin, Phenolics Content and Antioxidant Activity of Germinated Selected Legumes and Their Fractions. *J. Saudi Soc. Agric. Sci.* **2019**, *18*, 294–301. [CrossRef]
17. Setia, R.; Dai, Z.; Nickerson, M.T.; Sopiwnyk, E.; Malcolmson, L.; Ai, Y. Impacts of Short-Term Germination on the Chemical Compositions, Technological Characteristics and Nutritional Quality of Yellow Pea and Faba Bean Flours. *Food Res. Int.* **2019**, *122*, 263–272. [CrossRef] [PubMed]
18. Brasil Ministério da Saúde; Agência Nacional De Vigilância Sanitária. *Métodos Químicos E Físico-Químicos Para Análise de Alimentos*; Ministério da Saúde: Brasilia, Brazil, 2005; p. 1017f.
19. Lewis, W.K. The rate of drying of solid materials. *J. Ind. Eng. Chem.* **1921**, *13*, 427–432. [CrossRef]
20. Page, G.E. Factors Influencing the Maximum Rate of Air Drying Shelled Corn in Thin-Layers. Master's Thesis, Purdue University, West Lafayette, IN, USA, 1949.

21. Henderson, S.M.; Pabis, S. Grain drying theory I: Temperature effect on drying coefficient. *J. Agric. Eng. Res.* **1961**, *6*, 169–174.
22. Karathanos, V.T. Determination of water content of dried fruits by drying kinetics. *J. Food Eng.* **1999**, *39*, 337–344. [CrossRef]
23. Thompson, T.L.; Peart, P.M.; Foster, G.H. Mathematical simulation of corn drying: A new model. *Trans. ASAE* **1968**, *11*, 582–586. [CrossRef]
24. Yagcioglu, A.; Degirmencioglu, A.; Cagatay, F. Drying characteristics of laurel leaves under different conditions. In Proceedings of the 7th International Congress on Agricultural Mechanization and Energy, Adana, Turkey, 26–27 May 1999; pp. 565–569.
25. Henderson, S.M. Progress in developing the thin layer drying equation. *Trans. ASAE* **1974**, *17*, 1167–1168. [CrossRef]
26. Midilli, A.; Kucuk, H.; Yapar, Z. A New Model for Single-Layer Drying. *Dry. Technol.* **2002**, *20*, 1503–1513. [CrossRef]
27. Sharaf-Eldeen, Y.I.; Blaisdell, J.L.; Hamdy, M.Y. A model for ear corn drying. *Trans. ASAE* **1980**, *23*, 1261–1265. [CrossRef]
28. Verma, L.R.; Bucklin, R.A.; Endan, J.B.; Wratten, F.T. Effects of Drying Air Parameters on Rice Drying Models. *Trans. ASAE* **1985**, *28*, 296–301. [CrossRef]
29. Crank, J. *The Mathematics of Diffusion*, 1st ed.; Clarendon Press: Oxford, UK, 1975.
30. Da Silva, H.W.; Rodovalho, R.S.; Velasco, M.F.; Silva, C.F.; Vale, L.S.R. Kinetics and Thermodynamic Properties Related to the Drying of "Cabacinha" Pepper Fruits. *Rev. Bras. Eng. Agrícola Ambient.* **2016**, *20*, 174–180. [CrossRef]
31. Miller, G.L. Use of Dinitrosalicylic Acid Reagent for Determination of Reducing Sugar. *Anal. Chem.* **1959**, *31*, 426–428. [CrossRef]
32. Yemm, E.W.; Willis, A.J. The Estimation of Carbohydrates in Plant Extracts by Anthrone. *Biochem. J.* **1954**, *57*, 508–514. [CrossRef]
33. De Assis Santos e Silva, F.; de Azevedo, C.A.V. The Assistat Software Version 7.7 and Its Use in the Analysis of Experimental Data. *Afr. J. Agric. Res.* **2016**, *11*, 3733–3740. [CrossRef]
34. Da Silva, E.C.O.; da Silva, W.P.; Gomes, J.P.; Silva, C.M.D.P.S.; Alexandre, H.V.; Farias, V.S.O.; de Melo, B.A.; Queiroz, A.J.M.; de Figuiredo, R.M.F. Drying of Albedo and Whole Peel of Yellow Passion Fruit. *J. Agric. Sci.* **2019**, *11*, 501. [CrossRef]
35. Lisboa, H.M.; Araujo, H.; Paiva, G.; Oriente, S.; Pasquali, M.; Duarte, M.E.; Mata, M.E.R.M.C. Determination of Characteristic Properties of Mulatto Beans (*Phaseolus vulgaris* L.) during Convective Drying. *J. Agric. Food Res.* **2019**, *1*, 100003. [CrossRef]
36. Rahmanian-Koushkaki, H.; Nourmohamadi-Moghadami, A.; Zare, D.; Karimi, G. Experimental and Theoretical Investigation of Hot Air- Infrared Thin Layer Drying of Corn in a Fixed and Vibratory Bed Dryer. *Eng. Agric. Environ. Food* **2017**, *10*, 191–197. [CrossRef]
37. Zielinska, M.; Michalska, A. Microwave-Assisted Drying of Blueberry (*Vaccinium corymbosum* L.) Fruits: Drying Kinetics, Polyphenols, Anthocyanins, Antioxidant Capacity, Colour and Texture. *Food Chem.* **2016**, *212*, 671–680. [CrossRef] [PubMed]
38. Chielle, D.P.; Bertuol, D.A.; Meili, L.; Tanabe, E.H.; Dotto, G.L. Convective Drying of Papaya Seeds (*Carica papaya* L.) and Optimization of Oil Extraction. *Ind. Crops Prod.* **2016**, *85*, 221–228. [CrossRef]
39. Hasan, A.A.M.; Bala, B.K.; Rowshon, M.K. Thin Layer Drying of Hybrid Rice Seed. *Eng. Agric. Environ. Food* **2014**, *7*, 169–175. [CrossRef]
40. Chaji, H.; Hedayatizadeh, M. Quality Assessment and Kinetics of Dehydrated Watermelon Seeds: Part 1. *Eng. Agric. Environ. Food* **2017**, *10*, 178–185. [CrossRef]
41. Hemis, M.; Choudhary, R.; Gariépy, Y.; Raghavan, V.G.S. Experiments and Modelling of the Microwave Assisted Convective Drying of Canola Seeds. *Biosyst. Eng.* **2015**, *139*, 121–127. [CrossRef]
42. De Melo Silva, L.M.; de Sousa, F.C.; de Sousa, E.P.; Cavalcanti Mata, M.E.R.M.; Duarte, M.E.M. Modelos de Predição da Cinética de Secagem Dos Grãos de Guandu. *Braz. J. Food Technol.* **2014**, *17*, 310–318. [CrossRef]
43. De Lima Ferreira, J.P.; de Melo Queiroz, A.J.; de Figueirêdo, R.M.F.; da Silva, W.P.; Gomes, J.P.; da Costa Santos, D.; Paiva, Y.F.; do Nascimento Silva, S.; Amadeu, L.T.S.; de Lima, T.L.B.; et al. Mathematical Modelling of Drying Kinetics and Effective Diffusivity of Germinated Pumpkin Seeds. *Sylwan* **2021**, *165*, 347–376.
44. Madamba, P.S.; Driscoll, R.H.; Buckle, K.A. The Thin-Layer Drying Characteristics of Garlic Slices. *J. Food Eng.* **1996**, *29*, 75–97. [CrossRef]
45. Guimarães, R.M.; de Oliveira, D.E.C.; Resende, O.; de Santana Silva, J.; de Rezende, T.A.M.; Egea, M.B. Thermodynamic Properties and Drying Kinetics of 'Okara'. *Rev. Bras. Eng. Agrícola Ambient.* **2018**, *22*, 418–423. [CrossRef]
46. Resende, O.; de Oliveira, D.E.C.; Costa, L.M.; Ferreira, W.N., Jr. Drying Kinetics of Baru Fruits (*Dipteryx alata* Vogel). *Eng. Agrícola* **2018**, *38*, 103–109. [CrossRef]
47. Zogzas, N.P.; Maroulis, Z.B.; Marinos-Kouris, D. Moisture Diffusivity Data Compilation in Foodstuffs. *Dry. Technol.* **1996**, *14*, 2225–2253. [CrossRef]
48. De Morais, M.F.; dos Santos, J.R.O.; dos Santos, M.P.; da Costa Santos, D.; da Costa, T.N.; Lima, J.B. Modeling and Thermodynamic Properties of 'Bacaba' Pulp Drying. *Rev. Bras. Eng. Agrícola Ambient.* **2019**, *23*, 702–708. [CrossRef]
49. Shafaei, S.M.; Masoumi, A.A.; Roshan, H. Analysis of Water Absorption of Bean and Chickpea during Soaking Using Peleg Model. *J. Saudi Soc. Agric. Sci.* **2016**, *15*, 135–144. [CrossRef]
50. Carvalho Lago, C.; Noreña, C.P.Z. Thermodynamic Analysis of Sorption Isotherms of Dehydrated Yacon (*Smallanthus sonchifolius*) Bagasse. *Food Biosci.* **2015**, *12*, 26–33. [CrossRef]
51. Chen, J.; Wang, Y.; Lang, X.; Ren, X.; Fan, S. Evaluation of Agricultural Residues Pyrolysis under Non-Isothermal Conditions: Thermal Behaviors, Kinetics, and Thermodynamics. *Bioresour. Technol.* **2017**, *241*, 340–348. [CrossRef]
52. Corrêa, J.L.G.; Rasia, M.C.; Mulet, A.; Cárcel, J.A. Influence of Ultrasound Application on Both the Osmotic Pretreatment and Subsequent Convective Drying of Pineapple (*Ananas comosus*). *Innov. Food Sci. Emerg. Technol.* **2017**, *41*, 284–291. [CrossRef]

53. Silva, L.P.; dos Santos, S.G.F.; Queiroz, J.S.; Rodovalho, R.S.; Buso, W.H.D. Drying Kinetics of Soybean Grains. *Científica* **2020**, *48*, 99. [CrossRef]
54. Ministério da Agricultura, Pecuária e Abastecimento. *Regulamento Técnico Da Farinha De Mandioca*; Ministério da Agricultura, Pecuária e Abastecimento: Brasília, Brazil, 2011; p. 6f.
55. Ministério da Saúde. *Agência Nacional De Vigilância Sanitária, Regulamento Técnico Para Produtos De Cereais, Amidos, Farinhas E Farelos*; Ministério da Saúde: Brasília, Brazil, 2005; p. 6f.
56. Khouryieh, H.; Aramouni, F. Physical and Sensory Characteristics of Cookies Prepared with Flaxseed Flour. *J. Sci. Food Agric.* **2012**, *92*, 2366–2372. [CrossRef] [PubMed]
57. Santos, N.C.; da Silva, W.P.; Barros, S.L.; Almeida, R.L.J.; Brito Araújo, A.J.; da Silva Nascimento, A.P. Red Rice (*Oryza sativa* L.) Use in Flour Production: Convective Drying and Bioactive Quality. *J. Food Process Eng.* **2020**, *43*, e13490. [CrossRef]
58. Olagunju, O.; Mchunu, N.; Durand, N.; Alter, P.; Montet, D.; Ijabadeniyi, O. Effect of Milling, Fermentation or Roasting on Water Activity, Fungal Growth, and Aflatoxin Contamination of Bambara Groundnut (*Vigna subterranea* (L.) Verdc). *LWT* **2018**, *98*, 533–539. [CrossRef]
59. Dueñas, M.; Sarmento, T.; Aguilera, Y.; Benitez, V.; Mollá, E.; Esteban, R.M.; Martín-Cabrejas, M.A. Impact of Cooking and Germination on Phenolic Composition and Dietary Fibre Fractions in Dark Beans (*Phaseolus vulgaris* L.) and Lentils (*Lens culinaris* L.). *LWT—Food Sci. Technol.* **2016**, *66*, 72–78. [CrossRef]
60. Singh, A.; Sharma, S.; Singh, B. Effect of Germination Time and Temperature on the Functionality and Protein Solubility of Sorghum Flour. *J. Cereal Sci.* **2017**, *76*, 131–139. [CrossRef]
61. Xu, M.; Jin, Z.; Simsek, S.; Hall, C.; Rao, J.; Chen, B. Effect of Germination on the Chemical Composition, Thermal, Pasting, and Moisture Sorption Properties of Flours from Chickpea, Lentil, and Yellow Pea. *Food Chem.* **2019**, *295*, 579–587. [CrossRef] [PubMed]
62. Silva, M.L.T.; Brinques, G.B.; Gurak, P.D. Utilização de Farinha de Subproduto de Brotos Para Elaboração de Massa Alimentícia Fresca. *Braz. J. Food Technol.* **2019**, *22*, e2018063. [CrossRef]
63. Kadam, D.M.; Balasubramanian, S. Foam Mat Drying of Tomato Juice. *J. Food Processing Preserv.* **2011**, *35*, 488–495. [CrossRef]
64. De Araújo, C.S.P.; de Andrade, F.H.A.; Galdino, P.O.; de Caldas Pinto, M.d.S. Desidratação de Batata-Doce Para Fabricação de Farinha. *Agropecuária Cient. No Semiárido* **2015**, *11*, 33–41.
65. Reis, D.S.; Figueiredo Neto, A.; de Vasconcelos Ferraz, A.; de Freitas, S.T. Produção e Estabilidade de Conservação de Farinha de Acerola Desidratada Em Diferentes Temperaturas. *Braz. J. Food Technol.* **2017**, *20*. [CrossRef]
66. Santos, N.C.; Silva, W.P.; Barros, S.L.; Araújo, A.J.B.; Gomes, J.P.; Almeida, R.L.J.; Nascimento, A.P.S.; Almeida, R.D.; e Silva, C.M.D.P.S.; Queiroz, A.J.M.; et al. Study on Drying of Black Rice (*Oryza sativa* L.) Grains: Physical-Chemical and Bioactive Quality. *J. Agric. Sci.* **2019**, *11*, 203. [CrossRef]
67. Amadeu, L.T.S.; de Melo Queiroz, A.J.; de Figueirêdo, R.M.F.; Paiva, Y.F.; de Lima Ferreira, J.P.; dos Reis, C.G.; da Silva, R.C.; Araújo, K.T.A.; Coelho, N.O.; de Sá Carneiro, E.F. Farinha de Sementes Germinadas de Abóbora: Aspectos Físicos, Físico-Químicos e Colorimétricos. *Res. Soc. Dev.* **2021**, *10*, e18810313005. [CrossRef]
68. De Rezende Queiroz, E.; de Abreu, C.M.P.; dos Santos, C.M.; Simão, A.A. Composição Química e Fitoquímica das Farinhas Da Casca e Da Semente de Lichias (*Litchi chinensis* Sonn) Cultivar "Bengal". *Ciência Rural* **2015**, *45*, 329–334. [CrossRef]
69. Sikorski, Z.E.; Pokorny, J.; Damodaran, S. Interações Físicas e Químicas Dos Componentes Dos Alimentos. In *Química de Alimentos de Fennema*; Damodaran, S., Parkin, K.L., Fennema, O.R., Eds.; Artmed: Porto Alegre, Brazil, 2010; Volume 1, p. 900.
70. Moongngarm, A.; Saetung, N. Comparison of Chemical Compositions and Bioactive Compounds of Germinated Rough Rice and Brown Rice. *Food Chem.* **2010**, *122*, 782–788. [CrossRef]
71. Driscoll, R. Food Dehydration. In *Food Processing: Principles and Applications*; Smith, J.S., Hui, Y.H., Eds.; Blackwell Publishing Professional: Ames, IA, USA, 2004; Volume 1.
72. Núcleo de Estudos e Pesquisas em Alimentos. *Tabela Brasileira de Composição de Alimentos (TACO)*; Universidade Estadual de Campinas: Campinas, Brazil, 2011; 164p.
73. Cornejo, F.; Novillo, G.; Villacrés, E.; Rosell, C.M. Evaluation of the Physicochemical and Nutritional Changes in Two Amaranth Species (*Amaranthus quitensis* and *Amaranthus caudatus*) after Germination. *Food Res. Int.* **2019**, *121*, 933–939. [CrossRef] [PubMed]

processes

MDPI

*Article*

# Hawthorn Drying: An Exploration of Ultrasound Treatment and Microwave–Hot Air Drying

**Mohammad Kaveh [1,*], Małgorzata Nowacka [2,*], Esmail Khalife [3], Kamal Imanian [4], Yousef Abbaspour-Gilandeh [5], Maryam Sabouri [6] and Safoura Zadhossein [5]**

1 Department of Petroleum Engineering, College of Engineering, Knowledge University, Erbil 44001, Iraq
2 Department of Food Engineering and Process Management, Institute of Food Sciences, Warsaw University of Life Sciences—SGGW, 02-776 Warsaw, Poland
3 Department of Civil Engineering, Cihan University-Erbil, Kurdistan Region, Erbil 44001, Iraq
4 Agricultural Engineering Research Department, West Azarbaijan Agricultural and Natural Resources Research and Education Center, Urmia 57169-63963, Iran
5 Department of Biosystems Engineering, College of Agriculture and Natural Resources, University of Mohaghegh Ardabili, Ardabil 56199-11376, Iran
6 Scientific Research Center, Erbil Polytechnic University, Erbil 44001, Iraq
* Correspondence: sirwan.kaweh@knu.edu.iq (M.K.); malgorzata_nowacka@sggw.edu.pl (M.N.)

**Abstract:** Drying is one of the methods used for preserving fruits and vegetables. However, due to the lengthy process and elevated temperature of convective drying, other pretreatment and drying methods are studied to shorten the drying time and obtain high-quality products. This study aimed to examine the effect of ultrasonic (US) pretreatment and microwave–hot air drying (MW–HA) on the drying time, specific energy (SEC), qualitative properties (e.g., color, shrinkage, and rehydration ratio), and bioactive compound properties (e.g., antioxidant activity, phenolic, and flavonoid contents) of hawthorn fruit. Experiments were performed using ultrasound pretreatment and a microwave dryer (microwave power: 180, 360, and 540 W) at air temperatures of 40, 55, and 70 °C. Drying of hawthorn lasts from 35 min for the ultrasound-treated sample (dried at 540 W and 70 °C) to 180 min (dried at 180 W and 40 °C without US treatment). The lowest amount of SEC (24.11 MJ/kg) was obtained using the US–MW–HA air drying method (dried at 540 W and 70 °C). The lowest values in total color change (13.37) and shrinkage (22.47%) were recorded for the sample dried with a MW power of 360 W and air temperature at 55 °C with US pretreatment prior to drying. Generally, the use of US and MW–HA air drying reduces the antioxidant activity (AC), total phenolic content (TPC), and total flavonoid content (TFC) during processing compared to fresh samples. The highest values for AA (28.01%), TPC (69.44 mg GAE/g d.m.), and TFC (64.38 mg QE/g) obtained at 360 W and 55 °C with US pretreatment for hawthorn fruit dried.

**Keywords:** hawthorn fruit; microwave drying; ultrasound treatment; specific energy; drying characteristics

**Citation:** Kaveh, M.; Nowacka, M.; Khalife, E.; Imanian, K.; Abbaspour-Gilandeh, Y.; Sabouri, M.; Zadhossein, S. Hawthorn Drying: An Exploration of Ultrasound Treatment and Microwave–Hot Air Drying. *Processes* **2023**, *11*, 978. https://doi.org/10.3390/pr11040978

Academic Editor: Jan Havlík

Received: 25 February 2023
Revised: 18 March 2023
Accepted: 20 March 2023
Published: 23 March 2023



## 1. Introduction

Fruits and vegetables are potential sources of plant chemical compounds, especially phenolic compounds, which are actually bioactive compounds and natural antioxidants [1]. Hawthorn (*Crataegus pinnatifida*) belongs to the rose family, which is mostly distributed in the Northern Hemisphere and mainly in areas of China, Iran, Europe, and North America [2]. There are different types of ripe hawthorns, the color of the fully ripe fruit is yellow, red, dark red, and dark purple. It has high medicinal and nutritional value and is used to treat various human diseases, including heart diseases [3], high blood pressure, chest pain, and the hardening of the arteries [4,5]. The chemical compounds of hawthorn are flavonoids, terpenic acids, proanthospanins, and organic acids [6].

One of the most important processes to increase the shelf life of foods with high humidity is the drying process. The purpose of the drying process is to evaporate a certain amount of water in the product. Also, by using the drying process, access to dried food is easier and the storage time of food is increased [7]. The drying of solid materials takes place by the simultaneous transfer of mass and energy between a drying fluid and solid samples [8]. There are many common methods for drying, including solar [9], vacuum [10], hot air [11], freezing [12], etc. Most of these drying methods have high energy consumption and long drying time, which leads to the low quality of dried products under the mentioned methods. In recent studies, there have been significant advances in the use of new methods, which include microwave [13], infrared [14], ultrasound [15], pulsed electric field [16], blanching, and plasma [17]. These methods will increase the efficiency of the process and increase the quality of the final dried products. Some of these methods can be used as pretreatment or used in combination with common methods to reduce the initial humidity or change the texture of crops (in a way that shortens the drying time).

New technologies in the food industry always seek to produce high quality products with increased efficiency and reduced energy consumption. The use of ultrasound (US) waves is one of these new and non-thermal technologies that is used today as a pretreatment before the drying process [18]. Ultrasound consists of sound waves with a frequency beyond the range of human hearing. By adjusting the frequency and power, US can be used in many industrial applications, including food. The use of ultrasonic waves as a pretreatment is a suitable non-thermal method to increase productivity, and during the application of this process, the physicochemical and quality characteristics of the food are less damaged [19]. US pretreatment includes immersing the fruit in distilled water or hypertonic aqueous solution simultaneously with the application of ultrasound waves. US waves cause rapid alternating contractions and expansions (sponge effect) and maintain the moisture inside the capillary tubes by creating a difference in suction pressure of the capillary tube [15]. In addition, US creates cavitation (cavity) inside the food material, which may be useful for separating water [20]. Pretreatment with the help of US has been carried out before using different dryers. For example, Zhang et al. [21], Sledz et al. [22], and Dehghannya et al. [23] used US pretreatment before MW–HA dryer to study the TPC, TFC, color, energy, rehydration rate, and shrinkage for Chinese hickory, parsley leaves, and potato, respectively. Chouaibi et al. [24] used US bath pretreatment and then MW, infrared, freeze, and hot-air (HA) dryers to obtain the qualitative and bioactive properties of Tunisian eggplant. Rybak et al. [16] investigated the effect of thermal and non-thermal pretreatments on the quality and bioactive properties of red bell pepper in hot-air and MW–HA dryers.

In addition, most of the new methods produce better quality products compared to common drying methods such as hot-air, and also reduce the duration of the process and consume less energy. Meanwhile, the use of MWs in combination with other methods is an effective proposed method that will be discussed in this study. Among the methods combined with the MW method are hot-air [25], vacuum [26], infrared [27], electrohydrodynamic [28], and freeze drying [29], of which hot air is the most common. The combined method of MW and HA has been used to dry different agricultural products such as bitter melon [30], apple slices [31], edamame [32], onion [33], potato [23], and raspberries [34]. In general, it can be noted that microwaves cause heat in the material by moving water molecules, and due to the condensation of surface moisture, burns occur on the surface of the final product, which reduces its quality. However, in the combined method, air flow causes the evaporation of a part of the condensed moisture on the surface of the sample and prevents the loss of physical and chemical characteristics of the dried sample.

Therefore, due to the lack of studies in the field of drying hawthorn, in the present study, the effect of US pretreatment and MW–HA drying of hawthorn slices with the application of different parameters (MW power: 180, 360 and 540 W; air temperature: 40, 55, and 70 °C) were studied. Additionally, the kinetics of the drying process, drying time, specific energy consumption, product quality in terms of color, shrinkage, rehydration ratio, and bioactive properties including total phenol content, total flavonoids content as well as antioxidants were analyzed.

## 2. Materials and Methods

### 2.1. Material

Hawthorn samples prepared for drying experiments were obtained from a local garden located in Sardasht city in West Azerbaijan province and were kept in a cold room at a temperature of 4 ± 1 °C until the end of the experiments. In order to balance the temperature of the samples with the ambient temperature, the samples were transferred from the cold room to the laboratory about 60 min before the start of each experiment. To carry out this research, at first, hawthorn slices with a thickness of 4 mm were carefully cut by a sharp knife, and after weighing, they were quickly placed in the US bath machine. The used hawthorn slices had an average initial moisture content (MC) of 2.27 on a dry basis (d.b.). The MC of the samples was measured according to the AOAC method [35] by placing the samples (three samples of 30 g) in an oven (Memmert, UFB 500, Schwabach, Germany) and the temperature of 70 °C for 24 h until reaching a constant weight [5].

### 2.2. Processing: Ultrasound Treatment and Microwave—Hot Air Dryer (US–MW–HA)

Hawthorn samples were divided into two groups: (1) samples without US pretreatment; (2) samples with US pretreatment. To investigate the effect of using US pretreatment, an US bath Model Parsonic 7500S with a frequency of 28 kHz, power of 70 W, and a time period of 15 min was used. Before starting the drying process, the samples were treated with US waves under ambient temperature. After a certain period of time, the samples were taken out of the US bath and the excess moisture on the surface of the samples was dried with absorbent paper. Then the samples were placed in microwave–hot air dryer made at Mohaghegh Ardabili University, Ardebil, Iran [36]. About 30 min before the drying process started, the dryer was turned on. According to the specifications of each test, the MW power and air temperature inside the dryer were adjusted to the desired values (air temperature of 40, 55, and 70 °C and MW power of 180, 360, and 540 W) until the air temperature inside the dryer was stable. Samples were weighed every three minutes using a digital scale. The samples were dried from the initial MC of 2.27 d.b. until reaching the moisture content of 0.11 d.b. Drying experiments were performed in 3 replications, which resulted in a total of 54 variants of treatments.

The kinetics of the drying process were analyzed. Moisture ratio was calculated using Equation (1) where $M_t$ is the MC on a dry basis at any time $t$ (grams of water per gram of dry matter), $M_0$ is the initial MC on a dry basis and $M_e$ is the equilibrium MC. $M_e$ is neglected because it is insignificant compared to $M_t$ and $M_0$ [25,37]:

$$MR = \frac{M_t - M_e}{M_o - M_e} \quad (1)$$

### 2.3. Specific Energy Consumption (SEC)

#### 2.3.1. Microwave Dryer

Energy consumption in the drying method using the MW method was obtained as follows [38]:

$$E_{mic} = P_{mic} t \quad (2)$$

where $E_{mic}$ is SEC in MW (KJ), $P_{mic}$ is the MW power (W), $t$ is the drying time (s).

#### 2.3.2. Hot-Air Dryer

SEC in the drying method using hot-air (HA) was calculated as follows [33]:

$$E_{con} = AV_a\rho_a\Delta Ht \quad (3)$$

where $E_{con}$ is the energy consumed in the HA dryer (KJ), $A$ is the area of the sample container (m$^2$), $V_a$ is the inlet air velocity (m/s), $\rho_a$ is the air mass density (kg/m$^3$), $\Delta H$ is the air enthalpy (kJ/kg of dry air), and $t$ is the time (min).

2.3.3. Ultrasound

Energy consumption in US pretreatment is calculated as follows [23]:

$$E_{US} = W \cdot V \cdot t \quad (4)$$

where $E_{US}$, $W$, and $V$ are the energy used in US pretreatment (KJ), *US* volumetric power (W/L) and water volume (L), respectively, and $t$ is the time (min).

2.3.4. US–MW–HA

The SEC in the combined MW–HA dryer with US pretreatment for drying hawthorn was calculated as follows [5]:

$$SEC = \frac{E_{mic} + E_{con} + E_{US}}{M_w} \quad (5)$$

where *SEC* is the specific energy consumption of the whole system (KJ/kg) and $M_w$ is the amount of moisture removed from the sample (kg).

*2.4. Proprties of the US–MW–HA Dried Hawthorn*

2.4.1. Shrinkage Assessment

The amount of shrinkage of hawthorn samples was checked by measuring the volume change before and after drying. To measure the volume, the toluene displacement method was used according to the method of Dehghannya et al. [23]. Thus, shrinkage is defined as the percentage change in the volume of the processed sample compared to the raw sample (Equation (6)) [39]:

$$\%S = \frac{V_i - V_j}{V_j} \times 100 \quad (6)$$

where $V_i$ is the volume of hawthorn samples before drying (g), $V_j$ is the volume of hawthorn samples after drying (g), and $\%S$ is the shrinkage percentage.

2.4.2. Color of the Samples

Samples color was measured in $L^*a^*b$ space. The color parameter $L^*$ represents the lightness, $a^*$ represents the redness/greenness and $b^*$ represents the yellowness/blueness of the sample [10,40]. The color changes of dried hawthorn were measured using a colorimeter (HP-200, China). Then, the color changes of the dried product compared to the fresh product ($\Delta E$) were calculated using Equation (7), [15].

$$\Delta E = \sqrt{(\Delta L^*)^2 + (\Delta a^*)^2 + (\Delta b^*)^2} \quad (7)$$

2.4.3. Rehydration Ratio

To calculate the RR parameter, dry samples were weighed and immersed in water with a temperature of 20 °C. Then, after 60 min [41], the samples were taken out of the water and weighed using a digital scale (AND, GF-6000, Japan), with an accuracy of ±0.01 g. All treatments were performed in three repetitions and their average was reported. The RR was calculated by Equation (8) [9]:

RR = (weight of dehydrated sample)/(drained weight of the rehydrated sample) (8)

2.4.4. Bioactive Compounds

Extract Preparation

To the 5 g of powdered dried samples, 100 mL of 80% methanol solution was added and shaken at room temperature for 24 h. Then, the supernatant of the centrifuged solutions was used to measure the total phenol content (TPC) and total flavonoid content (TFC) as well as antioxidant activity (AA).

Total Phenol Content (TPC)

The amount of TPC was determined by the Folin-Ciocalteu method described by Shahidi and Naczk, [42]. The amount of 1 mL of the extract of the sample was mixed with 2.5 mL of 10% Folin-Ciocalto solution and after 3 min, 2 mL of 7.5% sodium carbonate solution (75 g/L) was added into it. The samples were placed in the dark for 90 min, then the absorbance of the sample was read at 765 nm by a spectrophotometer (UV/Vis BIO-RAD-USA).

Total Flavonoid Content (TFC)

A 20 μL portion of the extract was mixed with 1 mL of distilled water and then 0.075 mL of sodium nitrite (5%) was added to it. After 5 min, 0.15 mL of $AlCl_3$ solution (10%) was added and after 6 min, 0.5 mL of NaOH (1 M) was added. The final volume of the solution was brought to 3 mL with distilled water. The absorbance of the resulting solution was immediately read at 510 nm by a spectrophotometer (UV/Vis BIO-RAD-USA).

Antioxidant Activity (AA)

The AA was measured using 2,2-diphenyl-1-picrylhydrazyl (DPPH). A 150 μL portion of the extract from hawthorn sample was completely mixed with 2 mL of 80% methanol and centrifuged for 10 min at a speed of 3500 rpm. Finally, its optical absorption intensity at the wavelength of 517 nm was read by a spectrophotometer (UV/Vis BIO-RAD-USA) and DPPH radical inhibitory capacity were calculated using Equation (9) [43]:

$$\%DPPH = [(A_0 - A_i)/A_0] \cdot 100 \quad (9)$$

where $A_i$ is the absorption of the control sample and $A_0$ is the absorption of the tested sample read using a spectrophotometer.

### 2.5. Statistical Analysis

In order to investigate the effect of US, drying air temperature and microwave power on the properties of hawthorn samples (drying time, SEC, color, shrinkage, RR, TPC, TFC, and AA), a factorial experiment in the form of a completely randomized design with three replications was used. The independent variables included US time at one level of 15 min, drying air temperature at three levels of 40, 55, and 70 °C, and MW power of the dryer at three levels of 180, 360, and 540 W were used (Table 1). Analysis of variance and the presence of significant differences between treatments were performed using Duncan's multi-range test at the probability level of 5% ($p < 0.05$) using SPSS version 21 statistical software.

**Table 1.** Design of experiment properties.

| Type of Experiment | Independent Variable | Level | Dependant Variable |
|---|---|---|---|
| Complete factorial design of experiments | Microwave Power | 180 W | Drying time |
| | | 360 W | SEC |
| | | 540 W | Color |
| | Temperature | 40 °C | Shrinkage |
| | | 55 °C | RR |
| | | 70 °C | TPC |
| | Ultrasound | 15 min | TFC |
| | | | AA |

## 3. Results and Discussion

### 3.1. Design of Experiments and Analysis

The aim of the investigation was to evaluate the effect of US pretreatment before drying hawthorn slices using a MW–HA dryer. Accordingly, a complete factorial design of experiments was conducted to judge the effect of US, drying air temperature, and MW power on drying time, SEC, total color change, shrinkage, RR, AA, TPC, and TFC was plotted. ANOVA results for the main effects and interaction of air temperature and MW power on the studied parameters are presented in Table 2.

**Table 2.** ANOVA analysis results of the drying time, color, energy, shrinkage, RR, TPC, TFC and AA at different MW power, temperature and MW power *temperature.

| Parameter | MW Power | | | Temperature | | | MW Power × Temperature | | | |
|---|---|---|---|---|---|---|---|---|---|---|
| | Sum of Squares | Mean Square | Sig. | Sum of Squares | Mean Square | Sig. | Sum of Squares | Mean Square | Sig. | C.V * |
| Drying time | 24,033.333 | 12,016.667 | 0.000 | 52,533.333 | 10,506.667 | 0.000 | 466.667 | 46.667 | 0.278 | 5.544 |
| SEC | 6797.737 | 3398.868 | 0.000 | 5640.968 | 1128.193 | 0.000 | 570.659 | 57.066 | 0.000 | 4.705 |
| Color | 2918.686 | 1459.343 | 0.000 | 291.084570 | 58.216914 | 0.000 | 39.844 | 3.984 | 0.001 | 3.874 |
| Shrinkage | 1289.128 | 644.564 | 0.000 | 2649.188 | 529.838 | 0.000 | 38.928 | 3.893 | 0.123 | 3.675 |
| RR | 8.929 | 2.978 | 0.000 | 2.660 | 0.443 | 0.000 | 1.889 | 0.105 | 0.000 | 2.586 |
| TPC | 3451.115 | 1150.372 | 0.000 | 1234.116 | 205.686 | 0.000 | 588.697 | 32.705 | 0.000 | 2.061 |
| TFC | 4830.773 | 1610.257 | 0.000 | 1803.648 | 300.608 | 0.000 | 820.722 | 45.596 | 0.000 | 1.580 |
| AA | 7798.419 | 2599.473 | 0.000 | 3579.063 | 596.510 | 0.000 | 1308.834 | 72.713 | 0.000 | 1.519 |

* The coefficient of variation (CV) is a statistical measure of the relative dispersion of data points in a data series about the mean.

A significant correlation between the average parameters (drying time, SEC, color, shrinkage, RR, TPC, TFC, and AA) was observed at the level of 1%. In addition to the main effects, the results showed that the mutual effects of temperature and MW power, except for the drying time and shrinkage parameters, also caused a significant difference between the average parameters at the level of 1%.

### 3.2. *Drying Time and Specific Energy Consumption*

The effect of US pretreatment, drying air temperature and MW power on the drying time of hawthorn samples is presented in Figure 1. The drying process lasted between 35 and 180 min. The longest drying time (180 min) was obtained when drying (temperature of 40 °C and MW power of 180 W) without US pretreatment was applied, whereas the shortest drying time (35 min) was obtained when samples were subjected to US treatment and dried at a temperature of 70 °C and MW power of 540 W. Using ultrasound at the MW power of 360 W and a temperature of 70 °C reduced the drying time by 25%. This value was 12% at the MW power of 540 W and a temperature of 40 °C. The positive effect of US on reducing drying time during the process has been reported in many studies [21,23,28,44]. The use of US can improve water removal from hawthorn samples, especially during the fall rate period when water evaporation becomes more challenging. The use of ultrasound creates a sponge effect in the product, and also the sponge effect created by US causes the formation of micro-channels on the skin layer of the samples. In fact, US weakens the tight bond with of tissue and facilitates the outflow of water [45,46]. Also, the cavitation factor leads to the formation of high-intensity turbulence and causes the subsequent decrease in the resistance of the boundary layer [22].

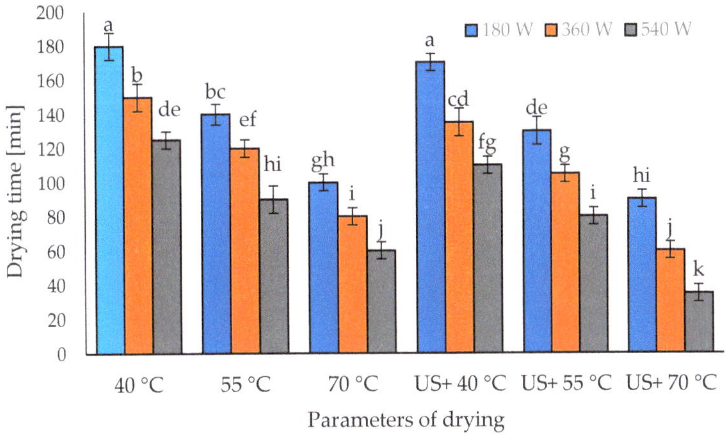


**Figure 1.** Drying time during different drying schemes. Data followed by different letters (a–k) in each column are significantly different at $p < 0.01$. Columns with the same letters are not significantly different.

Figure 1 shows that drying time decreased with increasing MW power. In the early stages of drying, the samples have high humidity and the use of microwave power causes molecules to be polarized, so the transfer of moisture to the surface of the product happens faster and the samples dry faster. The flow of water vapor from the inner parts to the surface of the product creates a porous structure, and as a result, the moisture evaporates faster, and finally the drying speed increases [23,47]. As it is evident in Figure 1, the increase in air temperature reduced the drying time. It is believed that the reason is the increase in the drying temperature increases the temperature of the drying product. After increasing the temperature of the product, a change in the latent heat of evaporation and an increase in the movement of water molecules in the product (substance) will occur due to the heat, and finally, the rate of evaporation and the rate of mass transfer in the drying substance will increase [48]. Similarly, other researchers also reported the simultaneous effect of increasing MW power and air temperature on reducing the drying time of agricultural products, including dragon fruit [47], white mushrooms [49] and carrot [50].

The drying process is known as one of the processes with high energy consumption [51]. Thus, in our study of hawthorn the specific energy consumption (SEC) was evaluated. The lowest amount of SEC was obtained under the conditions of 70 °C and 540 W of MW power with the use of US pretreatment. As seen in the results shown in Table 2, it was observed that the linear effect of the variables of air temperature, MW power, and the use of US as well as the interaction effect of air temperature and MW power on the SEC for drying were significant. Figure 2 shows the interaction effect of air temperature and MW power and US pretreatment on the SEC of drying hawthorn. As expected, with the increase in air temperature and MW power, the amount of SEC for drying decreased due to the reduction in drying time. Increasing MW power decreased the amount of energy consumed at all the tested treatments. However, increasing MW power had a greater effect on reducing drying energy than increasing air temperature. The SEC depends on various factors such as air temperature, air velocity, specific heat of air, and latent heat of water evaporation [52]. Increasing air temperature and MW power accelerates the evaporation of free water in the product, significantly reducing drying time and total energy consumption. [53]. Similar results were achieved by Szadzinska and Mierzwa [49] for mushrooms drying, Maftoonazad et al. [33] for onion drying, and Kaveh and Abbaspour- Gilandeh [36] for drying green peas.

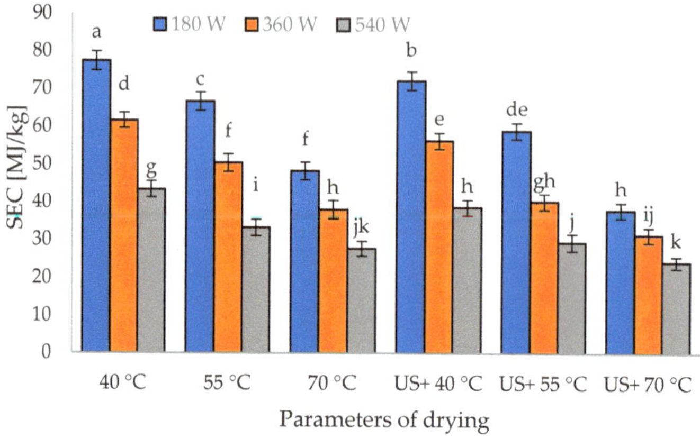


**Figure 2.** Specific energy consumption during different drying schemes. Data followed by different letters (a–k) in each column are significantly different at $p < 0.01$. Columns with the same letters are not significantly different.

Also, as seen in Figure 2, it can be noted that the use of US pretreatment has reduced the SEC. Products such as hawthorn form a hard surface layer when the moisture is removed from them, and then the moisture is removed from the product slowly. The application of US pretreatments reduces the formation of this hard surface layer, and as a result, the moisture removal from the surface of the product increases and the amount of SEC decreases [5,50]. The results obtained by Mirzaei-Baktash et al. [28] for button mushroom, Szadzińska et al. [54] for raspberries, and Mierzwa et al. [34] for raspberries, regarding the reduction in SEC when applying US pretreatment, are in line to the current research findings.

### 3.3. *Effect of Drying on Physical Property Changes*

#### 3.3.1. Color

The results of investigating the effect of temperature, MW power and US on total color changes (ΔE), regarding fresh raw material, are shown in Figure 3. The results showed that the color parameter has the highest value when the highest temperature and the highest microwave power were used: at a temperature of 70 °C and a power of 540 W. On the other hand, application of US treatment before the drying results in a lower value of the ΔE. US treatment followed by drying at 70 °C and 540 W MW power resulted in a significant decrease in total color difference compared to drying at the same parameters without US treatment. This trend was also observed at other temperature and microwave power levels. Such changes can be connected to the shorter drying time of about 5–42%, when US was applied before drying. Compared to the other parameters, the best color (the lowest ΔE) was obtained for the sample subjected to US treatment and dried at a temperature of 55 °C with the MW power at 360 W. The higher MW power and air temperature showed the highest rate of total color changes. In this method, although the samples are placed inside the dryer for a short period of time, the temperature used for drying is so high that the non-enzymatic browning reaction is carried out with high intensity [48]. This reaction causes burns on the surface of the sample and increases the final color changes of the samples. As can be seen in Figure 3, with the increase in air temperature from 40 to 55 °C and MW power from 180 to 360 W, the overall color changes decreased. One of the reasons is the low temperature affects the properties of the food due to the prolonged drying time by creating free radicals and sonochemicals as a result of cavitation, and the color changes of hawthorn samples increase at low temperature [50,55]. In addition, by increasing the air temperature from 60 to 70 °C and MW power from 360 to 540 W, enzymatic reactions and non-enzymatic brown reactions can occur and result in higher values of total color change. This can be linked with the pigment decomposition which happens at high temperature [56]. Also, it can be seen that the use of US treatment in all drying conditions reduced the color changes. Similar results have been reported by other researchers for agricultural products such as hawthorn [50], sunflower [44], carrot [57], and raspberries [54].

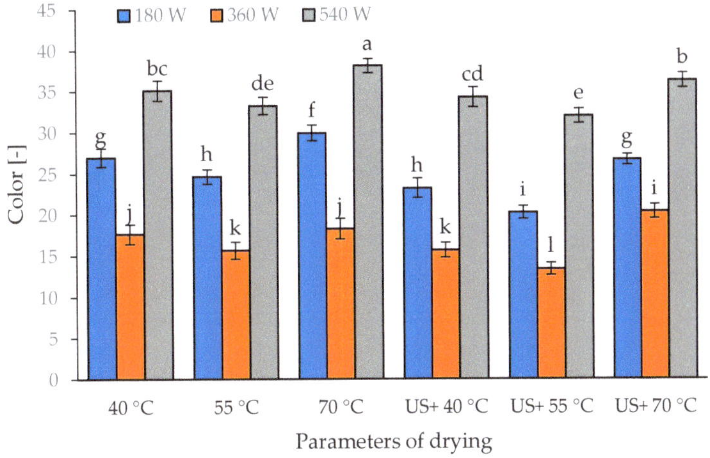


**Figure 3.** Effects of US pretreatment, MW power and air temperature during drying on total color difference (ΔE) in hawthorn. Data followed by different letters (a–l) in each column are significantly different at $p < 0.01$. Columns with the same letters are not significantly different.

3.3.2. Shrinkage

The application of microwave during drying results in porous plant material in comparison to traditional hot-air drying [58]. As the results in Table 2 show, the effect of MW power, air temperature and US on the amount of shrinkage was significant (0.01%), but the mutual effects on shrinkage were not significant (0.01%). The highest amount of shrinkage (57.69%) was obtained under drying conditions of air temperature of 70 °C, MW power 540 W, and without the application of US pretreatment. The lowest amount of shrinkage (22.47%) was obtained under the conditions of air temperature equal to 55 °C with the MW power of 360 W, and with the use of US pretreatment. As shown in Figure 4, by increasing the air temperature from 40 to 55 °C and MW power from 180 to 360 W, the shrinkage decreased. Due to the lengthening of the drying process, the volume of the product decreased significantly due to the creation of viscoelastic stresses in the pores, which leads to an increase in shrinkage [39]. In addition, shrinkage increased with increasing temperature from 55 to 70 °C and MW power from 360 to 540 W. This is due to the production of extensive internal heat, which accelerates the removal of water from the tissue in hawthorn samples at high power and temperature [50]. During the drying process, water removal from the tissue of the product by applying tension to its cell wall causes shrinkage. When the water in the intercellular wall of the product evaporates, air replaces it; as a result, the tissue is not able to maintain the structural network and the outer structure of the cell collapses and shrinkage results [32]. Joudi-Sarighayeh et al. [59] observed the amount of shrinkage in pumpkin slices dried by the MW/HA method. These results are in agreement with the findings of the studies reported by Bhat et al. [60], and Wang et al. [61].

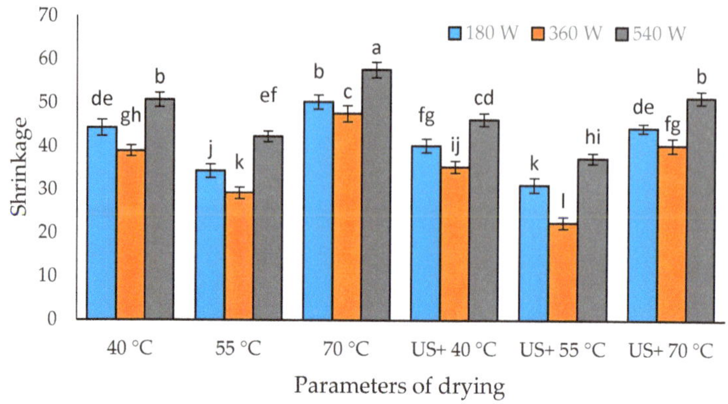


**Figure 4.** Effects of US pretreatment, MW power and air temperature during drying on shrinkage of hawthorn. Data followed by different letters (a–l) in each column are significantly different at $p < 0.01$. Columns with the same letters are not significantly different.

Also, the use of US pretreatment in all conditions of hawthorn drying reduced shrinkage compared to the methods of US-free pretreatment. During the drying process, the water leaving the cell causes an increase in the tension applied by the liquid on the cell wall. This increase in tension causes the fabric of the substance to shrink. Applying US pretreatment reduced the tension in the cell wall of the product [62]. Dehghannya et al. [50] showed that the use of US pretreatment reduces shrinkage for drying carrot in MW–CV. Jahanbakhshi et al. [63] also obtained similar results for drying nectarine in a hot air dryer under US pretreatment.

3.3.3. Rehydration Ratio (RR)

According to the results listed in Table 2, the independent effect of temperature, MW power, and the use of US pretreatment and the mutual effects of the desired parameters on the RR of dried hawthorn slices were significant at the probability level of 1%. As shown in Figure 5, the highest percentage of water reabsorption was observed in samples dried at 55 °C and MW power of 540 W with the use of US pretreatment. These results showed that a structure with less shrinkage has a higher water RR [30]. On the other hand, the lowest amount of RR was also observed in the samples without US pretreatment and at the MW power of 360 W dried at the temperature of 70 °C. This may be due to the changes in the structure or texture degradation of the samples during MW–HA air drying. Due to the increases in the internal temperature of the samples and the migration of sugars from the inside of the sample to the surface of the sample and because of the reduction of the pores during this drying stage, the RR decreased [64]. The results of Souza et al. [65] also showed that the RR of dried products was strongly dependent on the drying process, so that the carrots samples that were dried using higher MW powers had the lowest RR compared to other treatments. The results obtained in the present research are in agreement with the previous literature [59,66,67].

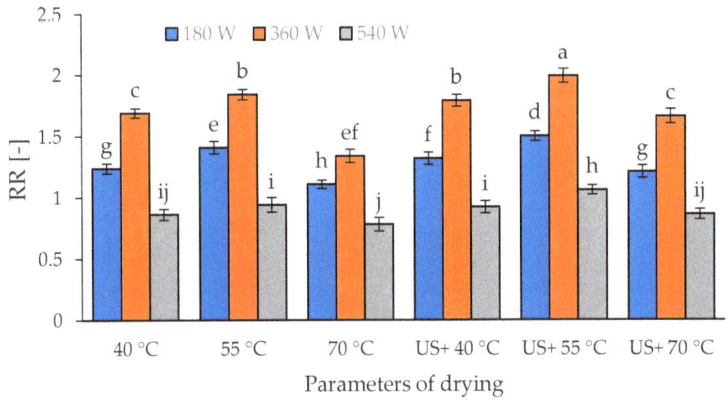


**Figure 5.** Effects of US pretreatment, MW power, and air temperature during drying on rehydration ratio (RR) in hawthorn. Data followed by different letters (a–j) in each column are significantly different at $p < 0.01$. Columns with the same letters are not significantly different.

Furthermore, the use of US pretreatment in all treatments improved the RR. The use of US can enlarge the capillaries and loosen the internal structure of the product by cavitation effect and mechanical effect, and this phenomenon is beneficial to improve the RR of the sample [68]. Tao et al. [69] also found that the use of US during the drying of white cabbage improved the RR of dried samples, which could be due to the microstructural changes produced by US during drying. Also, the study of Horuz et al. [70] proved that the use of US pretreatment before MW–HA drying can accelerate the RR process of dried tomato.

### 3.4. *Effect of Drying on Chemical Property Changes*

#### 3.4.1. Total Phenol Content (TPC) and Total Flavonoid Content (TFC)

The content of bioactive compounds (TPC, TFC) of hawthorn slices before and after drying is presented in Figures 6 and 7. Hawthorn is a great source of TPC and TFC, but processing conditions such as exposure to heat, oxygen, and light can affect its preservation. After drying, TPC and TFC decreased significantly compared to the fresh slices. The results showed that the increase in temperature and MW power had an adverse effect on the TPC and TFC. In fact, the increase in MW power and temperature caused the TPC and TFC content to decrease compared to the fresh sample. On the other hand, using US pretreatment before drying at the temperatures at all three MW power levels did not have much effect on the TPC and TFC values. As shown in Figures 4 and 5, the TPC and TFC content at all the temperature levels was higher when the MW power of 360 W was used, regardless of the application of US pretreatment. The lowest content of the TPC and TFC is related to the treatment at the air temperature of 70 °C and microwave power at 540 W without US. The difference in the amount of TPC and TFC at different temperatures and MW powers of drying is mostly due to the duration of exposure of the samples to temperature [67]. Since polyphenols are destroyed during drying or are attached to other compounds (e.g., proteins), any drying can partially reduce the amount of TPC and TFC as a result [61]. Also, as the drying time increases, due to the increased exposure of TPC and TFC to heat; the destruction of these compounds increases and finally the amount of phenol decreases [71]. Since US pretreatment is a suitable non-thermal method to increase TPC and TFC, the use of this pretreatment before drying increases the TPC and TFC compared to non-pretreated samples [17]. This could be due to the fact US destroys the cell wall of the fruit and causes the release of TPC compounds, and this causes an increase in the amount of TPC and TFC [57]. A similar observation was highlighted in sonicated cranberries [72], carrots [57], and turmeric [73].

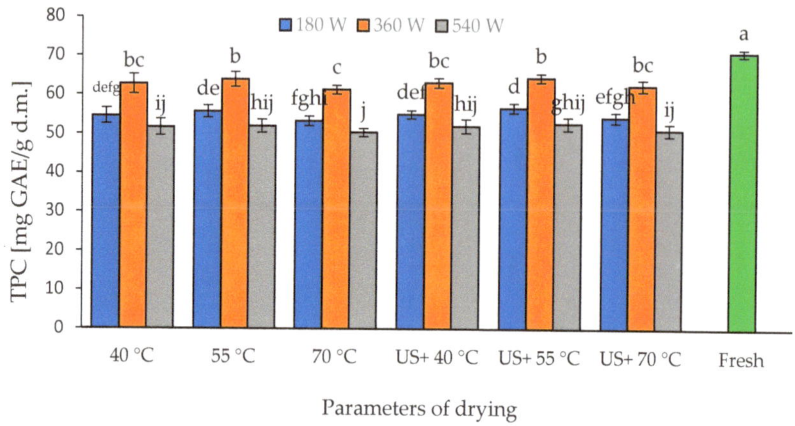


**Figure 6.** Effects of US pretreatment, MW power and air temperature during drying on total polyphenol content (TPC) in hawthorn. Data followed by different letters (a–j) in each column are significantly different at $p < 0.01$. Columns with the same letters are not significantly different.

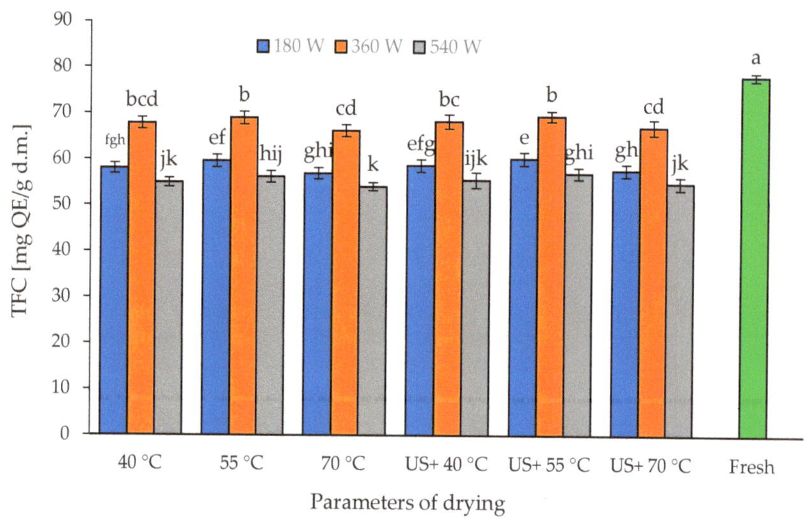


**Figure 7.** Effects of US pretreatment, MW power and air temperature during drying on total flavonoid content (TFC) in hawthorn. Data followed by different letters (a–k) in each column are significantly different at $p < 0.01$. Columns with the same letters are not significantly different.

3.4.2. Antioxidant Activity (AA)

The AA of hawthorn was evaluated using DPPH scavenging. According to Table 2, the effect of independent parameters (MW power and air temperature) and the mutual effect of MW power and air temperature on AA were significant at the 0.01% level. Regarding the AA, the results are similar to the TFC and TPC parameters; in other words, using air temperature and MW power had an adverse effect on the AA and decreased the AA of hawthorn compared to the fresh sample. However, the reduction rate was more significant compared to TFC and TPC. The decreasing trend in all temperature levels was more evident in the MW power of 540 W; therefore, the lowest value of the AA related to the treatment when 70 °C with the MW power of 540 W with and without US pretreatment was applied. On the other hand, the application of US pretreatment before drying in all three power levels and air temperatures did not have much effect on the AA. The results of the current research presented in Figure 8 show that the highest amount of AA was observed in the sample when US was applied before drying, with the use of the MW power of 360 W at a temperature of 55 °C. These four samples had a statistically significant difference at the level of 0.01% with other samples. The lowest amount of AA was obtained by samples dried at a temperature of 70 °C and MW power of 540 W ($p < 0.01\%$). In addition, these results could be related to the degradation of bioactive compounds such as TPC, TFC, and carotenoid due to higher thermal load [24,73]. US pretreatment may cause inhibition of cell respiration, inactivation of enzymes, and reduction in drying time, so the hawthorn slices retained more polyphenols and showed higher AA [28].

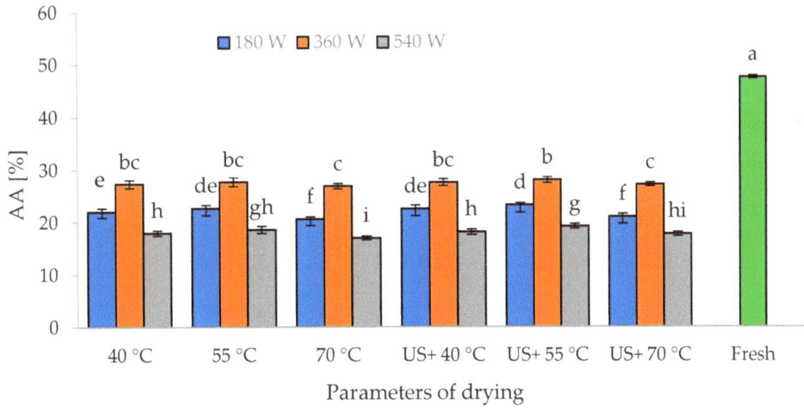


**Figure 8.** Effects of US pretreatment, MW power and air temperature during drying on antioxidant activity (AA) in hawthorn. Data followed by different letters (a–i) in each column are significantly different at $p < 0.01$. Columns with the same letters are not significantly different.

## 4. Conclusions

This research evaluated the effect of using US pretreatment and microwave–hot air (US–MW–HA) drying conditions on chosen physical and chemical properties of hawthorn pieces. The results indicated that US pretreatment could decrease drying time and specific energy consumption by enhancing moisture removal during the microwave–hot air drying process. Comparison between the use of US pretreatment and US-free pretreatment for drying time and specific energy consumption at the temperature of 70 °C and MW power of 540 W was reduced by 41.6% and 12.8%, respectively. However, it is important to evaluate the quality of the food product and the different parameters used for drying (temperature, microwave power) have an effect on the properties of plant tissue. Thus, results showed that the sample treated with US and dried at 60 °C and microwave power of 360 W had the highest RR amount (1.99) and the lowest amount of shrinkage and total color difference. Moreover, exposing raw material to the US pretreatment improves retention of total phenolic compounds, total flavonoids compounds and antioxidant activity due to the faster drying process. A microwave power level of 360 W and temperature of 55 °C with US pretreatment were the best drying conditions for hawthorn, which was characterized by the highest amount of bioactive compounds. Considering the medicinal value of hawthorn and also the use of this product in traditional medicine, it is necessary to conduct scientific studies in order to achieve the maximum quantitative and qualitative yield after harvesting by using the appropriate drying method. Thus, the use of US as a pretreatment technology resulted in better quality dried products than similar products without pretreatment.

**Author Contributions:** Conceptualization, M.K. and E.K.; methodology, M.K. and Y.A.-G.; software, E.K. and M.S.; validation, K.I., S.Z. and Y.A.-G.; formal analysis, M.S., M.N. and K.I.; investigation, M.K., M.N. and Y.A.-G.; resources, M.K., M.N. and S.Z.; data curation, M.K., S.Z. and K.I.; writing—original draft preparation, M.N., M.K. and E.K.; writing—review and editing, M.K., Y.A-G. and M.N.; visualization, M.K. and M.N.; supervision, M.N. All authors have read and agreed to the published version of the manuscript.

**Funding:** This research received no external funding.

**Institutional Review Board Statement:** Not applicable.

**Informed Consent Statement:** Not applicable.

**Data Availability Statement:** Not applicable.

**Acknowledgments:** Authors would like to thank their universities for support.

**Conflicts of Interest:** The authors declare no conflict of interest.

## References

1. Belwal, T.; Cravotto, C.; Prieto, M.A.; Venskutonis, P.R.; Daglia, M.; Devkota, H.P.; Baldi, A.; Ezzat, S.M.; Gomez-Gomez, L.; Salama, M.M.; et al. Effects of different drying techniques on the quality and bioactive compounds of plant-based products: A critical review on current trends. *Dry. Technol.* **2022**, *40*, 1539–1561. [CrossRef]
2. Duan, X.; Liu, W.C.; Ren, G.Y.; Yang, X. Effects of different drying methods on the physical characteristics and flavor of dried hawthorns (*Crataegus* spp.). *Dry. Technol.* **2017**, *35*, 1412–1421. [CrossRef]
3. Li, J.; Li, Z.; Raghavan, G.S.V.; Song, F.; Song, C.; Liu, M.; Pei, Y.; Fu, W.; Ning, W. Fuzzy logic control of relative humidity in microwave drying of hawthorn. *J. Food Eng.* **2021**, *310*, 11070. [CrossRef]
4. Aral, S.; Bese, A.V. Convective drying of hawthorn fruit (*Crataegus* spp.): Effect of experimental parameters on drying kinetics, color, shrinkage, and rehydration capacity. *Food Chem.* **2016**, *210*, 577–584. [CrossRef] [PubMed]
5. Abbaspour-Gilandeh, Y.; Kaveh, M.; Fatemi, H.; Aziz, M. Combined hot air, microwave, and infrared drying of hawthorn fruit: Effects of ultrasonic pretreatment on drying time, energy, qualitative, and bioactive compounds' properties. *Foods* **2021**, *10*, 1006. [CrossRef] [PubMed]
6. Li, Y.; Wang, X.; Wu, Z.; Wan, N.; Yang, M. Dehydration of hawthorn fruit juices using ultrasound-assisted vacuum drying. *Ultrason. Sonochemistry* **2020**, *68*, 10521. [CrossRef]
7. Reis, F.R.; Marques, C.; de Moraes, A.C.S.; Masson, M.L. Trends in quality assessment and drying methods used for fruits and vegetables. *Food Control* **2022**, *142*, 10925.
8. Majumder, P.; Sinha, A.; Gupta, R.; Sablani, S.S. Drying of Selected Major Spices: Characteristics and Influencing Parameters, Drying Technologies, Quality Retention and Energy Saving, and Mathematical Models. *Food Bioprocess Technol.* **2021**, *14*, 1028–1054. [CrossRef]
9. Mohammed, H.H.; Tola, Y.B.; Taye, A.H.; Abdisa, Z.K. Effect of pretreatments and solar tunnel dryer zones on functional properties, proximate composition, and bioactive components of pumpkin (*Cucurbita maxima*) pulp powder. *Heliyon* **2022**, *8*, e10747. [CrossRef]
10. Zhou, Y.-H.; Pei, Y.-P.; Sutar, P.P.; Liu, D.-H.; Deng, L.-Z.; Duan, X.; Liu, Z.-L.; Xiao, H.-W. Pulsed vacuum drying of banana: Effects of ripeness on drying kinetics and physicochemical properties and related mechanism. *LWT Food Sci. Technol.* **2022**, *161*, 1133. [CrossRef]
11. Pei, Y.; Li, Z.; Song, C.; Li, J.; Xu, W.; Zhu, G. Analysis and modelling of temperature and moisture gradient for ginger slices in hot air drying. *J. Food Eng.* **2022**, *323*, 11100. [CrossRef]
12. Karwacka, M.; Ciurzyńska, A.; Galus, S.; Janowicz, M. Freeze-dried snacks obtained from frozen vegetable by-products and apple pomace—Selected properties, energy consumption and carbon footprint. *Innov. Food Sci. Emerg. Technol.* **2022**, *77*, 102949. [CrossRef]
13. Tepe, F.B.; Tepe, T.K.; Ekinci, A. Drying kinetics and energy efficiency of microwave-dried lemon slices. *Chem. Ind. Chem. Eng. Q.* **2022**, *28*, 297–304. [CrossRef]
14. Sadeghi, E.; Movagharnejad, K.; Asl, A.H. Parameters optimization and quality evaluation of mechanical properties of infrared radiation thin layer drying of pumpkin samples. *J. Food Process Eng.* **2019**, *43*, e13309. [CrossRef]
15. Dadan, M.; Nowacka, M. The assessment of the possibility of using ethanol and ultrasound to design the properties of dried carrot tissue. *Appl. Sci.* **2021**, *11*, 689. [CrossRef]
16. Rybak, K.; Wiktor, A.; Kaveh, M.; Dadan, M.; Witrowa-Rajchert, D.; Nowacka, M. Effect of thermal and non-thermal technologies on kinetics and the main quality parameters of red bell pepper dried with convective and microwave–convective methods. *Molecules* **2022**, *27*, 2164. [CrossRef]
17. Bao, T.; Hao, X.; Shishir, M.R.S.; Karim, N.; Chen, W. Green alternative methods for pretreatment of whole jujube before drying process. *J. Sci. Food Agric.* **2022**, *102*, 1030–1039. [CrossRef]

18. Deng, Y.; Zhao, Y. Effects of pulsed-vacuum and ultrasound on the osmodehydration kinetics and microstructure of apples (Fuji). *J. Food Eng.* **2008**, *85*, 84–93. [CrossRef]
19. Awad, T.S.; Moharram, H.A.; Shaltout, O.E.; Asker, D.; Youssef, M.M. Applications of ultrasound in analysis, processing and quality control of food: A review. *Food Res. Int.* **2012**, *48*, 410–427. [CrossRef]
20. Rostamabadi, H.; Rohit, T.; Karaca, A.C.; Nowacka, M.; Colussi, R.; Frasson, S.F.; Aaliya, B.; Sunooj, K.V.; Falsafi, S.R. How non-thermal processing treatments affect physicochemical and structural attributes of tuber and root starches? *Trends Food Sci. Technol.* **2022**, *128*, 217–237. [CrossRef]
21. Zhang, J.; Li, M.; Ding, Z.; Wang, C.; Cheng, J. Evaluation of ultrasound-assisted microwave hot air convective drying Chinese hickory—Drying kinetics and product's quality properties. *J. Food Process Eng.* **2021**, *44*, e13842. [CrossRef]
22. Sledz, M.; Wiktor, A.; Rybak, K.; Nowacka, M.; Witrowa-Rajchert, D. The impact of ultrasound and steam blanching pre-treatments on the drying kinetics, energy consumption and selected properties of parsley leaves. *Appl. Acoust.* **2016**, *103*, 148–156. [CrossRef]
23. Dehghannya, J.; Kadkhodaei, S.; Heshmati, M.K.; Ghanbarzadeh, B. Ultrasound-assisted intensification of a hybrid intermittent microwave—Hot air drying process of potato: Quality aspects and energy consumption. *Ultrasonics* **2019**, *96*, 104–122. [CrossRef]
24. Chouaibi, M.; Snoussi, A.; Attouchi, S.; Ferrari, G. Influence of drying processes on bioactive compounds profiles, hydroxymethylfurfural, color parameters, and antioxidant activities of Tunisian eggplant (*Solanum melongena* L.). *J. Food Process. Preserv.* **2021**, *45*, e15460. [CrossRef]
25. Zeng, S.; Wang, B.; Lv, W.; Wu, Y. Effects of microwave power and hot air temperature on the physicochemical properties of dried ginger (*Zingiber officinale*) using microwave hot-air rolling drying. *Food Chem.* **2023**, *404*, 134741. [CrossRef] [PubMed]
26. Ando, Y.; Nei, D. Comparison of potato void structures dried by air-drying, freeze-drying, and microwave-vacuum-drying, and the physical properties of powders after grinding. *Food Bioprocess Technol.* **2023**, *16*, 447–458. [CrossRef]
27. Nanvakenari, S.; Movagharnejad, K.; Latifi, A. Modelling and experimental analysis of rice drying in new fluidized bed assisted hybrid infrared-microwave dryer. *Food Res. Int.* **2022**, *159*, 111617. [CrossRef]
28. Mirzaei-Baktash, H.; Hamdami, N.; Torabi, P.; Fallah-Joshaqani, S.; Dalvi-Isfahan, M. Impact of different pretreatments on drying kinetics and quality of button mushroom slices dried by hot-air or electrohydrodynamic drying. *LWT Food Sci. Technol.* **2022**, *155*, 112894. [CrossRef]
29. Li, L.; Zhang, M.; Wang, W. Ultrasound-assisted osmotic dehydration pretreatment before pulsed fluidized bed microwave freeze-drying (PFBMFD) of Chinese yam. *Food Biosci.* **2020**, *35*, 100548. [CrossRef]
30. Nguyen, T.-V.-L.; Nguyen, P.-B.-D.; Tran, T.T.V.; Tran, B.-L.; Huynh, T.-P. Low-temperature microwave-assisted drying of sliced bitter melon: Drying kinetics and rehydration characteristics. *J. Food Process Eng.* **2022**, *45*, e14177. [CrossRef]
31. Tepe, F.B. Impact of pretreatments and hybrid microwave assisting on drying characteristics and bioactive properties of apple slices. *J. Food Process. Preserv.* **2022**, *46*, e17067. [CrossRef]
32. An, N.; Sun, W.; Li, B.; Wang, Y.; Shang, N.; Lv, W.; Li, D.; Wang, L. Effect of different drying techniques on drying kinetics, nutritional components, antioxidant capacity, physical properties and microstructure of edamame. *Food Chem.* **2022**, *373*, 131412. [CrossRef]
33. Maftoonazad, N.; Dehghani, M.R.; Ramaswamy, H.S. Hybrid microwave-hot air tunnel drying of onion slices: Drying kinetics, energy efficiency, product rehydration, color, and flavor characteristics. *Dry. Technol.* **2022**, *40*, 966–986. [CrossRef]
34. Mierzwa, D.; Szadzińska, J.; Pawłowski, A.; Pashminehazar, R.; Kharaghani, A. Nonstationary convective drying of raspberries, assisted by microwaves and ultrasound. *Dry. Technol.* **2019**, *37*, 988–1001. [CrossRef]
35. AOAC. *Official Methods of Analysis*, 16th ed.; Association of Official Analytical Chemists: Washington, DC, USA, 2010.
36. Kaveh, M.; Abbaspour-Gilandeh, Y. Drying characteristics, specific energy consumption, qualitative properties, total phenol compounds, and antioxidant activity during hybrid hot air-microwave- rotary drum drying of green pea. *Iran. J. Chem. Chem. Eng.* **2022**, *40*, 655–672.
37. Sharifian, F.; Motlagh, A.M.; Nikbakht, A.M. Pulsed microwave drying kinetics of fig fruit (*Ficus carica* L.). *Aust. J. Crop Sci.* **2012**, *6*, 1441–1444.
38. Çetin, N.; Sağlam, C. Effects of ultrasound pre-treatment assisted drying methods on drying characteristics, physical and bioactive properties of windfall apples. *J. Sci. Food Agric.* **2022**, *103*, 534–547. [CrossRef]
39. Geng, Z.; Torki, M.; Kaveh, M.; Beigi, M.; Yang, X. Characteristics and multi-objective optimization of carrot dehydration in a hybrid infrared /hot air dryer. *LWT Food Sci. Technol.* **2022**, *172*, 114229. [CrossRef]
40. Sharifian, F.; Modarres-Motlagh, A.; Komarizade, M.H.; Nikbakht, A.M. Colour change analysis of fig fruit during microwave drying. *Int. J. Food Eng.* **2013**, *9*, 107–114. [CrossRef]
41. Wiktor, A.; Landfeld, A.; Matys, A.; Novotná, P.; Dadan, M.; Kováříková, E.; Nowacka, M.; Mulenko, M.; Witrowa-Rajchert, D.; Strohalm, J.; et al. Selected Quality Parameters of Air-Dried Apples Pretreated by High Pressure, Ultrasounds and Pulsed Electric Field—A Comparison Study. *Foods* **2021**, *10*, 1943. [CrossRef]
42. Shahidi, F.; Naczk, M. *Phenolics in Food and Nutraceuticals*; CRC Press: Boca Raton, FL, USA, 2004.
43. Kaveh, M.; Abbaspour-Gilandeh, Y.; Nowacka, M. Optimisation of microwave-rotary drying process and quality parameters of terebinth. *Biosyst. Eng.* **2021**, *208*, 113–130. [CrossRef]
44. Dibagar, N.; Kowalski, S.J.; Chayjan, R.A.; Figiel, A. Accelerated convective drying of sunflower seeds by high-power ultrasound: Experimental assessment and optimization approach. *Food Bioprod. Process.* **2020**, *123*, 42–59. [CrossRef]

45. Huang, Y.; Zhang, M.; Mujumdar, A.S.; Luo, Z.; Fang, Z. Dehydrated fruits and vegetables using low temperature drying technologies and their application in functional beverages: A review. *Dry. Technol.* 2022, *in press*. [CrossRef]
46. Dehsheikh, F.N.; Dinani, S.T. Coating pretreatment of banana slices using carboxymethyl cellulose in an ultrasonic system before convective drying. *Ultrason. Sonochemistry* **2019**, *52*, 401–413. [CrossRef]
47. Raj, G.V.S.B.; Dash, K.K. Effect of intermittent microwave convective drying on physicochemical properties of dragon fruit. *Food Sci. Biotechnol.* **2022**, *31*, 549–560. [CrossRef]
48. Pham, N.D.; Karim, M.A. Investigation of nutritional quality evolution of papaya during intermittent microwave convective drying. *Dry. Technol.* **2022**, *40*, 3694–3707. [CrossRef]
49. Szadzinska, J.; Mierzwa, D. The influence of hybrid drying (microwave-convective) on drying kinetics and quality of white mushrooms. *Chem. Eng. Process. Process Intensif.* **2021**, *16*, 108532. [CrossRef]
50. Dehghannya, J.; Seyed-Tabatabaei, S.-R.; Heshmati, M.K.; Ghanbarzadeh, B. Influence of three stage ultrasound—Intermittent microwave—Hot air drying of carrot on physical properties and energy consumption. *Heat Mass Transf.* **2021**, *57*, 1893–1907. [CrossRef]
51. Witrowa-Rajchert, D.; Wiktor, A.; Sledz, M.; Nowacka, M. Selected emerging technologies to enhance the drying process: A review. *Dry. Technol.* **2014**, *32*, 1386–1396. [CrossRef]
52. Motevali, A.; Minaei, S.; Banakar, A.; Ghobadian, B.; Khoshtaghaza, M.H. Comparison of energy parameters in various dryers. *Energy Convers. Manag.* **2014**, *87*, 711–725. [CrossRef]
53. Kaveh, M.; Abbaspour-Gilandeh, Y.; Nowacka, M. Comparison of different drying techniques and their carbon emissions in green peas. *Chem. Eng. Process. Process Intensif.* **2021**, *160*, 108274. [CrossRef]
54. Szadzińska, J.; Łechtańska, J.; Pashminehazar, R.; Kharaghani, A.; Tsotsas, E. Microwave- and ultrasound-assisted convective drying of raspberries: Drying kinetics and microstructural changes. *Dry. Technol.* **2019**, *37*, 1–12. [CrossRef]
55. İlter, I.; Akyıl, S.; Devseren, E.; Okut, D.; Koç, M.; Ertekin, F.K. Microwave and hot air drying of garlic puree: Drying kinetics and quality characteristics. *Heat Mass Transf.* **2018**, *54*, 2101–2112. [CrossRef]
56. Zeng, Y.; Liu, Y.; Zhang, J.; Xi, H.; Duan, X. Effects of far-infrared radiation temperature on drying characteristics, water status, microstructure and quality of kiwifruit slices. *J. Food Meas. Charact.* **2019**, *13*, 3086–3096. [CrossRef]
57. Kroehnke, J.; Szadzińska, J.; Stasiak, M.; Radziejewska-Kubzdela, E.; Biegańska-Marecik, R.; Musielak, G. Ultrasound- and microwave-assisted convective drying of carrots—Process kinetics and product's quality analysis. *Ultrason. Sonochemistry* **2018**, *48*, 249–258. [CrossRef]
58. Witrowa-Rajchert, D.; Rzaca, M. Effect of drying method on the microstructure and physical properties of dried apples. *Dry. Technol.* **2009**, *27*, 903–909. [CrossRef]
59. Joudi-Sarighayeh, F.; Abbaspour-Gilandeh, Y.; Kaveh, M.; Hernández-Hernández, J.L. The optimization of the physical–thermal and bioactive properties of pumpkin slices dried in a hybrid microwave– convective dryer using the response surface method. *Agronomy* **2022**, *12*, 2291. [CrossRef]
60. Bhat, T.A.; Hussain, S.Z.; Wani, S.M.; Rather, M.A.; Reshi, M.; Naseer, B.; Qadri, T.; Khalil, A. The impact of different drying methods on antioxidant activity, polyphenols, vitamin C and rehydration characteristics of Kiwifruit. *Food Biosci.* **2022**, *48*, 101821. [CrossRef]
61. Wang, Y.; Li, X.; Chen, X.; Li, B.; Mao, X.; Miao, J.; Zhao, C.; Huang, L.; Gao, W. Effects of hot air and microwave-assisted *48*drying on drying kinetics, physicochemical properties, and energy consumption of chrysanthemum. *Chem. Eng. Process. Process Intensif.* **2018**, *129*, 84–94. [CrossRef]
62. Gharkhloo, Z.R.; Sharifian, F.; Rahimi, A.; Yamchi, A.A. Influence of high wave sound pretreatment on drying quality parameters of Echinacea root with infrared drying. *J. Sci. Food Agric.* **2022**, *102*, 2153–2164. [CrossRef]
63. Jahanbakhshi, A.; Yeganeh, R.; Momeny, M. Influence of ultrasound pre-treatment and temperature on the quality and thermodynamic properties in the drying process of nectarine slices in a hot air dryer. *J. Food Process. Preserv.* **2020**, *44*, e14818. [CrossRef]
64. Çetin, N. Comparative assessment of energy analysis, drying kinetics, and biochemical composition of tomato waste under different drying conditions. *Sci. Hortic.* **2022**, *305*, 111405. [CrossRef]
65. Souza, A.U.; Correa, J.L.G.; Tanikawa, D.H.; Abrahao, F.R.; Junqueira, J.R.J.; Jimenez, E.C. Hybrid microwave-hot air drying of the osmotically treated carrots. *LWT Food Sci. Technol.* **2022**, *156*, 113046. [CrossRef]
66. Darıcı, M.; Süfer, O.; Simsek, M. Determination of microwave drying and rehydration kinetics of green peppers with the bioactive and textural properties. *J. Food Process Eng.* **2021**, *44*, e13755. [CrossRef]
67. Zahoor, I.; Khan, M.A. Microwave assisted fluidized bed drying of red bell pepper: Drying kinetics and optimization of process conditions using statistical models and response surface methodology. *Sci. Hortic.* **2021**, *286*, 11020. [CrossRef]
68. Song, Y.; Tao, Y.; Zhu, X.; Han, Y.; Show, P.L.; Song, C.; Zaid, H.F.M. Ultrasound-Enhanced Hot Air Drying of Germinated Highland Barley Seeds: Drying Characteristics, Microstructure, and Bioactive Profile. *Agri Eng.* **2019**, *1*, 496–510. [CrossRef]
69. Tao, Y.; Han, M.; Gao, X.; Han, Y.; Show, P.-L.; Liu, C.; Ye, X.; Xie, G. Applications of water blanching, surface contacting ultrasound-assisted air drying, and their combination for dehydration of white cabbage: Drying mechanism, bioactive profile, color and rehydration property. *Ultrason. Sonochemistry* **2019**, *53*, 192–201. [CrossRef]
70. Horuz, E.; Jaafar, H.J.; Maskan, M. Ultrasonication as pretreatment for drying of tomato slices in a hot air–microwave hybrid oven. *Dry. Technol.* **2017**, *35*, 849–859. [CrossRef]

71. Li, M.; Wang, B.; Wang, Y.; Liu, J.; Zhang, M. Evaluation of the uniformity, quality and energy cost of four types of vegetables and fruits after pilot-scale pulse-spouted bed microwave (915 MHz) freeze-drying. *Dry. Technol.* **2023**, *41*, 290–370. [CrossRef]
72. Zhou, Y.-H.; Staniszewska, I.; Liu, Z.-L.; Zielinska, D.; Xiao, H.-W.; Pan, Z.; Nowak, K.W.; Zielinska, M. Microwave-vacuum-assisted drying of pretreated cranberries: Drying kinetics, bioactive compounds and antioxidant activity. *LWT Food Sci. Technol.* **2021**, *146*, 111464. [CrossRef]
73. An, N.; Shang, N.; Lv, W.; Li, D.; Wang, L.; Wang, Y. Effects of carboxymethyl cellulose/pectin coating combined with ultrasound pretreatment before drying on quality of turmeric (*Curcuma longa* L.). *Int. J. Biol. Macromol.* **2022**, *202*, 354–365. [CrossRef] [PubMed]

processes

Article

# Effect of Drying Pretreatment on Cellulolytic Enzymatic Hydrolysis of Lignin from Napier Grass

**Syazmi Zul Arif Hakimi Saadon [1] and Noridah Binti Osman [1,2,*]**

[1] Department of Chemical Engineering, Universiti Teknologi PETRONAS, Bandar Seri Iskandar 32610, Perak, Malaysia
[2] Higher Education Center of Excellence–Center for Biofuel and Biochemical Research, Institute of Self-Sustainable Building, Universiti Teknologi PETRONAS, Bandar Seri Iskandar 32610, Perak, Malaysia
* Correspondence: noridah.osman@utp.edu.my

**Abstract:** Biomass can be a viable supplement and alternative to non-renewable sources of fuel and chemicals. Lignin is an important part of biomass sources which can be used in various chemical and fuel industries. This study explores the pretreatment of lignin from Napier grass using thermal and physical means, as well as extraction of lignin via cellulolytic enzymatic hydrolysis to determine the optimum condition for feedstock pretreatment. Napier grass parts under various drying conditions and particle sizes were treated with enzymes. Moisture analysis, FTIR spectroscopy, UV–Vis analysis, and Klason lignin were carried out to analyze the moisture, functional group, and yield of lignin. Moisture content of the samples were inversely proportional to the drying conditions. The FTIR result showed lower peak intensity for higher drying conditions, while ball-milling showed less reduction in peak intensity. More Klason lignin was extracted under higher drying conditions. The yield of cellulolytic enzymatic lignin (CEL) was found to be more than actual lignin content, suggesting cellulose was not fully degraded. The FTIR spectra of CEL was found to be closer to that of lignin, but purification was still needed. Optimization was carried out by evaluating the statistical significance of each pretreatment effect of the pretreatments.

**Keywords:** cellulase; enzymatic hydrolysis; lignin; Napier grass; pretreatment

**Citation:** Saadon, S.Z.A.H.; Osman, N.B. Effect of Drying Pretreatment on Cellulolytic Enzymatic Hydrolysis of Lignin from Napier Grass. *Processes* **2023**, *11*, 1092. https://doi.org/10.3390/pr11041092

Academic Editor: Jan Havlík

Received: 27 January 2023
Revised: 5 February 2023
Accepted: 7 February 2023
Published: 4 April 2023



## 1. Introduction

Biomass is considered to be a potential source of renewable and sustainable energy going forward. Its usage is gaining momentum because of its wide availability and eco-friendly nature. In Malaysia, Energy Statistics published in 2020 show that fossil fuel is still heavily relied on while biomass-based energy only accommodated 0.2% of the primary fuel production for the year 2018, retaining the same percentage as three years prior [1,2]. Biomass can be categorized into six main sources: (1) dedicated energy crops; (2) agricultural crop residue; (3) forestry residues; (4) algae; (5) wood processing residues; (6) sorted municipal waste; and (7) wet waste [3–5]. Each of these biomass sources can be turned into value-added products for energy consumption, chemical manufacturing, or valuable material production. Napier grass is currently classed as agricultural crop residue, but it has the potential to become an energy crop since it can be turned into solid fuel [6], bioethanol [7–9] and also biogas [10,11].

These valuable products are derived from the lignocellulosic material, which comprises mostly cellulose, lignin hemicellulose and extractives. Lignin is the second most abundant naturally occurring polymer and the most abundant aromatic compound in nature. Its purposes in woody biomass are to provide biological and chemical protection from degradation, as well as structural strength and rigidity to the plant [12]. It can be found in all types of plant, including hardwood, softwood, and herbaceous plants as a major constituent in the structural cell wall. Lignin is the only non-carbohydrate polymer of the three components, and it comprises about one-third of the mass of the lignocellulose. As it is commonly found in biomass products, it is needless to mention that it can be exploited and applied in various sectors. The research on lignin has increased over the years. As interest in cellulose rises, it has resulted in an incremental increase in lignin research since lignin is considered a by-product of cellulose extraction. According to Web of Science, the number of research publications of cellulose and lignin has increased steadily by an average of 11% each year. Due to the low reproducibility of woody biomass, much research is focusing on herbaceous plants to supplement lignin production from biomass. Although lignin content in herbaceous plants is lower than hardwood and softwood, herbaceous plants have high annual renewability and the largest annual biomass stock [13].

Pretreatment of feedstock is an important step in all biochemical conversions. It is a necessary step to prepare the feedstock for further processes, and functions to enhance the digestibility of lignocellulosic components, increase accessibility to the targeted component, and also to ease extraction. As mentioned by Kumar and Sharma [14], the goal of all pretreatments is to avoid a reduction in size, preserve the saccharide fractions, limit formation of degradation products, and minimize energy and cost. During the pre-treatment process, the recalcitrance of the lignocellulose structure is disrupted when the lignin sheath is broken down, degradation of hemicellulose occurs, and there is a reduction in both crystallinity and degree of cellulose polymerization [15,16].

The aim of this study is to optimize the drying pretreatment process to maximize the production of lignin from the Klason method and cellulolytic enzymatic hydrolysis. In this study, we compared different parts of Napier grass (NG), which were then pretreated with variations in drying time, drying temperature, and milling process. Subsequent enzymatic hydrolysis was performed on the pretreated samples with additional variables of incubation temperature and day. The effect of the variables was analyzed using moisture content analysis, Fourier Transformed Infrared (FTIR) Spectroscopy Analysis and Ultraviolet–visible (UV-Vis) spectrophotometry, as well as by measuring the lignin yield upon pretreatment and enzymatic hydrolysis incubation.

## 2. Materials and Methods

### 2.1. Materials

The Napier grass was collected from Lambor Kanan, Perak, Malaysia, where it had been cultivated for about 3 months. The sample was then separated into stems and leaves and cleaned using tap water to remove any dirt or impurities. The Napier grass was left to sun-dry for 5 h. The sample was then placed in a cold room at a temperature of −4 °C to minimize rotting or degradation. Prior to pretreatment, the samples were thoroughly checked for degradation.

Acetic acid, ethanol, sodium bisulfite, sodium hydroxide pellets, nitric acid, sodium chlorite, toluene, and cellulase from Aspergillus Niger were of analytical grade and purchased from Avantis Laboratory Supply, Malaysia. Pure microcrystalline cellulose (MCC) and Kraft lignin (KrL) were also obtained from the same supplier to compare with extracted lignin. Crystalline nanocellulose (CNC) was obtained from PowerNano Malaysia, also to compare with extracted lignin.

### 2.2. Pretreatments

For thermal pretreatment, the feedstock was dried in the oven at 45, 75, 105 and 135 °C. The drying process was done for 5, 15 and 25 h. In each batch, the Napier grass was weighed to be around 30g, and its weight was recorded again after the drying process.

For physical pretreatment, the feedstock was then shredded using a laboratory blender and was sieved to ensure a uniform size of 250 μm using a sieve shaker. Some of the particles larger than 250 μm were then further milled using a planetary ball mill for 10 min at 500 rpm. The grinded feedstock was kept in an air-tight container to prevent moisture absorption. The pretreatment parameter is listed in Table 1 below.

**Table 1.** Pretreatment parameters.

| Sample | Thermal Pretreatment | Physical Pretreatment |
|---|---|---|
| Type<br>• Leaf<br>• Stem | Drying temperature<br>• 45 °C<br>• 75 °C<br>• 105 °C<br>• 135 °C<br>Drying time<br>• 5 h<br>• 15 h<br>• 25 h | Particle size<br>• 250 μm<br>• Ball-milled |

### 2.3. *Klason Lignin*

Klason lignin is the insoluble residue of acid hydrolysis. This method employs the ASTM D1105 Standard Test Method for Preparation of Extractive-Free Wood [17], the ASTM D1107 Standard Test Method for Ethanol-Toluene Solubility of Wood [18], and the ASTM D1106 Standard Test Method for Acid-Insoluble Lignin in Wood [19]. In general, 1 g of prepared sample was extracted for 4 h with 95% ethanol in a Soxhlet extraction apparatus. It was then extracted with ethanol–toluene mixture for another 8 h. After extraction, the sample was dried in the oven at 105 °C for 1 h and then was left to cool in a desiccator overnight.

The sample was digested in a hot water bath at approximately 100 °C for 3 h. The sample was filtered, washed with hot water and ethanol, and left to dry in air overnight. The air-dried sample was mixed with 15 mL of cold 72% sulfuric acid for 2 h at a temperature between 18 and 20 °C with frequent stirring. After that, the mixture was diluted to 3% concentration sulfuric acid and left to boil for 4 h. The sample was then filtered under suction and washed with hot water. The insoluble residue was dried in the oven for 2 h at 105 °C and left to cool in a desiccator overnight before it was weighed. The dilute acid solvent was collected and stored for analysis. The acid-insoluble lignin was calculated using Equation (1) as follows:

$$\%\ IL = \frac{W_{solid\ extract}}{W_{sample}} \times 100\% \quad (1)$$

### 2.4. *Cellulolytic Enzymatic Hydrolysis*

For the enzymatic hydrolysis, the samples were treated with cellulase from Aspergillus Niger. The treatment was based on the procedure described by NREL (ABBREVIATION) [20] and Chang et al. [21] and modified according to Rencoret et al. [22]. In brief, 0.02 M of pH 4.0 acetate buffer was prepared by dissolving sodium acetate trihydrate in acetic acid and water. Then, 1.5 g of sample was left to suspend in acetate buffer and 60 mg of cellulase was added to the suspension to maintain a 1:40 cellulose-to-sample ratio. The suspension was incubated at 250 rpm in an incubator shaker at varying temperatures and number of days. The incubation parameters are listed in Table 2 below. After the incubation, the solid was separated from the solution, where the solution contained soluble lignin. The solid was extracted with 96% dioxane for 4 h and subsequently extracted with 50% dioxane. After the extraction, the solid was washed with acetic acid and water. The solid was dried, weighed, and stored in an air-tight container. The CEL is calculated using Equation (1).

**Table 2.** Enzymatic incubation parameters.

| Incubation Temperature | Incubation Time |
|---|---|
| • 30 °C<br>• 40 °C<br>• 50 °C | • 1 day<br>• 3 days<br>• 5 days |

### 2.5. Moisture Analysis

Each sample was then evaluated after thermal and physical pretreatment using PRECISA XM60/XM60-HR moisture analyzing equipment. Each sample was weighed to around 0.5 g and placed on a heating plate. The test was repeated three times and the results were averaged. Moisture content was calculated using Equation (2).

$$Moisture\ content\ (\%) = \frac{Initial\ weight - Final\ weight}{Initial\ weight} \times 100\% \quad (2)$$

### 2.6. Fourier Transformed Infrared (FTIR) Spectroscopy Analysis

FTIR with Attenuated Total Reflectance (ATR) was used to evaluate the pretreated sample. Changes in the intensity of the functional group of the pretreated sample were determined and analyzed based on the FTIR spectrum. The spectra were recorded in the frequency range of 4000–400 $cm^{-1}$ with a resolution of 1 $cm^{-1}$. The conversion from generated percent transmittance (%T) to absorbance (A) was performed using Equation (3).

$$A = 2 - log(\%T) \quad (3)$$

### 2.7. Ultraviolet–Visible (UV–Vis) Spectrophotometry

After incubation, the dilute acid solvent and enzymatic solutions were assessed with UV–Vis for soluble lignin content. Acid-soluble lignin content was determined using the UV absorbance measurement of the pretreated samples at 280 nm and 320 nm according to the NREL procedure. Acid-soluble lignin content was calculated according to the following Equation (5):

$$\%\ ASL = \frac{UV_{abs} \times Vol_{filtrate} \times Dilution}{\varepsilon \times Weight_{sample} \times Pathlength} \times 100\% \quad (4)$$

where UVabs is the mean UV–Vis absorbance at 205 and 280 nm, Volfiltrate represents the volume of filtrate, ε refers to the molar absorptivity of biomass, Wsample represents the sample weight in milligrams, and Pathlength is the UV–Vis cell pathlength in cm. Dilution was set to 1 since no other diluting solvent was used.

### 2.8. Statistical Analysis

SAS statistical software was used to carry out statistical analysis of moisture content, FTIR spectrum peak intensity, and extracted lignin using Analysis of Variance (ANOVA). A significance level (denoted as α) of 0.05 indicated that a significant difference exists. If the *p*-value is less than or equal to the significance level, the null hypothesis can be rejected which means there is a significant difference between the changes in the set of data. However, when the *p*-value is greater than the significance level, it indicates that there is not enough evidence to reject the null hypothesis which means that the results do not have a significant difference between them.

## 3. Results and Discussion

### 3.1. Moisture Analysis

The moisture content of each datapoint was plotted as shown in Figure 1. As expected, a declining trend was observed with increasing time and temperature. Drying at 135 °C after 15 and 25 h resulted in the lowest moisture content, while a drying temperature of 45 °C showed the lowest loss of moisture content from the sample. A sharp decrease can be seen between 45 °C and 75 °C. At temperatures of 75 °C and above, it can be observed that the moisture content reached below 10%. This is in agreement with Houghton et al., who suggested that higher temperature is used to reduce moisture content [23]. It is also seen that the difference between stems and leaves was very significant, with stems containing and retaining more moisture. Higher moisture content results in less lignocellulose content per weight of sample, since much of the weight is contributed to by the weight of water molecules. In general, a moisture content of lower than 10 % is desired in all extractions so that the presence of water molecules does not interrupt or hinder the process of extraction and isolation.

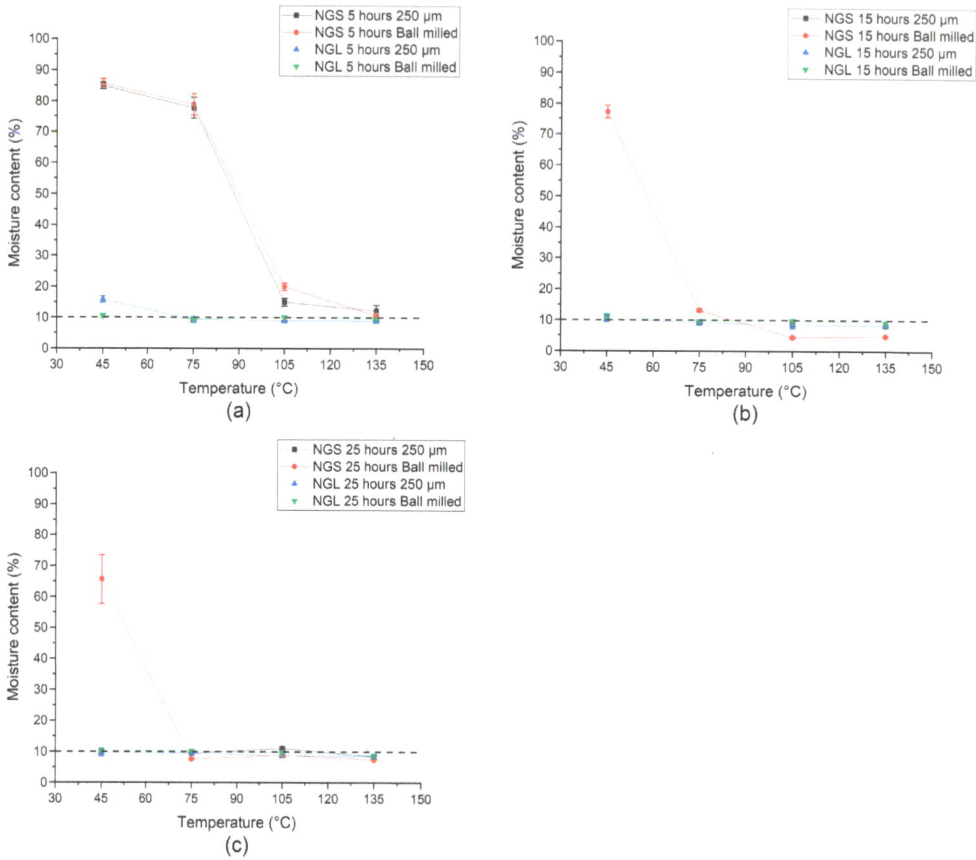


**Figure 1.** Moisture content against temperature (**a**) 5 h, (**b**) 15 h and (**c**) 25 h.

After statistical analysis, it was found that all effects had a significant impact on the moisture content as shown in Tables 3 and 4. Since both *p*-values for both effects as well as their interactions were less than the α-value (*p*-value < α), they were considered to be statistically significant. When further analyzed using the least mean square method, any increment above 15 h and 75 °C was found to not have a significant effect on the moisture content. The changes between 105 and 135 °C and 15 and 25 h were very insignificant, since the interaction between the two points had a *p*-value greater than 5% and the changes were not noticeable. The sample type was also significant, until 105 °C when the moisture started to congregate, rendering similar moisture content for both types. The size effect was significant, especially towards stem samples where ball-milled stems had a higher moisture content as compared to 250 μm. Since one of the main objectives of pretreatment is to reduce energy, it can be concluded that pretreatment at 15 h, 105 °C and ball-milled particle onto leaf samples are the optimal conditions, as opposed to the 103 °C for 24 h recommended by ASAE Standards.

**Table 3.** ANOVA table for moisture content.

| ANOVA | | | | Alpha | 0.05 |
|---|---|---|---|---|---|
| Source | DF | Sum of Squares | Mean Square | F-Value | Pr > F |
| Model | 47 | 75,479.11 | 1605.938 | 675.03 | <0.0001 |
| Error | 96 | 228.3904 | 2.37907 | | |
| Total | 143 | 75,707.5 | | | |

**Table 4.** ANOVA table of each effect for moisture content.

| Source | DF | Type I SS | Mean Square | F Value | Pr > F |
|---|---|---|---|---|---|
| Type | 1 | 10,758.53 | 10,758.53 | 4522.16 | <0.0001 |
| Time | 2 | 8408.365 | 4204.182 | 1767.16 | <0.0001 |
| Temp | 3 | 14,117.24 | 4705.748 | 1977.98 | <0.0001 |
| Size | 1 | 958.4184 | 958.4184 | 402.85 | <0.0001 |
| Type*Time | 2 | 7571.9 | 3785.95 | 1591.36 | <0.0001 |
| Type*Temp | 3 | 11,730.29 | 3910.095 | 1643.54 | <0.0001 |
| Type*Size | 1 | 854.2955 | 854.2955 | 359.09 | <0.0001 |
| Time*Temp | 6 | 5682.803 | 947.1338 | 398.11 | <0.0001 |
| Time*Size | 2 | 438.0065 | 219.0032 | 92.05 | <0.0001 |
| Temp*Size | 3 | 2648.561 | 882.8538 | 371.09 | <0.0001 |
| Type*Time*Temp | 6 | 5616.94 | 936.1566 | 393.5 | <0.0001 |
| Type*Time*Size | 2 | 268.1544 | 134.0772 | 56.36 | <0.0001 |
| Type*Temp*Size | 3 | 3228.422 | 1076.141 | 452.34 | <0.0001 |
| Time*Temp*Size | 6 | 1875.295 | 312.5491 | 131.37 | <0.0001 |
| Type*Time*Temp* Size | 6 | 1321.887 | 220.3145 | 92.61 | <0.0001 |

### 3.2. Pretreatment FTIR Analysis

Functional groups and chemical composition were characterized using FTIR analysis. For lignin, there are several functional groups that should be observed, such as hydroxyl, carbonyl, methoxyl, carboxyl, aromatic and phenolic. Assignment of functional groups to FTIR spectrum wavenumbers are listed in Table 5 below.

**Table 5.** Assignment of functional groups to FTIR spectrum wavenumbers [24,25].

| Wavenumber ($cm^{-1}$) | Assignments | Band Assignment |
|---|---|---|
| 3600–3100 | O-H | Stretching vibration of alcoholic and phenolic OH groups involved in hydrogen bonds |
| 2960–2820 | C-H | $-CH_2$, $-CH_3$ |
| 1770–1685<br>1680–1650 | C=O | Conjugated p-substituent carbonyl and carboxyl |
| 1600–1500,<br>1430–1420 | Aromatic skeletal | Aromatic ring vibrations |
| 1515–1511 | C=C | Aromatic skeletal breathing with C-O stretching |
| 1470–1450,<br>1370–1360 | C-H | C–H deformations methyl and methylene |
| 1427–1423 | C-H | Aromatic skeletal vibrations combined with C–H in-plane deformation |
| 1375–1397 | O-H<br>C-H | Phenolic OH and aliphatic C-H in methyl groups |
| 1170–1150 | C-H | Aromatic C-H in-plane deformation in the guaiacyl ring |
| 1145–1140 | C-H | Aromatic C-H in-plane deformation in the syringyl ring |
| 1035–1025 | C-O, C-H | Aromatic ring and primary alcohol |

The spectra at the fingerprint area are shown in Figures 2 and 3. All FTIR spectra of samples after thermal and physical pretreatments showed a similar pattern to the untreated samples. Some differences in terms of absorption intensity were detected between samples in comparison with each other. When varying the 250 µm sample, not much change could be seen except for that dried at 105 °C for 15 h. The sample dried at 75 °C for 5 h seemed to retain most of its bonds, most evidently from its high absorbance at 1098 $cm^{-1}$. For ball-milled particles, the effect of temperature and time was more noticeable. Higher temperatures reduced the intensity of absorbance, suggesting that higher temperatures broke some of the chemical bonds. Reductions in intensity of 3700–3000 $cm^{-1}$ and 1640 $cm^{-1}$ suggested a loss of water molecules in terms of water and moisture content. According to the literature, the bands at 1240 $cm^{-1}$ (asymmetric stretching vibrations of C–O–C bonds in G-lignin), 1510 $cm^{-1}$ (phenyl ring skeletal vibrations) and 1732 $cm^{-1}$ (carbonyl) were indicative of delignification [25]. When comparing between 250 µm and ball-milled particles, the ball-milled particles had lower intensity at peaks associated with 3314, 1732, 1485, 1244 and 1159 $cm^{-1}$, and higher peaks at 2920, 2853, 1633, 1316 and 1035 $cm^{-1}$. This means that ball-milling causes partial delignification to occur. Most of the changes in intensity were not very significant, since most of the changes were of less than 50%.

ANOVA analyses were carried out on peaks at 3314, 2920, 2853, 1732, 1633, 1517, 1485, 1316, 1244, 1159, 1142 and 1035 $cm^{-1}$. These wavenumbers were chosen since they represent the lignin fingerprint wavenumbers, changes in moisture presence, and also a significant reduction or improvement in intensity compared to the pretreated sample. A summary of the statistical significance is tabulated in Tables 6 and 7. It was found that the interaction between temperature and time with size is not significant for any of the spectra wavenumbers.

**Table 6.** Summary of F-value significance based on ANOVA for physical and thermal pretreatment at 3314, 2920, 2853, 1732, 1633 and 1517 $cm^{-1}$.

| Source | O-H 3323 | C-H 2923 | C-H 2855 | C=O 1730 | Aromatics 1628 | Aromatics 1518 |
|---|---|---|---|---|---|---|
| Type | <0.0001 | 0.6167 | 0.1123 | 0.2045 | 0.1351 | 0.0028 |
| Size | 0.4327 | 0.5907 | 0.9691 | 0.1388 | 0.1830 | 0.0966 |
| Time | <0.0001 | 0.0014 | 0.0142 | 0.5556 | 0.001 | 0.2233 |
| Temp | 0.2076 | 0.1796 | 0.2215 | 0.3114 | 0.7933 | 0.8583 |
| Type*Size | 0.4952 | 0.1576 | 0.3258 | 0.3725 | 0.2543 | 0.816 |
| Type*Time | <0.0001 | 0.2055 | 0.9864 | 0.4693 | 0.002 | 0.2516 |
| Type*Temp | 0.0682 | 0.1164 | 0.1051 | 0.0898 | 0.1776 | 0.4277 |
| Size*Time | 0.8083 | 0.0791 | 0.0723 | 0.3141 | 0.3674 | 0.2864 |
| Size*Temp | 0.5814 | 0.3147 | 0.3846 | 0.7167 | 0.4108 | 0.3785 |
| Time*Temp | 0.1882 | 0.0378 | 0.0680 | 0.1713 | 0.3216 | 0.4365 |
| Type*Size*Time | 0.8950 | 0.3432 | 0.3269 | 0.5186 | 0.2168 | 0.2085 |
| Type*Size*Temp | 0.4862 | 0.2653 | 0.3086 | 0.3193 | 0.341 | 0.6868 |
| Type*Time*Temp | 0.1151 | 0.0268 | 0.0392 | 0.1582 | 0.2182 | 0.4483 |
| Size*Time*Temp | 0.1882 | 0.0275 | 0.0486 | 0.1402 | 0.0709 | 0.1140 |

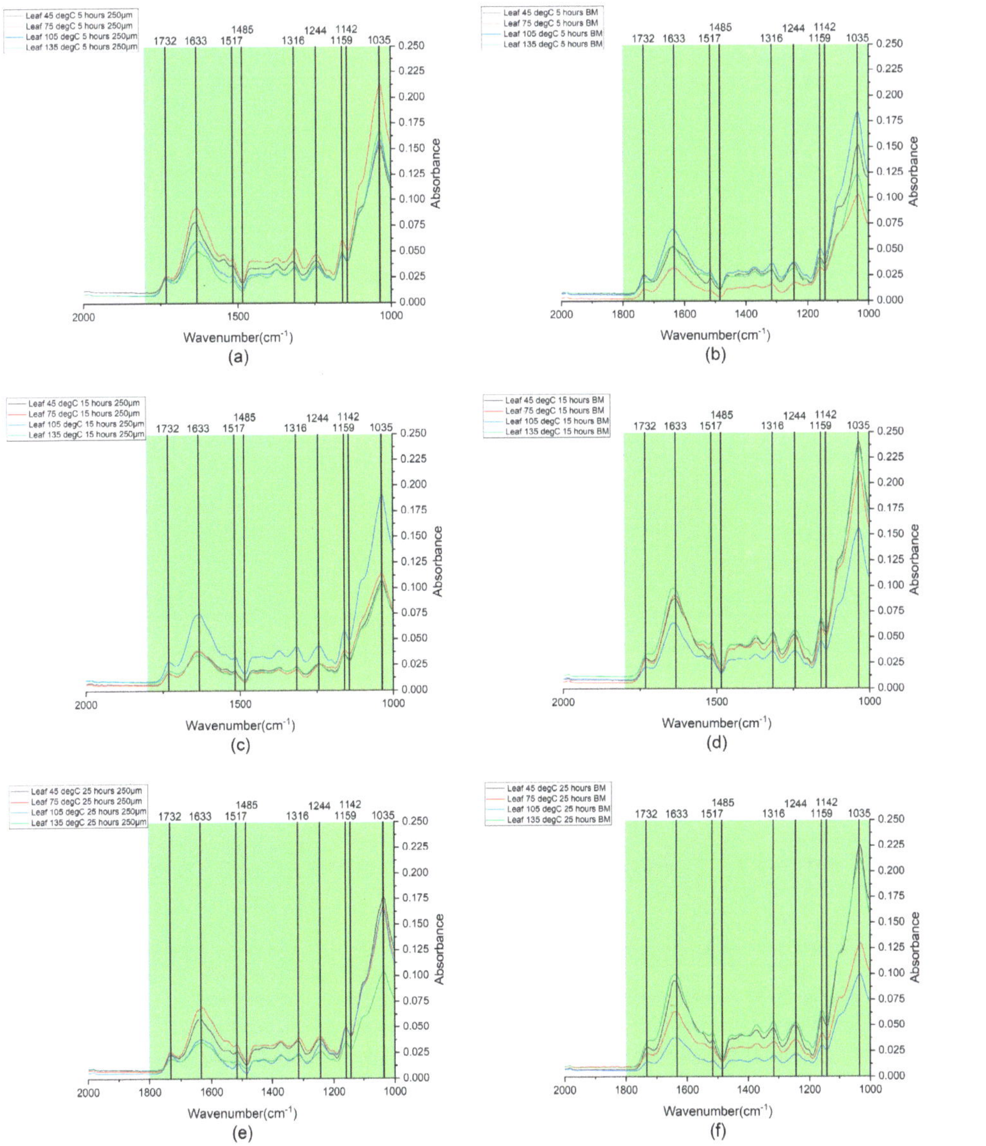


**Figure 2.** FTIR spectra for leaf particles at varying sizes, times, and temperatures; (**a**) 250 µm for 5 h, (**b**) 250 µm for 15 h, (**c**) 250 µm for 25 h, (**d**) ball-milled for 5 h, € ball-milled for 15 h and (**f**) ball-milled for 25 h.

**Figure 4.** Klason lignin for leaf samples after pretreatment.

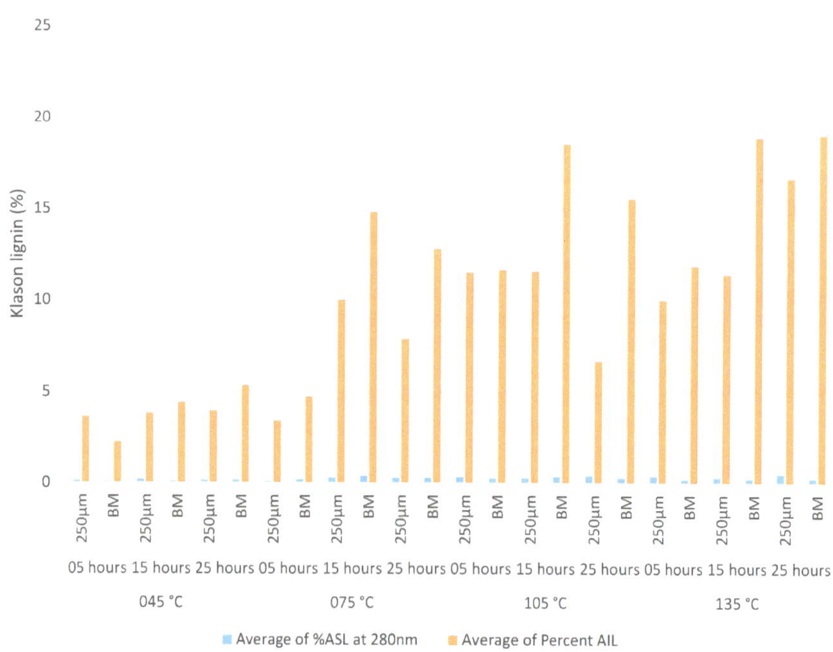


**Figure 5.** Klason lignin for stem samples after pretreatment.

**Table 8.** ANOVA table of acid-insoluble and acid-soluble lignin for each effect after physical and thermal pretreatment.

| | Acid-Insoluble Lignin | | | | | Acid-Soluble Lignin | | | | |
|---|---|---|---|---|---|---|---|---|---|---|
| Source | DF | Type I SS | Mean Square | F Value | Pr > F | DF | Type I SS | Mean Square | F Value | Pr > F |
| Type | 1 | 32.847 | 32.847 | 8.760 | 0.025 | 1 | 0.351 | 0.351 | 12.550 | 0.012 |
| Time | 3 | 320.572 | 106.857 | 28.510 | 0.001 | 3 | 0.185 | 0.062 | 2.200 | 0.188 |
| Temp | 2 | 62.566 | 31.283 | 8.350 | 0.019 | 2 | 0.036 | 0.018 | 0.650 | 0.557 |
| Size | 1 | 16.253 | 16.253 | 4.340 | 0.083 | 1 | 0.000 | 0.000 | 0.000 | 0.988 |
| Type*Time | 3 | 116.797 | 38.932 | 10.390 | 0.009 | 3 | 0.331 | 0.110 | 3.930 | 0.072 |
| Type*Temp | 2 | 27.170 | 13.585 | 3.620 | 0.093 | 2 | 0.012 | 0.006 | 0.210 | 0.819 |
| Type*Size | 1 | 53.075 | 53.075 | 14.160 | 0.009 | 1 | 0.022 | 0.022 | 0.770 | 0.414 |
| Time*Temp | 6 | 82.059 | 13.677 | 3.650 | 0.070 | 6 | 0.190 | 0.032 | 1.130 | 0.442 |
| Time*Size | 3 | 18.129 | 6.043 | 1.610 | 0.283 | 3 | 0.025 | 0.008 | 0.300 | 0.825 |
| Temp*Size | 2 | 11.966 | 5.983 | 1.600 | 0.278 | 2 | 0.026 | 0.013 | 0.460 | 0.654 |
| Type*Time*Temp | 6 | 28.075 | 4.679 | 1.250 | 0.397 | 6 | 0.151 | 0.025 | 0.900 | 0.550 |
| Type*Time*Size | 3 | 9.742 | 3.247 | 0.870 | 0.508 | 3 | 0.061 | 0.020 | 0.730 | 0.572 |
| Type*Temp*Size | 2 | 21.853 | 10.927 | 2.920 | 0.131 | 2 | 0.002 | 0.001 | 0.040 | 0.958 |
| Time*Temp*Size | 6 | 33.178 | 5.530 | 1.480 | 0.324 | 6 | 0.244 | 0.041 | 1.450 | 0.331 |

### 3.4. Cellulolytic Enzymatic Lignin (CEL) Yield

Based on the results, the pretreatment parameters were reduced to only 75 and 105 °C for the temperature while for time, it was reduced to 15 and 25 h only. The yields of CEL with regard to the pretreatment and incubation parameters are shown in Figures 6 and 7. From all the CEL, it was found that the yield was much more than the Klason lignin, which suggests that the solids obtained are not entirely lignin. Cellulose was not fully removed; instead it is inferred that the cellulose was broken down into nanocellulose. Although the cellulase is selective towards the cellulose component, the reaction might not be as severe and the cellulose might not be completely disintegrated into simple sugars such as glucose. This is also because of the presence of xylan within the lignocellulose complex, which acts as the limiting factor in enzymatic hydrolysis [30].

A further look into the statistical analysis of the pretreatment conditions reveals that the changes in the factors did not all have a significant effect on the final yield of CEL. The summary of the ANOVA and the $p$-values of each factor interaction are shown in Tables 9 and 10. It is seen that NGL produced a more consistent CEL when compared to the stem samples. For the leaf samples, the CEL extracted after ball-milling was slightly less, whilst more CEL was obtained from the stem after ball-milling. The sample type provided the most significance when interacting with other factors. Leaf samples seemed to produce a higher amount and more consistent yield of CEL as compared to stem, which is more susceptible to pretreatment conditions.

**Table 9.** ANOVA table for CEL.

| ANOVA | | | | Alpha | 0.05 |
|---|---|---|---|---|---|
| Source | DF | Sum of Squares | Mean Square | F-Value | Pr > F |
| Model | 139 | 7603.004 | 54.698 | 2.58 | 0.183 |
| Error | 4 | 84.931 | 21.233 | | |
| Total | 143 | 7687.935 | | | |

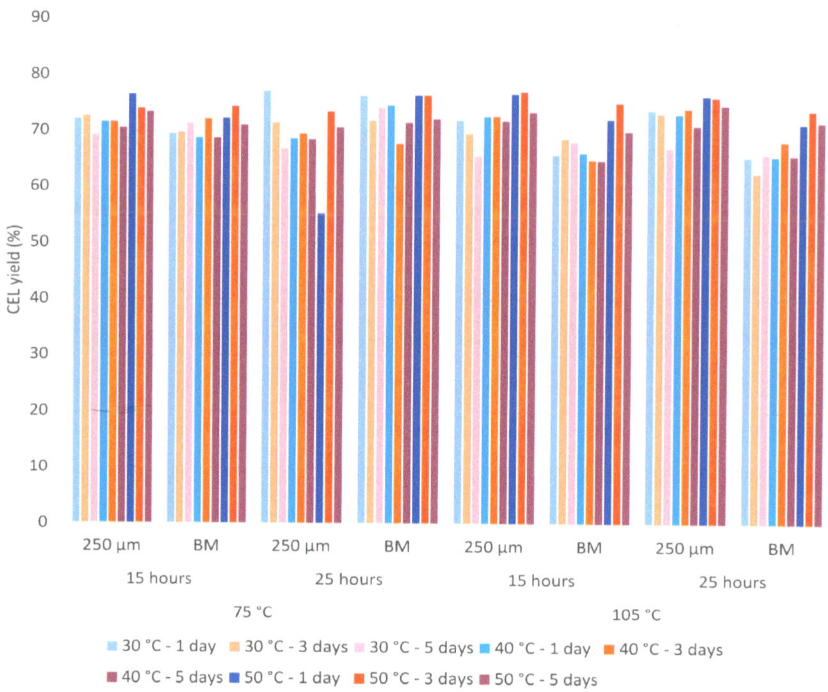


**Figure 6.** CEL yield for leaf samples.

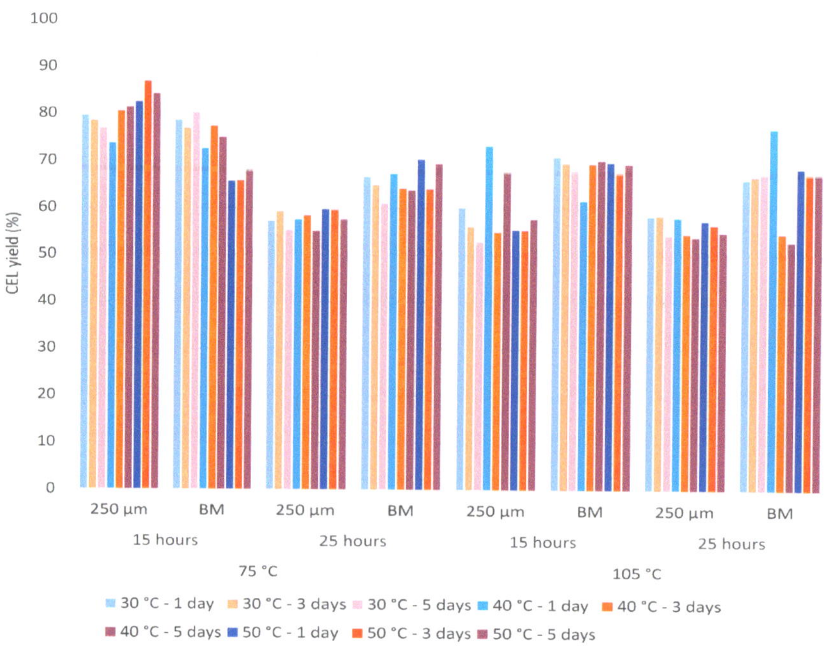


**Figure 7.** CEL yield for stem samples.

**Table 10.** *p*-value summary for CEL yield ANOVA.

| Source | Pr > F | Source | Pr > F | Source | Pr > F |
|---|---|---|---|---|---|
| Type | 0.0026 | Type*Pretreatment Temperature *Pretreatment Time | 0.0199 | Type*Pretreatment Temperature*Pretreatment Time*Incubation Temperature | 0.3135 |
| Pretreatment Temperature | 0.0068 | Type*Pretreatment Temperature*Size | 0.0084 | Type*Pretreatment Temperature*Pretreatment Time*Incubation Day | 0.9583 |
| Pretreatment Time | 0.0039 | Type*Pretreatment Temperature*Incubation Temperature | 0.4052 | Type*Pretreatment Temperature*Size*Incubation Temperature | 0.2545 |
| Size | 0.1086 | Type*Pretreatment Temperature*Incubation Day | 0.5976 | Type*Pretreatment Temperature*Size*Incubation Day | 0.8537 |
| Incubation Temperature | 0.2495 | Type*Pretreatment Time*Size | 0.1639 | Type*Pretreatment Temperature*Incubation Temperature*Incubation Day | 0.6087 |
| Incubation Day | 0.3419 | Type*Pretreatment Time*Incubation Temperature | 0.2597 | Type*Pretreatment Time*Size*Incubation Temperature | 0.7927 |
| Type*Pretreatment Temperature | 0.0175 | Type*Pretreatment Time*Incubation Day | 0.4462 | Type*Pretreatment Time*Size*Incubation Day | 0.9339 |
| Type*Pretreatment Time | 0.0043 | Type*Size*Incubation Temperature | 0.4247 | Type*Pretreatment Time*Incubation Temperature*Incubation Day | 0.5310 |
| Type*Size | 0.014 | Type*Size*Incubation Day | 0.9383 | Type*Size*Incubation Temperature*Incubation Day | 0.8234 |
| Type*Incubation Temperature | 0.266 | Type*Incubation Temperature*Incubation Day | 0.8406 | Pretreatment Temperature*Pretreatment Time*Size*Incubation Temperature | 0.2566 |
| Type*Incubation Day | 0.552 | Pretreatment Temperature*Pretreatment Time*Size | 0.0209 | Pretreatment Temperature*Pretreatment Time*Size*Incubation Day | 0.9723 |
| Pretreatment Temperature*Pretreatment Time | 0.0147 | Pretreatment Temperature*Pretreatment Time*Incubation Temperature | 0.7661 | Pretreatment Temperature*Pretreatment Time*Incubation Temperature*Incubation Day | 0.7745 |
| Pretreatment Temperature*Size | 0.4488 | Pretreatment Temperature*Pretreatment Time*Incubation Day | 0.9394 | Pretreatment Temperature*Size*Incubation Temperature*Incubation Day | 0.9792 |
| Pretreatment Temperature*Incubation Temperature | 0.4221 | Pretreatment Temperature*Size*Incubation Temperature | 0.2330 | Pretreatment Time*Size*Incubation Temperature*Incubation Day | 0.8115 |

**Table 10.** *Cont.*

| Source | Pr > F | Source | Pr > F | Source | Pr > F |
|---|---|---|---|---|---|
| Pretreatment Temperature*Incubation Day | 0.5145 | Pretreatment Temperature*Size*Incubation Day | 0.5347 | Type *Pretreatment Temperature*Pretreatment Time*Size*Incubation Temperature | 0.5699 |
| Pretreatment Time*Size | 0.0343 | Pretreatment Temperature*Incubation Temperature*Incubation Day | 0.8158 | Type*Pretreatment Temperature*Pretreatment Time*Size*Incubation Day | 0.3964 |
| Pretreatment Time*Incubation Temperature | 0.6025 | Pretreatment Time*Size*Incubation Temperature | 0.1792 | Type*Pretreatment Temperature*Pretreatment Time*Incubation Temperature*Incubation Day | 0.7400 |
| Pretreatment Time*Incubation Day | 0.5302 | Pretreatment Time*Size*Incubation Day | 0.2509 | Type*Pretreatment Temperature*Size*Incubation Temperature*Incubation Day | 0.9281 |
| Size*Incubation Temperature | 0.393 | Pretreatment Time*Incubation Temperature*Incubation Day | 0.6616 | Type*Pretreatment Time*Size*Incubation Temperature*Incubation Day | 0.5420 |
| Size*Incubation Day | 0.7657 | Size*Incubation Temperature*Incubation Day | 0.5843 | Pretreatment Temperature*Pretreatment Time*Size*Incubation Temperature*Incubation Day | 0.5709 |
| Incubation Temperature*Incubation Day | 0.6866 | Type*Pretreatment Temperature*Pretreatment Time*Size | 0.2655 | | |

An increase in the pretreatment temperature and time, as well as their interaction with other factors, decreased the overall solid yield. The pretreatment helps to remove the unwanted component from the materials, exposing the lignocellulosic component to be extracted. This is aligned with previous studies which indicated that exposing the sample at higher temperatures for a prolonged period of time could degrade the phenol content, decrease water content and may retard the extraction [31]. For the single effect of pretreatment the size effect was not significant, while both of the enzymatic hydrolysis parameters did not affect the CEL content. When looking at the incubation parameters, the changes are not obvious. This shows that the incubation effect did not have a significant effect. The change in incubation temperature is not prominent but shows the expected trend. At 40 °C, the production of CEL was lowest when compared to the results at 30 and 50 °C. This shows that the cellulase is most effective at breaking down the cellulose content, which is in agreement with the literature [32]. Above the optimum, the enzyme would be denatured, and less cellulose would be degraded, as is shown in the results. For incubation time, the longer the incubation takes, the more cellulose is degraded, and this translates into a lower solid yield.

### 3.5. CEL UV–Vis Analysis

The soluble lignin was tested at 205 and 280 nm, and the results are shown in Figures 8 and 9. It was found that the percent of soluble lignin obtained from the buffer solution after enzymatic hydrolysis was extremely small. Lignin was not dissolved in the solution and remained as a solid. This is good for the whole process since no separation of lignin from the solution is needed.

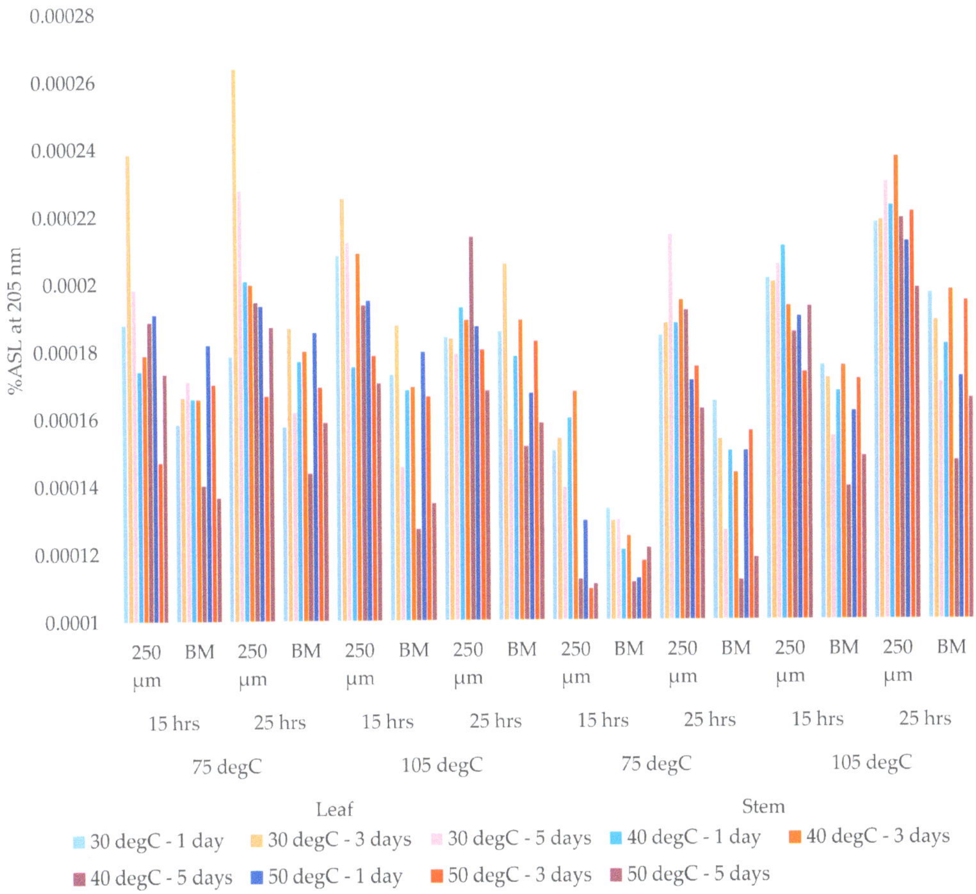


**Figure 8.** Soluble lignin content after enzymatic hydrolysis at 205 nm.

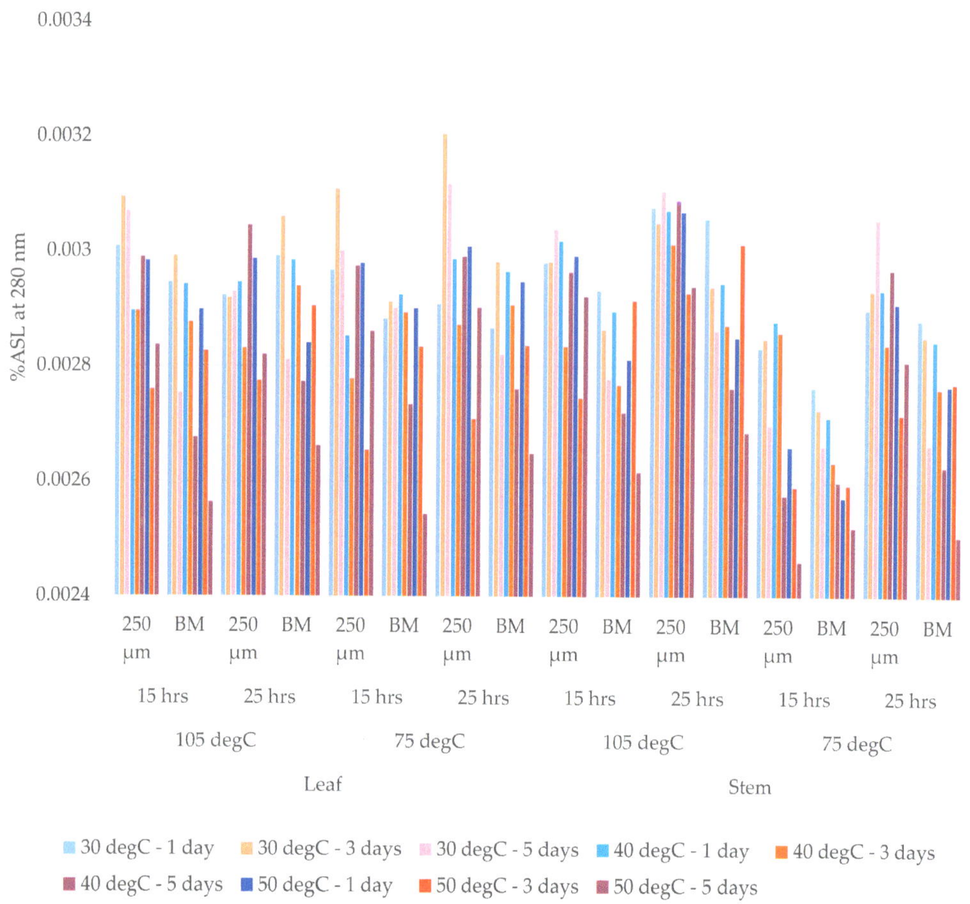


**Figure 9.** Soluble lignin content after enzymatic hydrolysis at 280 nm.

The statistical analyses for soluble lignin indicated that most factors had a significant effect, as shown in the summary in Table 11. Leaf samples and higher drying conditions produced higher soluble lignin content while ball-milling only decreased the soluble lignin content. Higher incubation temperatures and longer times also produced lower amounts of soluble lignin. Overall, since the percentage of soluble lignin was very little, it is of low concern which means that lignin does not have to be separated from the solution and can be disregarded from the total lignin extracted.

**Table 11.** *p*-value summary for soluble lignin yield ANOVA.

| Factors | 205 nm | 280 nm | Factors | 205 nm | 280 nm |
|---|---|---|---|---|---|
| Type | <0.0001 | <0.0001 | Pretreatment Temperature*Pretreatment Time*Size | 0.0005 | 0.0012 |
| Pretreatment Temperature | <0.0001 | <0.0001 | Pretreatment Temperature*Pretreatment Time*Incubation Temperature | 0.0054 | 0.0119 |
| Pretreatment Time | <0.0001 | <0.0001 | Pretreatment Temperature*Pretreatment Time*Incubation Day | 0.0241 | 0.0032 |
| Size | <0.0001 | <0.0001 | Pretreatment Temperature*Size*Incubation Temperature | 0.003 | 0.004 |
| Incubation Temperature | <0.0001 | <0.0001 | Pretreatment Temperature*Size*Incubation Day | 0.2397 | 0.406 |
| Incubation Day | 0.0118 | <0.0001 | Pretreatment Temperature*Incubation Temperature*Incubation Day | 0.0077 | 0.0094 |
| Type*Pretreatment Temperature | <0.0001 | <0.0001 | Pretreatment Time*Size*Incubation Temperature | 0.0254 | 0.0184 |
| Type*Pretreatment Time | <0.0001 | 0.0001 | Pretreatment Time*Size*Incubation Day | 0.0661 | 0.0367 |
| Type*Size | 0.0059 | 0.0015 | Pretreatment Time*Incubation Temperature*Incubation Day | 0.0776 | 0.2712 |
| Type*Incubation Temperature | 0.0318 | 0.0426 | Size*Incubation Temperature*Incubation Day | 0.0235 | 0.0017 |
| Type*Incubation Day | 0.0291 | 0.0546 | Type*Pretreatment Temperature*Pretreatment Time*Size | 0.1019 | 0.0562 |
| Pretreatment Temperature*Pretreatment Time | 0.0009 | 0.0018 | Type*Pretreatment Temperature*Pretreatment Time*Incubation Temperature | 0.3378 | 0.1633 |
| Pretreatment Temperature*Size | 0.585 | 0.4558 | Type*Pretreatment Temperature*Pretreatment Time*Incubation Day | 0.0183 | 0.0057 |
| Pretreatment Temperature*Incubation Temperature | 0.008 | 0.0389 | Type*Pretreatment Temperature*Size*Incubation Temperature | 0.0457 | 0.0961 |
| Pretreatment Temperature*Incubation Day | 0.1851 | 0.0119 | Type*Pretreatment Temperature*Size*Incubation Day | 0.578 | 0.1493 |
| Pretreatment Time*Size | 0.0128 | 0.1254 | Type*Pretreatment Temperature*Incubation Temperature*Incubation Day | 0.0122 | 0.0172 |
| Pretreatment Time*Incubation Temperature | 0.0109 | 0.0234 | Type*Pretreatment Time*Size*Incubation Temperature | 0.0351 | 0.018 |
| Pretreatment Time*Incubation Day | 0.0299 | 0.013 | Type*Pretreatment Time*Size*Incubation Day | 0.0649 | 0.0233 |
| Size*Incubation Temperature | 0.003 | 0.001 | Type*Pretreatment Time*Incubation Temperature*Incubation Day | 0.0137 | 0.0223 |
| Size*Incubation Day | 0.0029 | 0.0002 | Type*Size*Incubation Temperature*Incubation Day | 0.0023 | 0.002 |
| Incubation Temperature*Incubation Day | 0.0014 | 0.0002 | Pretreatment Temperature*Pretreatment Time*Size*Incubation Temperature | 0.0505 | 0.038 |
| Type*Pretreatment Temperature*Pretreatment Time | 0.1902 | 0.1306 | Pretreatment Temperature*Pretreatment Time*Size*Incubation Day | 0.2941 | 0.0353 |
| Type*Pretreatment Temperature*Size | 0.009 | 0.0153 | Pretreatment Temperature*Pretreatment Time*Incubation Temperature*Incubation Day | 0.0752 | 0.0261 |

**Table 11.** *Cont.*

| Factors | 205 nm | 280 nm | Factors | 205 nm | 280 nm |
|---|---|---|---|---|---|
| Type*Pretreatment Temperature*Incubation Temperature | 0.0514 | 0.0208 | Pretreatment Temperature*Size*Incubation Temperature*Incubation Day | 0.0206 | 0.0828 |
| Type*Pretreatment Temperature*Incubation Day | 0.0311 | 0.0041 | Pretreatment Time*Size*Incubation Temperature*Incubation Day | 0.0197 | 0.0672 |
| Type*Pretreatment Time*Size | 0.0005 | 0.0023 | Type*Pretreatment Temperature*Pretreatment Time*Size*Incubation Temperature | 0.0358 | 0.0205 |
| Type*Pretreatment Time*Incubation Temperature | 0.0103 | 0.024 | Type*Pretreatment Temperature*Pretreatment Time*Size*Incubation Day | 0.016 | 0.0032 |
| Type*Pretreatment Time*Incubation Day | 0.1095 | 0.0284 | Type*Pretreatment Temperature*Pretreatment Time*Incubation Temperature*Incubation Day | 0.0349 | 0.0693 |
| Type*Size*Incubation Temperature | 0.0183 | 0.0062 | Type*Pretreatment Temperature*Size*Incubation Temperature*Incubation Day | 0.0178 | 0.0176 |
| Type*Size*Incubation Day | 0.1845 | 0.0065 | Type*Pretreatment Time*Size*Incubation Temperature*Incubation Day | 0.3709 | 0.4726 |
| Type*Incubation Temperature*Incubation Day | 0.0037 | 0.0045 | Pretreatment Temperature*Pretreatment Time*Size*Incubation Temperature*Incubation Day | 0.8026 | 0.1518 |

### 3.6. CEL FTIR Analysis

The FTIR peaks obtained from each solid yield were compared with those from pure MCC and KrL. The spectra for full FTIR fingerprint wavenumber area for leaf and stem samples are shown in Figures 10 and 11 and Figures 12 and 13, respectively. All spectra lie around the same absorbance, indicating that the pretreatment and incubation parameters do not change the chemical bonding. The spectra also follow the same trend as each other. This means that the process is not selective towards any chemical bonding and removal of bonding happens only in the form of degradation of the sample. When compared to pure MCC and KrL, it was found that the spectra of extracted CEL were lower than MCC but higher than KrL. This suggests that the CEL is not entirely lignin and that microcellulose is present. Much of the literature suggests that nanocellulose can be produced when cellulase breaks down the lignocellulosic complex, but as indicated from the FTIR spectra of samples and pure CNC, this is not evident from the sample. Nanocellulose might also be produced in smaller amounts, but this cannot be determined from FTIR and requires other analyses to confirm. Ting et al. reported that nanocellulose spectra were lower than MCC, similar to our results which support this inference [33]. It can be said that lignin is partly extracted since all the spectra stayed close to those of pure KrL, even though some cellulose peaks can still be observed. Further purification is needed to remove the cellulose component still present in the sample.

Looking into the statistical analysis of the FTIR spectra, the significance of each factor and its interactions can be examined. The summary of *p*-values of each interaction is shown in Table 12. Characteristic peaks at 1639 and 1516 $cm^{-1}$ indicating conjugated carbonyl groups and aromatic skeletal vibrations, respectively, showed significant changes only when type and size were varied. This confirms that lignin is present much more in the stem, while the ball-milling process improves the lignin aromatic skeletal bonding within the sample. The non-conjugated carbonyl groups peak at 1729 $cm^{-1}$, G ring breathing with carbonyl stretching peak at 1251 $cm^{-1}$, aromatic C-H in-plane deformation in the guaiacyl ring peak at 1165 $cm^{-1}$, and the aromatic C-H in-plane deformation in the Syringyl ring peak at 1146 $cm^{-1}$, also show the same trend with only type, size and several of their interactions. The 1059 $cm^{-1}$ peak, which only present in cellulose and is absent in lignin, is seen to be greatly reduced in lignin. When comparing between leaf and stem samples, it is seen that stem samples produce a steadier result very close to pure KrL spectra, suggesting that the lignin extracted from stem samples is less affected by the variables and was able to be extracted more consistently.

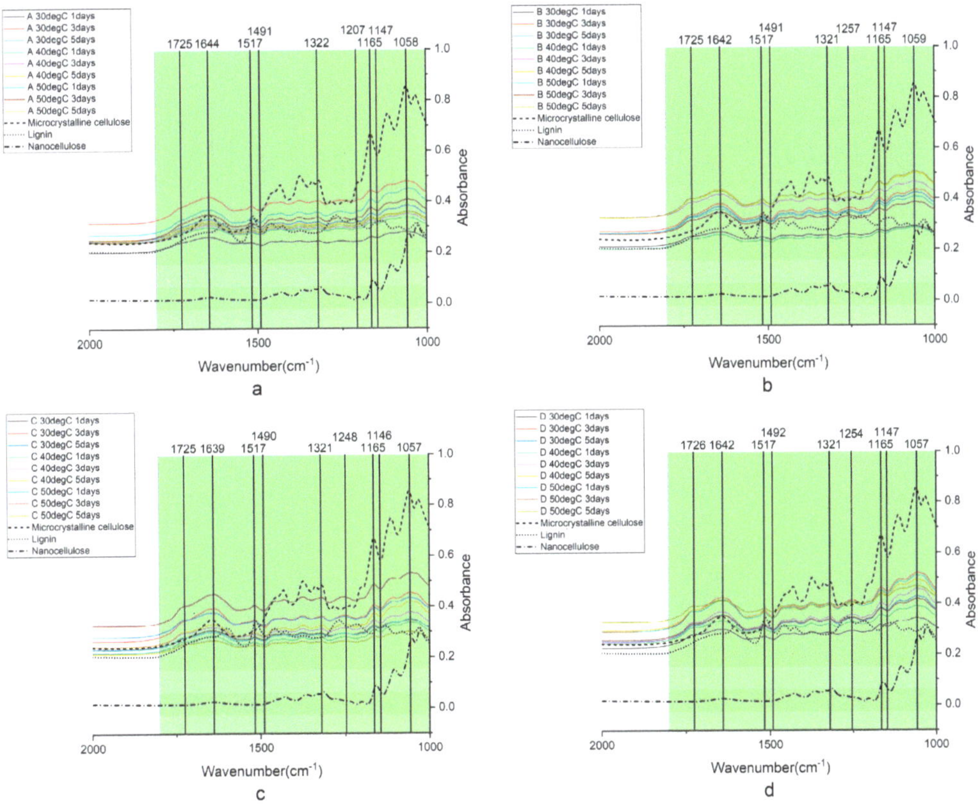


**Figure 10.** FTIR spectra of leaf samples pretreated at 75 °C at varying incubation temperatures and times; (**a**) 15 h 250 µm, (**b**) 15 h ball-milled, (**c**) 25 h 250 µm, (**d**) 25 h ball-milled.

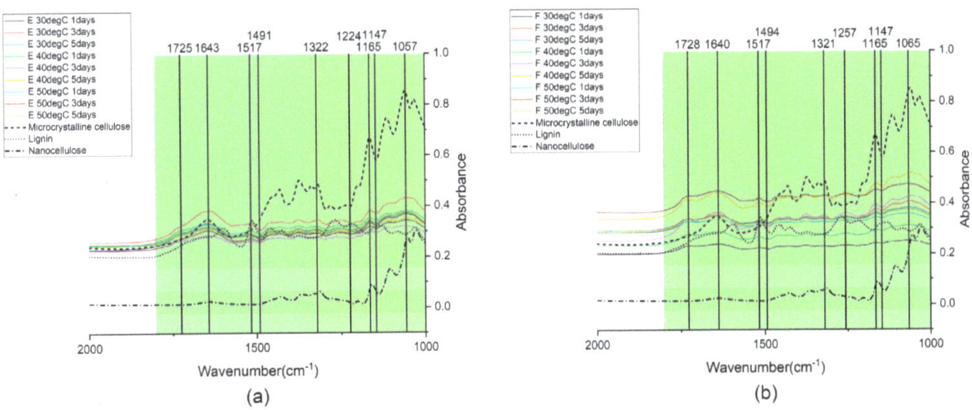


**Figure 11.** *Cont.*

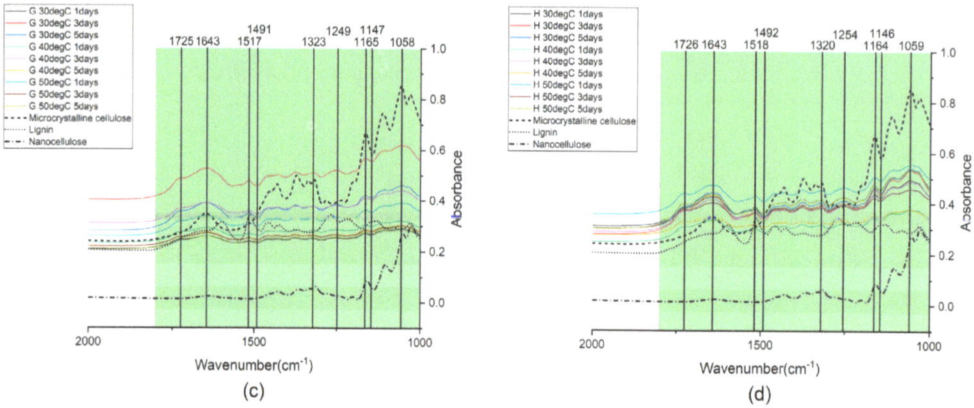


**Figure 11.** FTIR spectra of leaf samples pretreated at 105 °C at varying incubation temperatures and times; (**a**) 15 h 250 μm, (**b**) 15 h ball-milled, (**c**) 25 h 250 μm, (**d**) 25 h ball-milled.

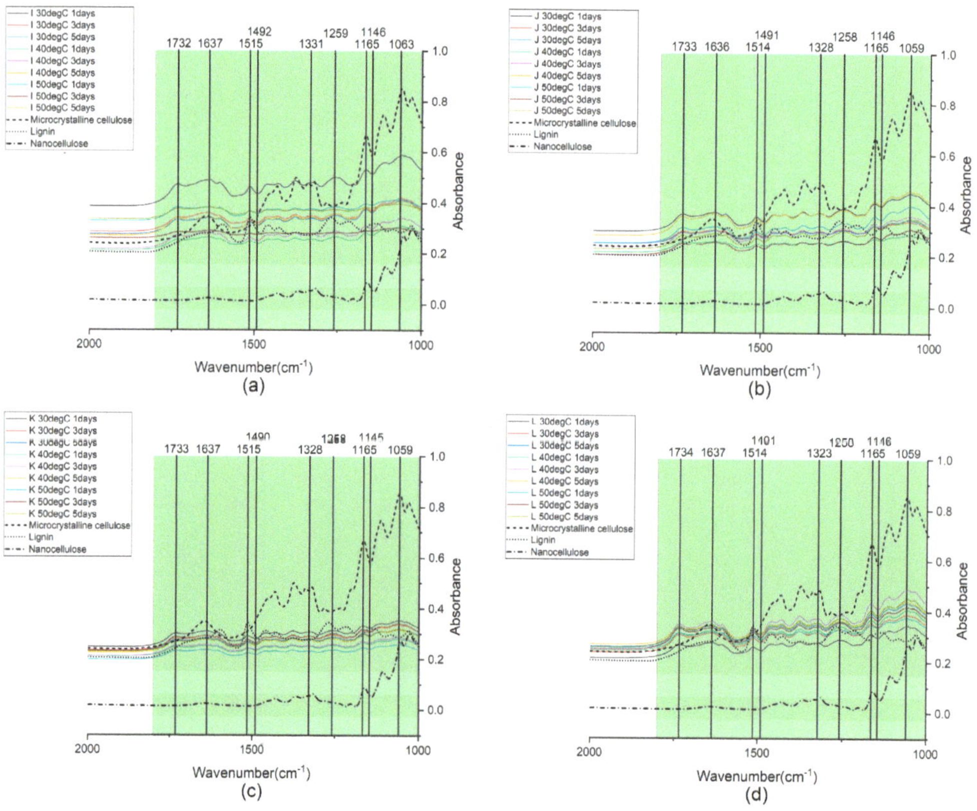


**Figure 12.** FTIR spectra of stem samples pretreated at 75 °C at varying incubation temperatures and times; (**a**) 15 h 250 μm, (**b**) 15 h ball-milled, (**c**) 25 h 250 μm, (**d**) 25 h ball-milled.

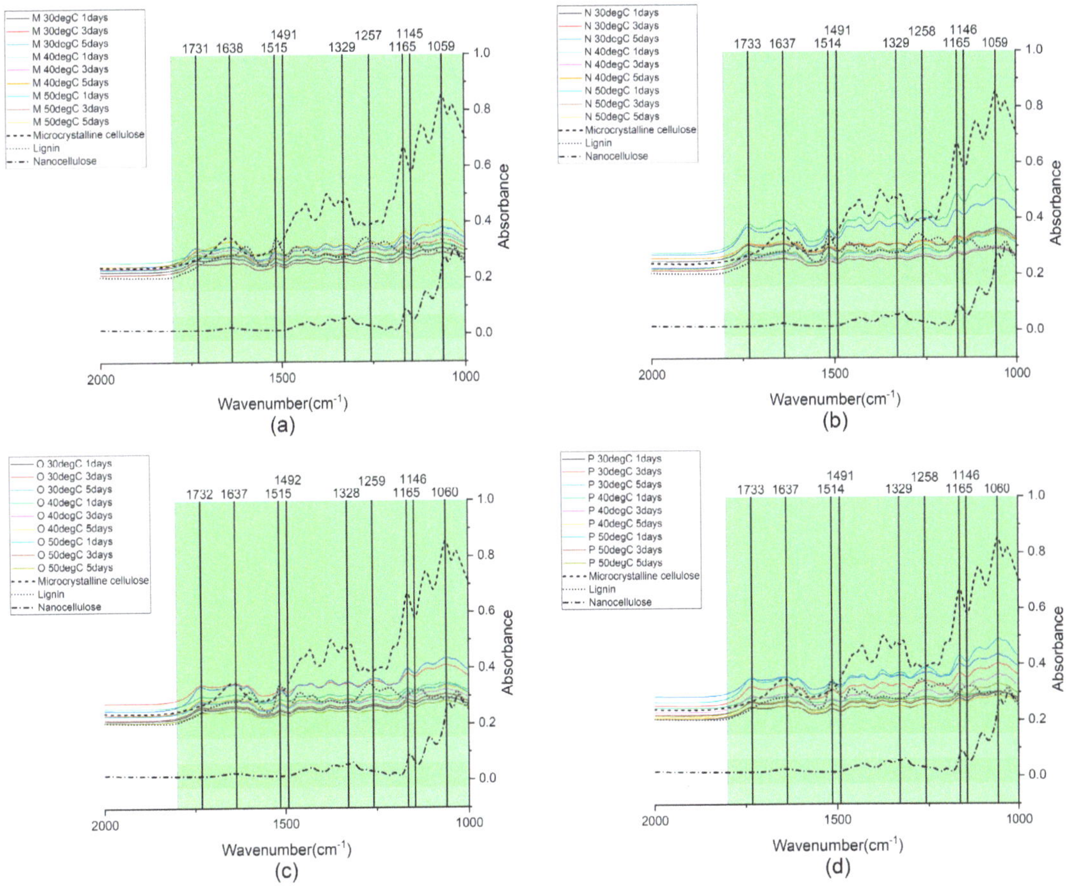


**Figure 13.** FTIR spectra of stem samples pretreated at 105 °C at varying incubation temperatures and times; (**a**) 15 h 250 μm, (**b**) 15 h ball-milled, (**c**) 25 h 250 μm, (**d**) 25 h ball-milled.

**Table 12.** *p*-value summary for CEL FTIR ANOVA.

| Factor | 3600–3100 | 2960–2820 | 2860–2840 | 1720 | 1670 | 1510 | 1490 | 1330 | 1270 | 1140 | 1125 | 1030 |
|---|---|---|---|---|---|---|---|---|---|---|---|---|
| Type | 0.0149 | 0.0061 | 0.0068 | 0.0104 | 0.0014 | 0.0050 | 0.0060 | 0.0068 | 0.0159 | 0.0103 | 0.0073 | 0.0073 |
| Pretreatment Temperature | 0.3148 | 0.3647 | 0.3967 | 0.5300 | 0.2863 | 0.3101 | 0.4058 | 0.3328 | 0.3267 | 0.3092 | 0.3149 | 0.3149 |
| Pretreatment Time | 0.3499 | 0.9165 | 0.8464 | 0.8614 | 0.6569 | 0.9195 | 0.6818 | 0.6488 | 0.5888 | 0.3735 | 0.4603 | 0.4603 |
| Size | 0.0154 | 0.0169 | 0.0203 | 0.0081 | 0.0104 | 0.0129 | 0.0201 | 0.0130 | 0.0108 | 0.0141 | 0.0142 | 0.0142 |
| Incubation Temperature | 0.2093 | 0.2258 | 0.2370 | 0.1658 | 0.1179 | 0.1524 | 0.2162 | 0.2197 | 0.2425 | 0.2769 | 0.2672 | 0.2672 |
| Incubation Day | 0.0958 | 0.0617 | 0.0670 | 0.0576 | 0.0462 | 0.0566 | 0.0720 | 0.0666 | 0.0822 | 0.0963 | 0.0862 | 0.0862 |
| Type*Pretreatment Temperature | 0.3134 | 0.1327 | 0.1172 | 0.0752 | 0.1223 | 0.1042 | 0.0949 | 0.1770 | 0.1759 | 0.3243 | 0.2755 | 0.2755 |
| Type*Pretreatment Time | 0.0674 | 0.0594 | 0.0648 | 0.0309 | 0.0256 | 0.0385 | 0.0507 | 0.0502 | 0.0496 | 0.0634 | 0.0555 | 0.0555 |
| Type*Size | 0.2492 | 0.0801 | 0.0707 | 0.0401 | 0.0347 | 0.0560 | 0.0517 | 0.1002 | 0.1321 | 0.2239 | 0.1614 | 0.1614 |
| Type*Incubation Temperature | 0.3131 | 0.1588 | 0.1518 | 0.1455 | 0.1793 | 0.1559 | 0.1508 | 0.1745 | 0.1837 | 0.2432 | 0.2121 | 0.2121 |
| Type*Incubation Day | 0.0448 | 0.0287 | 0.0304 | 0.0248 | 0.0186 | 0.0245 | 0.0294 | 0.0293 | 0.0355 | 0.0424 | 0.0381 | 0.0381 |
| Pretreatment Temperature*Pretreatment Time | 0.3632 | 0.1509 | 0.1269 | 0.1279 | 0.1137 | 0.1139 | 0.1019 | 0.1982 | 0.2306 | 0.3140 | 0.2615 | 0.2615 |
| Pretreatment Temperature*Size | 0.7040 | 0.3898 | 0.3275 | 0.2888 | 0.3823 | 0.3006 | 0.2624 | 0.5150 | 0.5801 | 0.7830 | 0.7174 | 0.7174 |
| Pretreatment Temperature*Incubation Temperature | 0.6238 | 0.4359 | 0.4276 | 0.3925 | 0.3374 | 0.3998 | 0.4125 | 0.4399 | 0.5098 | 0.5474 | 0.5217 | 0.5217 |
| Pretreatment Temperature*Incubation Day | 0.7832 | 0.5684 | 0.5376 | 0.5795 | 0.6252 | 0.5745 | 0.5105 | 0.6161 | 0.6776 | 0.7045 | 0.6776 | 0.6776 |
| Pretreatment Time*Size | 0.0779 | 0.0975 | 0.1315 | 0.0858 | 0.0430 | 0.0739 | 0.1140 | 0.0757 | 0.0798 | 0.0680 | 0.0694 | 0.0694 |
| Pretreatment Time*Incubation Temperature | 0.4411 | 0.5064 | 0.5368 | 0.5409 | 0.4226 | 0.5380 | 0.6242 | 0.4904 | 0.4915 | 0.4569 | 0.4704 | 0.4704 |
| Pretreatment Time*Incubation Day | 0.2547 | 0.1866 | 0.2011 | 0.1305 | 0.1455 | 0.1542 | 0.1931 | 0.1741 | 0.1705 | 0.2061 | 0.1953 | 0.1953 |
| Size*Incubation Temperature | 0.1391 | 0.0956 | 0.1058 | 0.0684 | 0.0671 | 0.0776 | 0.0975 | 0.0952 | 0.1008 | 0.1449 | 0.1283 | 0.1283 |
| Size*Incubation Day | 0.3959 | 0.2001 | 0.1909 | 0.1814 | 0.1880 | 0.1766 | 0.1776 | 0.2383 | 0.2662 | 0.3353 | 0.2926 | 0.2926 |
| Incubation Temperature*Incubation Day | 0.1247 | 0.0885 | 0.0931 | 0.0755 | 0.0619 | 0.0757 | 0.0888 | 0.0853 | 0.0958 | 0.1073 | 0.0954 | 0.0954 |
| Type*Pretreatment Temperature*Pretreatment Time | 0.7798 | 0.7982 | 0.8453 | 0.9269 | 0.7199 | 0.8160 | 0.9868 | 0.6935 | 0.7314 | 0.6118 | 0.6874 | 0.6874 |
| Type*Pretreatment Temperature*Size | 0.5101 | 0.5020 | 0.5412 | 0.4435 | 0.5121 | 0.5039 | 0.5588 | 0.4885 | 0.5194 | 0.4916 | 0.5226 | 0.5226 |
| Type*Pretreatment Temperature*Incubation Temperature | 0.5891 | 0.5675 | 0.5568 | 0.4613 | 0.3656 | 0.4870 | 0.4140 | 0.5713 | 0.6367 | 0.6728 | 0.6316 | 0.6316 |
| Type*Pretreatment Temperature*Incubation Day | 0.9441 | 0.7779 | 0.7469 | 0.6570 | 0.7195 | 0.6531 | 0.7011 | 0.8108 | 0.7930 | 0.8918 | 0.8733 | 0.8733 |
| Type*Pretreatment Time*Size | 0.6590 | 0.3323 | 0.2636 | 0.2415 | 0.5562 | 0.2968 | 0.2175 | 0.3986 | 0.4043 | 0.6322 | 0.5707 | 0.5707 |
| Type*Pretreatment Time*Incubation Temperature | 0.2765 | 0.1883 | 0.1848 | 0.1526 | 0.1472 | 0.1533 | 0.1736 | 0.1944 | 0.2012 | 0.2438 | 0.2267 | 0.2267 |

**Table 12.** *Cont.*

| Factor | 3600–3100 | 2960–2820 | 2860–2840 | 1720 | 1670 | 1510 | 1490 | 1330 | 1270 | 1140 | 1125 | 1030 |
|---|---|---|---|---|---|---|---|---|---|---|---|---|
| Type*Pretreatment Time*Incubation Day | 0.3320 | 0.2020 | 0.2021 | 0.1860 | 0.2090 | 0.2000 | 0.2112 | 0.2164 | 0.2228 | 0.2731 | 0.2512 | 0.2512 |
| Type*Size*Incubation Temperature | 0.7287 | 0.2545 | 0.2116 | 0.1679 | 0.2307 | 0.1846 | 0.1500 | 0.3383 | 0.4046 | 0.6613 | 0.5099 | 0.5099 |
| Type*Size*Incubation Day | 0.4087 | 0.2862 | 0.2824 | 0.2485 | 0.2371 | 0.2556 | 0.2587 | 0.3074 | 0.3297 | 0.3806 | 0.3510 | 0.3510 |
| Type*Incubation Temperature*Incubation Day | 0.4449 | 0.4082 | 0.4286 | 0.4108 | 0.2644 | 0.3524 | 0.4342 | 0.3935 | 0.4169 | 0.4464 | 0.4252 | 0.4252 |
| Pretreatment Temperature*Pretreatment Time*Size | 0.2590 | 0.1225 | 0.1155 | 0.0932 | 0.2163 | 0.1297 | 0.1076 | 0.1461 | 0.1254 | 0.2208 | 0.1819 | 0.1819 |
| Pretreatment Temperature*Pretreatment Time*Incubation Temperature | 0.1337 | 0.1393 | 0.1547 | 0.1352 | 0.1083 | 0.1418 | 0.1657 | 0.1332 | 0.1416 | 0.1372 | 0.1345 | 0.1345 |
| Pretreatment Temperature*Pretreatment Time*Incubation Day | 0.7608 | 0.6672 | 0.6672 | 0.5691 | 0.6209 | 0.5972 | 0.6062 | 0.6946 | 0.6836 | 0.7862 | 0.7457 | 0.7457 |
| Pretreatment Temperature*Size*Incubation Temperature | 0.3439 | 0.1629 | 0.1527 | 0.1358 | 0.1480 | 0.1309 | 0.1296 | 0.1904 | 0.2310 | 0.2905 | 0.2480 | 0.2480 |
| Pretreatment Temperature*Size*Incubation Day | 0.5921 | 0.5322 | 0.5805 | 0.5831 | 0.5441 | 0.5821 | 0.6229 | 0.5243 | 0.5458 | 0.5219 | 0.5224 | 0.5224 |
| Pretreatment Temperature*Incubation Temperature*Incubation Day | 0.2043 | 0.1306 | 0.1393 | 0.1037 | 0.1196 | 0.1177 | 0.1403 | 0.1346 | 0.1389 | 0.1763 | 0.1634 | 0.1634 |
| Pretreatment Time*Size*Incubation Temperature | 0.3410 | 0.2253 | 0.2083 | 0.1768 | 0.2158 | 0.1976 | 0.1895 | 0.2412 | 0.2626 | 0.3459 | 0.3008 | 0.3008 |
| Pretreatment Time*Size*Incubation Day | 0.4295 | 0.1718 | 0.1495 | 0.1283 | 0.1398 | 0.1274 | 0.1252 | 0.2106 | 0.2383 | 0.3577 | 0.2948 | 0.2948 |
| Pretreatment Time*Incubation Temperature*Incubation Day | 0.4761 | 0.3539 | 0.3611 | 0.2930 | 0.2855 | 0.3073 | 0.3505 | 0.3863 | 0.4162 | 0.5004 | 0.4688 | 0.4688 |
| Size*Incubation Temperature*Incubation Day | 0.6888 | 0.7602 | 0.7888 | 0.7867 | 0.6536 | 0.7530 | 0.7674 | 0.7628 | 0.7924 | 0.7536 | 0.7516 | 0.7516 |
| Type*Pretreatment Temperature*Pretreatment Time*Size | 0.0472 | 0.0505 | 0.0594 | 0.0414 | 0.0304 | 0.0399 | 0.0668 | 0.0507 | 0.0581 | 0.0583 | 0.0616 | 0.0616 |
| Type*Pretreatment Temperature*Pretreatment Time*Incubation Temperature | 0.7992 | 0.6296 | 0.5763 | 0.6068 | 0.6550 | 0.6235 | 0.5131 | 0.7188 | 0.7930 | 0.8737 | 0.8299 | 0.8299 |
| Type*Pretreatment Temperature*Pretreatment Time*Incubation Day | 0.8352 | 0.7252 | 0.6731 | 0.6132 | 0.7497 | 0.6397 | 0.6187 | 0.7365 | 0.7180 | 0.8291 | 0.7976 | 0.7976 |
| Type*Pretreatment Temperature*Size*Incubation Temperature | 0.4940 | 0.3678 | 0.3863 | 0.3452 | 0.3420 | 0.3473 | 0.4063 | 0.3581 | 0.3604 | 0.4311 | 0.4076 | 0.4076 |
| Type*Pretreatment Temperature*Size*Incubation Day | 0.5379 | 0.5154 | 0.5600 | 0.4484 | 0.5164 | 0.5152 | 0.6201 | 0.4991 | 0.4947 | 0.5041 | 0.5286 | 0.5286 |
| Type*Pretreatment Temperature*Incubation Temperature*Incubation Day | 0.2936 | 0.2801 | 0.2985 | 0.2269 | 0.2644 | 0.2607 | 0.3012 | 0.2749 | 0.2733 | 0.3071 | 0.3034 | 0.3034 |
| Type*Pretreatment Time*Size*Incubation Temperature | 0.7681 | 0.8461 | 0.8103 | 0.7199 | 0.7454 | 0.7357 | 0.7482 | 0.8756 | 0.9080 | 0.8715 | 0.8959 | 0.8959 |

**Table 12.** *Cont.*

| Factor | 3600–3100 | 2960–2820 | 2860–2840 | 1720 | 1670 | 1510 | 1490 | 1330 | 1270 | 1140 | 1125 | 1030 |
|---|---|---|---|---|---|---|---|---|---|---|---|---|
| Type*Pretreatment Time*Size*Incubation Day | 0.2386 | 0.1419 | 0.1457 | 0.1208 | 0.1336 | 0.1263 | 0.1455 | 0.1422 | 0.1571 | 0.1835 | 0.1738 | 0.1738 |
| Type*Pretreatment Time*Incubation Temperature*Incubation Day | 0.9611 | 0.8023 | 0.7656 | 0.7042 | 0.7416 | 0.7105 | 0.6797 | 0.8285 | 0.8539 | 0.9362 | 0.8973 | 0.8973 |
| Type*Size*Incubation Temperature*Incubation Day | 0.5135 | 0.2975 | 0.3011 | 0.2390 | 0.2595 | 0.2644 | 0.2743 | 0.3185 | 0.3473 | 0.4278 | 0.3844 | 0.3844 |
| Pretreatment Temperature*Pretreatment Time*Size*Incubation Temperature | 0.6578 | 0.8091 | 0.8689 | 0.8346 | 0.8268 | 0.8946 | 0.9130 | 0.7462 | 0.6808 | 0.6363 | 0.6765 | 0.6765 |
| Pretreatment Temperature*Pretreatment Time*Size*Incubation Day | 0.9728 | 0.8484 | 0.8316 | 0.8280 | 0.8802 | 0.8666 | 0.8176 | 0.9083 | 0.9289 | 0.9528 | 0.9218 | 0.9218 |
| Pretreatment Temperature*Pretreatment Time*Incubation Temperature*Incubation Day | 0.2361 | 0.1348 | 0.1367 | 0.1047 | 0.1096 | 0.1114 | 0.1288 | 0.1292 | 0.1355 | 0.1694 | 0.1558 | 0.1558 |
| Pretreatment Temperature*Size*Incubation Temperature*Incubation Day | 0.6415 | 0.3512 | 0.3321 | 0.2964 | 0.3098 | 0.2959 | 0.2977 | 0.3928 | 0.4350 | 0.5786 | 0.5108 | 0.5108 |
| Pretreatment Time*Size*Incubation Temperature*Incubation Day | 0.6547 | 0.6076 | 0.6409 | 0.5564 | 0.5275 | 0.5725 | 0.6431 | 0.5922 | 0.6066 | 0.6390 | 0.6352 | 0.6352 |
| Type*Pretreatment Temperature*Pretreatment Time*Size*Incubation Temperature | 0.8285 | 0.8642 | 0.8943 | 0.7894 | 0.6337 | 0.7647 | 0.8315 | 0.8701 | 0.8982 | 0.8964 | 0.9016 | 0.9016 |
| Type*Pretreatment Temperature*Pretreatment Time*Size*Incubation Day | 0.6309 | 0.4455 | 0.4491 | 0.3646 | 0.4826 | 0.4089 | 0.4296 | 0.4408 | 0.4193 | 0.4989 | 0.4799 | 0.4799 |
| Type*Pretreatment Temperature*Pretreatment Time*Incubation Temperature*Incubation Day | 0.4897 | 0.4257 | 0.4426 | 0.3800 | 0.3826 | 0.3904 | 0.4677 | 0.3953 | 0.3760 | 0.4089 | 0.4138 | 0.4138 |
| Type*Pretreatment Temperature*Size*Incubation Temperature*Incubation Day | 0.6127 | 0.3346 | 0.3324 | 0.3002 | 0.3201 | 0.2974 | 0.3028 | 0.3700 | 0.4123 | 0.4989 | 0.4491 | 0.4491 |
| Type*Pretreatment Time*Size*Incubation Temperature*Incubation Day | 0.8532 | 0.7399 | 0.7522 | 0.6423 | 0.7510 | 0.7169 | 0.7383 | 0.7135 | 0.7213 | 0.7494 | 0.7523 | 0.7523 |
| Pretreatment Temperature*Pretreatment Time*Size*Incubation Temperature*Incubation Day | 0.7106 | 0.4441 | 0.4048 | 0.3571 | 0.4479 | 0.3932 | 0.3515 | 0.4392 | 0.4553 | 0.5412 | 0.4977 | 0.4977 |

### 3.7. Optimal Condition

The optimal condition for pretreatment was also applied to produce the most favorable CEL. As mentioned above, the main objectives for pretreatment are to avoid size reduction, preserve the saccharide fractions, limit formation of degradation products and minimize energy and cost. According to the moisture reduction, there were no significant decreases above 15 h and 75 °C. Ball-milling, longer incubation time and higher temperature used much more energy, and this translated into a higher cost of operation. When comparing the FTIR spectra, it is evident that higher temperature and time reduced the intensity of peaks. Ball-milling also reduced the intensity, but the changes were not too prominent within the lignin range. When comparing the overall time, the ball-milling process took a shorter time to prepare since it already can produce consistent size reduction without the need of a sieve, which could increase the energy consumption. Overall, it was found that the best conditions in which to pretreat the feedstock were at 15 h, 75 °C and using the ball mill on a leaf sample.

Stem sample FTIR spectra of CEL were much closer to KrL, indicating that more lignin was present. Increasing the temperature and the drying time caused the amount of CEL produced to be more consistent and closer to pure lignin spectra. Little change was seen with increasing incubation time and temperature but 40 °C appeared to obtain the lowest cellulose content, which is ideal. Longer incubation only improved cellulose breakdown by a tiny margin, such that choosing the middle point would be ideal to save energy and time while also producing a commendable amount of lignin. Therefore, using a stem sample, pretreated at 105 °C for 25 h and incubated at 40 °C for 3 days would be the optimal parameter to obtain lignin from Napier grass.

A comparison of Klason lignin and CEL with other studies is shown in Table 13.

**Table 13.** Comparison of lignin content from previous and current studies.

| Reference | Lignin Content |
|---|---|
| Manokhoon and Rangseesuriyachai [34] | Untreated: 16.7%<br>NaOH treated: 6.9–8.1% |
| Mohammed et al. [29] | Napier stem: 26.99 ± 1.29%<br>Napier leaf: 30.09 ± 1.30% |
| Phitsuwan et al. [35] | Untreated: 29.8%<br>NaOH treated: 9.1%<br>$CaOH_2$ treated: 20.1%<br>$NH_3$ treated: 12.0%<br>$aH_2O_2$ treated: 15.4% |
| Phitsuwan et al. [36] | 29.8–12.3% |
| Song et al. [37] | 5.7–6.2% |
| This study | Klason lignin: 4.48–38.2%<br>CEL: 52.9–86.9% |

## 4. Conclusions

Pretreatment has been successfully carried out on Napier grass leaf and stem samples through physical and thermal methods, and the pretreated samples were then incubated for cellulolytic enzymatic hydrolysis. The moisture content is directly affected by the drying temperature as seen in the result presented. A higher drying temperature and longer drying time will lead to higher moisture loss from the sample. From the observation of FTIR spectra after physical and thermal pretreatment, drying temperature does affect the composition of functional groups in pretreated samples, but it is evident that partial delignification occurs due to a reduction in the lignin fingerprint band. From the Klason method, ASL was found to differ significantly between the two types of samples. CEL was extracted from the pretreated sample after varying incubation parameters. The extracted CEL showed a higher solid yield than the actual lignin content, indicating that impure lignin was obtained and that cellulose was not fully disintegrated from the sample. Soluble lignin was detected in a very small amount, negating the need of separation from the solution. The FTIR results obtained for CEL were slightly higher than pure Kraft lignin which means that cellulose was still present in the sample. Optimization of parameters was carried out to ensure that an easier process can be performed while producing a better lignin product. The optimized conditions for pretreatment were found to be 75 °C, 15 h and balling onto a leaf sample. For cellulolytic enzymatic hydrolysis, incubation at 30 °C for 3 days is the optimum. These obtained data can be a reliable precursor for other studies on the extraction of lignin from Napier grass, as well as other grass-type biomass which optimize and improve the process of valorizing the biomass sources. The data also can be a good starting point for research on other pretreatment and extraction methods.

**Author Contributions:** Conceptualization, S.Z.A.H.S. and N.B.O.; methodology, S.Z.A.H.S. and N.B.O.; validation, S.Z.A.H.S. and N.B.O.; formal analysis, S.Z.A.H.S. and N.B.O.; investigation, S.Z.A.H.S.; resources, S.Z.A.H.S. and N.B.O.; data curation, S.Z.A.H.S.; writing—original draft preparation, S.Z.A.H.S.; writing—review and editing, S.Z.A.H.S. and N.B.O.; visualization, S.Z.A.H.S.; supervision, N.B.O.; project administration, N.B.O.; funding acquisition, N.B.O. All authors have read and agreed to the published version of the manuscript.

**Funding:** This research was funded by Yayasan Universiti Teknologi PETRONAS grant (015LC0-207) and Ministry of Higher Education (MOHE) Malaysia through the Higher Education Center of Excellence (HICoE) grant (grant number 015MA0-052).

**Data Availability Statement:** The data presented in this study are available on request from the corresponding author. The data are not publicly available as the data is under protection restrictions and university copyright policy.

**Acknowledgments:** The authors would like to acknowledge the funding support from Yayasan Universiti Teknologi PETRONAS grant (cost center: 015LC0-207) and the Ministry of Higher Education (MOHE) Malaysia through the Higher Education Center of Excellence (HICoE) grant to Center for Biofuel and Biochemical Research (cost center: 015MA0-052). The authors also would like to acknowledge the support by the Department of Chemical Engineering, Universiti Teknologi PETRONAS.

**Conflicts of Interest:** The authors declare no conflict of interest.

## References

1. Malaysia Energy Commission. *Malaysia Energy Statistics Handbook 2017*; Malaysia Energy Commission: Putrajaya, Malaysia, 2017.
2. Malaysia Energy Commission. *Malaysia Energy Statistics Handbook 2020*; Malaysia Energy Commission: Putrajaya, Malaysia, 2021.
3. Popp, J.; Lakner, Z.; Harangi-Rákos, M.; Fári, M. The effect of bioenergy expansion: Food, energy, and environment. *Renew. Sustain. Energy Rev.* **2014**, *32*, 559–578. [CrossRef]
4. Ambaye, T.C.; Vaccari, M.; Bonilla Petriciolet, A.; Prasad, S.; van Hullebusch, E.D.; Rtimi, S. Emerging technologies for biofuel production: A critical review on recent progress, challenges and perspectives. *J. Environ. Manag.* **2021**, *290*, 112627. [CrossRef] [PubMed]
5. Bioenergy Technologies Office. Biomass Resources. Available online: https://www.energy.gov/eere/bioenergy/biomass-resources (accessed on 23 November 2022).
6. Mohammed, I.Y.; Abakr, Y.A.; Yusup, S.; Kazi, F.K. Valorization of Napier grass via intermediate pyrolysis: Optimization using response surface methodology and pyrolysis products characterization. *J. Clean. Prod.* **2017**, *142*, 1848–1866. [CrossRef]
7. Ko, C.-H.; Yu, F.-C.; Chang, F.-C.; Yang, B.-Y.; Chen, W.-H.; Hwang, W.-S.; Tu, T.-C. Bioethanol production from recovered napier grass with heavy metals. *J. Environ. Manag.* **2017**, *203*, 1005–1010. [CrossRef] [PubMed]
8. Yasuda, M.; Ishii, Y.; Ohta, K. Napier grass (*Pennisetum purpureum* Schumach) as raw material for bioethanol production: Pretreatment, saccharification, and fermentation. *Biotechnol. Bioprocess Eng.* **2015**, *19*, 943–950. [CrossRef]
9. Tsai, M.-H.; Lee, W.-C.; Kuan, W.-C.; Sirisansaneeyakul, S.; Savarajara, A. Evaluation of different pretreatments of Napier grass for enzymatic saccharification and ethanol production. *Energy Sci. Eng.* **2018**, *6*, 683–692. [CrossRef]
10. Janejadkarn, A.; Chavalparit, O. Biogas Production from Napier Grass (Pak Chong 1) (*Pennisetum purpureum* × *Pennisetum americanum*). *Adv. Mater. Res.* **2014**, *856*, 327–332. [CrossRef]
11. Sawasdee, V.; Pisutpaisal, N. Feasibility of Biogas Production from Napier Grass. *Energy Procedia* **2014**, *61*, 1229–1233. [CrossRef]
12. Schoemaker, H.E.; Piontek, K. On the interaction of lignin peroxidase with lignin. *Pure Appl. Chem.* **1996**, *68*, 2089–2096. [CrossRef]
13. Buranov, A.U.; Mazza, G. Lignin in straw of herbaceous crops. *Ind. Crops Prod.* **2008**, *28*, 237–259. [CrossRef]
14. Kumar, A.K.; Sharma, S. Recent updates on different methods of pretreatment of lignocellulosic feedstocks: A review. *Bioresour. Bioprocess.* **2017**, *4*, 7. [CrossRef] [PubMed]
15. Baruah, J.; Nath, B.K.; Sharma, R.; Kumar, S.; Deka, R.C.; Baruah, D.C.; Kalita, E. Recent Trends in the Pretreatment of Lignocellulosic Biomass for Value-Added Products. *Front. Energy Res.* **2018**, *6*, 141. [CrossRef]

16. Botella, C.; Zhang, K.; Baugh, A.; Liang, Y.; Sivakumar, S.V. Reversible acid pretreatment scale up studies for the production of cellulosic ethanol from ensiled sweet sorghum. *Biochem. Eng. J.* **2019**, *150*, 107266. [CrossRef]
17. *ASTM D1105-21*; Standard Test Method for Preparation of Extractive-Free Wood. ASTM International: West Conshohocken, PA, USA, 2021.
18. *ASTM D1107-21*; Standard Test Method for Ethanol-Toluene Solubility of Wood. ASTM International: West Conshohocken, PA, USA, 2021.
19. *ASTM D1106-2*; Standard Test Method for Acid-Insoluble Lignin in Wood. ASTM International: West Conshohocken, PA, USA, 2021.
20. Decker, S.; Resch, M.; Baker, J. *Low Solids Enzymatic Saccharification of Lignocellulosic Biomass*; National Renewable Energy Laboratory: Golden CO, USA, 2015.
21. Chang, H.-m.; Cowling, E.B.; Brown, W. Comparative Studies on Cellulolytic Enzyme Lignin and Milled Wood Lignin of Sweetgum and Spruce. *Holzforsch* **1975**, *29*, 153–159. [CrossRef]
22. Rencoret, J.; Prinsen, P.; Gutierrez, A.; Martinez, A.T.; Del Rio, J.C. Isolation and structural characterization of the milled wood lignin, dioxane lignin, and cellulolytic lignin preparations from brewer's spent grain. *J. Agric. Food Chem.* **2015**, *63*, 603–613. [CrossRef] [PubMed]
23. Houghton, T.P.; Stevens, D.M.; Pryfogle, P.A.; Wright, C.T.; Radtke, C.W. The effect of drying temperature on the composition of biomass. *Appl. Biochem. Biotechnol.* **2009**, *153*, 4–10. [CrossRef] [PubMed]
24. Rashid, T.; Kait, C.F.; Murugesan, T. A "Fourier Transformed Infrared" Compound Study of Lignin Recovered from a Formic Acid Process. *Procedia Eng.* **2016**, *148*, 1312–1319. [CrossRef]
25. Lupoi, J.S.; Singh, S.; Parthasarathi, R.; Simmons, B.A.; Henry, R.J. Recent innovations in analytical methods for the qualitative and quantitative assessment of lignin. *Renew. Sustain. Energy Rev.* **2015**, *49*, 871–906. [CrossRef]
26. Tucker, M.P.; Kim, K.H.; Newman, M.M.; Nguyen, Q.A. Effects of temperature and moisture on dilute-acid steam explosion pretreatment of corn stover and cellulase enzyme digestibility. *Appl. Biochem. Biotechnol.* **2003**, *105*, 165–177. [CrossRef]
27. Mardawati, E.; Herliansah, H.; Suryadi, E.; Hanidah, I.I.; Siti Setiasih, I.; Andoyo, R.; Sukarminah, E.; Djali, M.; Rialita, T.; Cahyana, Y. Optimization of Particle Size, Moisture Content and Reaction Time of Oil Palm Empty Fruit Bunch through Ozonolysis Pretreatment. *J. Jpn. Inst. Energy* **2019**, *98*, 132–138. [CrossRef]
28. Brinkmann, K.; Blaschke, L.; Polle, A. Comparison of different methods for lignin determination as a basis for calibration of near-infrared reflectance spectroscopy and implications of lignoproteins. *J. Chem. Ecol.* **2002**, *28*, 2483–2501. [CrossRef] [PubMed]
29. Mohammed, I.; Abakr, Y.; Kazi, F.; Yusup, S.; Alshareef, I.; Chin, S. Comprehensive Characterization of Napier Grass as a Feedstock for Thermochemical Conversion. *Energies* **2015**, *8*, 3403–3417. [CrossRef]
30. Penttilä, P.A.; Várnai, A.; Pere, J.; Tammelin, T.; Salmén, L.; Siika-aho, M.; Viikari, L.; Serimaa, R. Xylan as limiting factor in enzymatic hydrolysis of nanocellulose. *Bioresour. Technol.* **2013**, *129*, 135–141. [CrossRef]
31. Muhamad, N.; Sahadan, W.; Hoon, H. Effect of drying temperatures and extraction solvents on total phenolic, flavonoid contents and antioxidant properties of immature Manis Terengganu Melon (Cucumis melo). *J. Agrobiotechnol.* **2018**, *9*, 114–121.
32. Michelin, M.; Gomes, D.G.; Romaní, A.; Polizeli, M.d.L.T.M.; Teixeira, J.A. Nanocellulose Production: Exploring the Enzymatic Route and Residues of Pulp and Paper Industry. *Molecules* **2020**, *25*, 3411. [CrossRef] [PubMed]
33. Ting, S.S. Comparative Properties Analysis between Microcrystalline Cellulose and Cellulose Nanocrystals Extracted From Rice Straw. *Malays. J. Microsc.* **2019**, *15*, 146–154.
34. Manokhoon, P.; Rangseesuriyachai, T. Effect of two-stage sodium hydroxide pretreatment on the composition and structure of Napier grass (Pakchong 1) (*Pennisetum purpureum*). *Int. J. Green Energy* **2020**, *17*, 864–871. [CrossRef]
35. Phitsuwan, P.; Sakka, K.; Ratanakhanokchai, K. Structural changes and enzymatic response of Napier grass (*Pennisetum purpureum*) stem induced by alkaline pretreatment. *Bioresour. Technol.* **2016**, *218*, 247–256. [CrossRef]
36. Phitsuwan, P.; Charupongrat, S.; Klednark, R.; Ratanakhanokchai, K. Structural features and enzymatic digestibility of Napier grass fibre treated with aqueous ammonia. *J. Ind. Eng. Chem.* **2015**, *32*, 360–364. [CrossRef]
37. Song, W.; Peng, L.; Bakhshyar, D.; He, L.; Zhang, J. Mild $O_2$-aided alkaline pretreatment effectively improves fractionated efficiency and enzymatic digestibility of Napier grass stem towards a sustainable biorefinery. *Bioresour. Technol.* **2021**, *319*, 124162. [CrossRef]

processes

Article

# Dried Droplets of Diluted Blood to Detect a High Concentration of Lipids

**Monserrat Ancheyta-Palacios, Iris G. Velasco-Terán, Yojana J. P. Carreón * and Jorge González-Gutiérrez ***

Facultad de Ciencias en Física y Matemáticas, Universidad Autónoma de Chiapas, Tuxtla Gutiérrez 29050, Chiapas, Mexico
* Correspondence: yojana.carreon@unach.mx (Y.J.P.C.); jorge.ggutierrez@unach.mx (J.G.-G.)

**Abstract:** Hyperlipidemia is the elevated concentration of lipids in the blood, and it increases the probability of arterial obstruction, infarctions, and other complications of the circulatory system. While there are indications that qualitative analysis of blood stains could potentially identify patients with this pathology, the efficacy of this method remains uncertain. In this paper, we report an experimental study that investigates the formation of patterns in dried blood droplets with varying concentrations of ultrapure water. Two blood samples, one healthy and one with moderate hyperlipidemia, were examined to determine the ideal water and blood mixtures for detecting high lipid concentrations. Numerous intricate patterns were observed throughout the central region and periphery of the dried droplet. These patterns encompass various forms, such as plaques, bump-like patterns, and a range of cracks including random, radial, and ortho-radial configurations. By calculating the entropy of the Gray Level Co-occurrence Matrix (GLCM) and analyzing ROC curves, we determined that solutions with 4% and 12% hematocrit (indicating a high percentage of ultrapure water) exhibit over 95% accuracy in differentiating high lipid concentrations. These findings provide a promising outlook for the development of diagnostic methods based on the study of diluted blood coatings.

**Keywords:** patterns; texture; drying; drop; blood; lipids

**Citation:** Ancheyta-Palacios, M.; Velasco-Terán, I.G.; Carreón, Y.J.P.; González-Gutiérrez, J. Dried Droplets of Diluted Blood to Detect a High Concentration of Lipids. *Processes* **2023**, *11*, 2047. https://doi.org/10.3390/pr11072047

Academic Editors: Dariusz Dziki and Jan ZawaÅa

Received: 9 February 2023
Revised: 2 July 2023
Accepted: 6 July 2023
Published: 9 July 2023



## 1. Introduction

The morphological characterization of structures resulting from droplet drying has garnered significant interest in various technological applications, particularly in the field of biosensors [1–5], quality control of liquid consumables [6–8], forensic applications [9,10], protein analysis [11–13] and diagnosis of pathologies [14,15]. Under controlled evaporation conditions, water removal from a sessile droplet gives rise to a deposit exhibiting intriguing morphological characteristics [16,17]. The ingredients responsible for the formation of a great diversity of patterns result from the competition between capillary flows and Marangoni flows, and complex aggregation processes [18]. Capillary flows are generated from inside the drop to move radially outward towards the edge. In cases where the sessile drop fluid contains nonvolatile particles, these are pushed outward by capillary flows, forming a ring-shaped structure known as a "coffee ring" [19,20]. On the other hand, the Marangoni flows depend on the concentration gradients and induce the fluid to recirculate towards the interior of the drop, generating a homogeneous stain [21–23].

The characterization of patterns in dried blood droplets is an interesting alternative to develop diagnostic strategies [24–26]. In general, these patterns are made up of a "central region" where plaques and random cracks are found. A large "crown" with wide white cracks and mobile plaques surrounds the central region, while a third region is formed by a thin peripheral belt of blood serum. Interestingly, the crown, which emerge by the accumulation of a high concentration of red cells, does not adhere to the substrate. In contrast, the central and periphery regions adhere more strongly due to the absence of erythrocytes. This occurs because the glycoproteins, contained in the red cells, prevent the covering of the walls [25,27]. The formation of the thin peripheral belt occurs due to the receding of the gel front induced by a change in internal flow that transports red blood cells from the inner to the periphery of the droplet [28].

Elevated blood lipid levels pose a risk for atherosclerosis and cardiovascular disease, potentially leading to fatal outcomes [29]. Recently, Brutin et al. [25] introduced a novel approach to identify hyperlipidemia by analyzing dried blood droplets. Interestingly, they found that the presence of high lipid concentrations correlates with the formation of small plaques in the central region. Although this study represents a significant advance in the search for alternatives to unveil hyperlipidemia, the effectiveness of dried droplet analysis as a diagnostic tool is unknown. Furthermore, despite intensive research on blood droplet drying, it is still unknown how blood lipid concentration affects the mechanisms of mass transport and aggregation processes involved in structure formation. A better understanding of such processes and the development of techniques for pattern identification could help to develop strategies that result in a highly effective methodology for the diagnosis of hyperlipidemia.

The gray level co-occurrence matrix (GLCM) is a measure of texture evaluation of an object and it is related to the spatial distribution of pixel intensities in a region of interest. The main characteristic of the GLCM parameters is that they act as indices of the frequency of gray level combinations in an image [30]. This metric is used to classify cancer types [31,32], characterize morphological changes in the organization of collagen tissues caused by some pathology [33], among other relevant systems [34,35]. On the other hand, different techniques have been used to evaluate classification algorithms such as receiver operating characteristic (ROC). ROC curves are used to quantify the accuracy of a test to discriminate between elements of two groups [36,37]. The use of GLCM in combination with ROC curve analysis has been successfully used to analyze patterns in dried droplets in the contexts such as drug quality control [6], protein folding detection [38], and biofluid characterization [39].

In this paper, we report an experimental study on the formation of patterns in dried blood droplets with different concentrations of ultrapure water. Two blood samples, one healthy and one with moderate hyperlipidemia, were investigated to identify optimal water and blood mixtures for detecting high lipid concentrations. Through the Entropy of the Gray Level Co-occurrence Matrix (GLCM), we performed texture analyses on two areas: one encompassing the central region and the corona (entire deposit), and the other focusing solely on the central region. By employing ROC curve analysis, we have ascertained that diluted solutions with 4% and 12% hematocrit (HCT) achieve an accuracy of over 95% in differentiating blood samples with a high lipid concentration.

## 2. Experimental Details

### 2.1. Sample Collection and Storage

Blood samples were collected from donors by medical personnel and informed consent was received from the volunteers. To determine specific parameters of their blood composition, volunteers underwent clinical laboratory (refer to Table 1 for details). All the samples were collected and stored inside a 4 mL vacutainer tube (EDTA.K2, Golden Vac). Ultrapure water was used to dilute the blood samples. Five dilutions (4%, 12%, 20%, 28% and 36% HCT) were made from whole blood samples (40% HCT).

**Table 1.** Complete blood count.

| Hematic Biometry | Healthy Patient | Dyslipidemic Patient | Unit |
|---|---|---|---|
| Leukocytes | 4.5 | 8.74 | $10^9/L$ |
| Erythrocytes | 4.3 | 4.63 | $10^6/mm^3$ |
| Hemoglobin | 129 | 138 | g/L |

**Table 1.** *Cont.*

| Hematic Biometry | Healthy Patient | Dyslipidemic Patient | Unit |
|---|---|---|---|
| Hematocrit | 40 | 39.3 | % |
| Mean Corpuscular Volume | 93 | 84.9 | fL |
| Mean Corpuscular Hb | 1.8617 | 1.8493 | fmol |
| Mean Corpuscular Hb Concentration | 20.0447 | 21.7823 | mmol/L |
| Platelet | 310 | 313 | $10^9$/L |
| Mean platelet volume | 7.1 | 9 | fL |
| Lymphocyte | 2.07 | 3.59 | $10^9$/L |
| Neutrophils | 1.89 | 4.01 | $10^9$/L |
| Monocytes | 0.54 | 0.8 | $10^9$/L |
| Eosinophils | 0 | 0.24 | $10^9$/L |
| Basophils | 0 | 0.05 | $10^9$/L |
| Lymphocyte | 46 | 41.1 | % |
| Neutrophils | 42 | 45.8 | % |
| Monocytes | 12 | 9.2 | % |
| Eosinophils | 0 | 2.7 | % |
| Basophils | 0 | 0.6 | % |
| Biochemistry | | | |
| Glucose | 5.1621 | 8.9809 | mmol/L |
| Urea | 4.995 | 4.8452 | mmol/L |
| Urea Nitrogen | 5.0286 | 4.8571 | mmol/L |
| Serum Creatinine | 0.0522 | 0.076 | mmol/L |
| Total Cholesterol | 3.3364 | 4.7899 | mmol/L |
| Triglyceride | 0.7458 | 2.765 | mmol/L |
| Cholesterol HDL | 1.2285 | 1.1018 | mmol/L |
| Cholesterol LDL | 1.7284 | 3.0958 | mmol/L |

### 2.2. Drop Evaporation

The evaporation substrate was brand new Poly(methyl methacrylate) (PMMA) which was thoroughly cleaned prior to being used. Eighteen droplets of the solution, with a volume of 3 μL, were placed on PMMA. The experiments were repeated three times at each experimental condition, i.e., at least 50 dried droplets in total were produced for each concentration (18 deposits in 3 different PMMA substrates). The drops were dried under controlled temperature at T = 21–27 °C and Relative Humidity (RH) of 30–35%. The relative humidity was controlled using the effect of water activity $a_w = \rho/\rho_0$, where $\rho$ is the vapor pressure of water in a substance and $\rho_0$ is the pressure of pure water vapor at the same temperature. The RH value was measured with a temperature and humidity sensor (Xiaomi NUN4126GL). The average evaporation times for the various erythrocyte concentrations were as follows: 47.15 min (4% HCT), 44.33 min (12% HCT), 43 min (20% HCT), 41.09 min (28% HCT), 40.3 min (35% HCT), and 38.3 min (40% HCT).

### 2.3. Image Acquisition

The deposits were observed after evaporation in ambient conditions using a microscope (Velab, VE-M4, 4×, and 10×) coupled with a Nikon camera (D3200). The resolution of the images was chosen to be 300 dpi, creating images of approximately 4000 pixels in length for the longest side.

## 2.4. Texture Analysis of Dried Drop Patterns

Gray Level Co-Occurrence Matrix (GLCM)

We use Entropy based on gray level co-occurrence matrix (GLCM) to measure the texture of patterns in dried droplets. Higher (lower) entropy values indicate large (small) heterogeneous regions in an image. Texture analysis based on GLCM has been used successfully in characterizing protein films containing complex crack patterns [40] and salt crystals [39].

A gray level co-occurrence matrix (GLCM) is a matrix where the number of rows and columns is equal to the number of gray levels $N_g$ in an image. Its analysis is based on the correlation among pixels in an image. Mathematically, the matrix element $p(i,j)$ is the probability values for changes between gray level $i$ and $j$ at a particular displacement distance ($d$) and angle ($\phi$) on an image. This probability can be defined as:

$$p(i,j) = \frac{C(i,j)}{\sum_{i=0}^{N_g-1}\sum_{j=0}^{N_g-1} C(i,j)}, \tag{1}$$

where $C(i,j)$ is the number of occurrences of gray levels $i$ and $j$ within the window, at a particular $(d,\phi)$ pair. The denominator is the total number of gray level pairs $(i,j)$ within the window and is bounded by an upper limit of $N_g \times N_g$. The mean and the standard deviation for the columns and rows of the matrix, using the above equation, can be defined as follows:

$$u_x = \sum_{i=0}^{N_g-1}\sum_{j=0}^{N_g-1} i \cdot p(i,j), \quad u_y = \sum_{i=0}^{N_g-1}\sum_{j=0}^{N_g-1} j \cdot p(i,j), \tag{2}$$

$$\sigma_x = \sum_{i=0}^{N_g-1}\sum_{j=0}^{N_g-1} (i-u_x)^2 \cdot p(i,j), \quad \sigma_y = \sum_{i=0}^{N_g-1}\sum_{j=0}^{N_g-1} (j-u_y)^2 \cdot p(i,j), \tag{3}$$

where $u_x$ and $u_y$ are the mean for the columns and rows, respectively; and $\sigma_x$ and $\sigma_y$ represent the standard deviation for the columns and rows, respectively. Now, using these equations, we can define the entropy $H$ as follows:

$$H = -\sum_{i=0}^{N_g-1}\sum_{j=0}^{N_g-1} p(i,j)log(p(i,j)). \tag{4}$$

## 2.5. The Receiver Operating Characteristic (ROC) Curve

A receiver operating characteristic (ROC) graph is a method used to organize and select classifiers based on their efficiency [36,37,41]. The ability of a classifier to provide a good relative instance score is measured by ROC plots. The ROC curve is obtained by plotting the True Positive rate (Sensitivity) as a function of the False Positive rate (1-Specificity) for each possible threshold value on the confidence score [36,42]. The Sensitivity represents the positive correctly classified samples to the total number of positive samples. Mathematically, this quantity is written as Equation (5). Whereas the Specificity (True Negative rate) represents the negative correctly classified samples to the total number of negative samples, this quantity is defined as Equation (5). The False Positive rate (1-Specificity) represents incorrectly classified negative samples to the total number of negative samples as in Equation (6). Specificity and 1-Specificity are complements since the first considers the actual negative samples and the second refers the negative samples that were incorrectly classified.

$$Sensibility = \frac{TP}{TP+FN}, \; Specificity = \frac{TN}{TN+FP} \tag{5}$$

$$1\text{-}Specificity = \frac{FP}{FP+TN} \tag{6}$$

A ROC curve is a two-dimensional depiction of classifier performance. Therefore, the ROC curve is a two-dimensional depiction of the relationship between the True Positive rate and the False Positive rate. Each point on the ROC curve is a cutoff point used to designate test-positive. Finally, to compare different classifiers in ROC curve is used the area under the curve ROC (AUC) metric. The AUC is the probability that the algorithm of a classifier ranks a randomly chosen positive instance higher than a randomly chosen negative instance. The detailed procedure of ROC curve analysis can be found in [36,42].

## 3. Results

### 3.1. Pattern Formation in Dried Blood Drops

At first glance, dried droplets of dilute blood with both normal and elevated lipid concentrations display apparent morphological similarities. However, closer scrutiny reveals significant distinctions in the interior structures that compose the patterns. Figure 1 shows the deposits with the diluted blood (4–36% HCT) and whole blood (40% HCT) samples. A close-up of the dried blood drops for the different concentrations of HCT is shown in Figure 2.

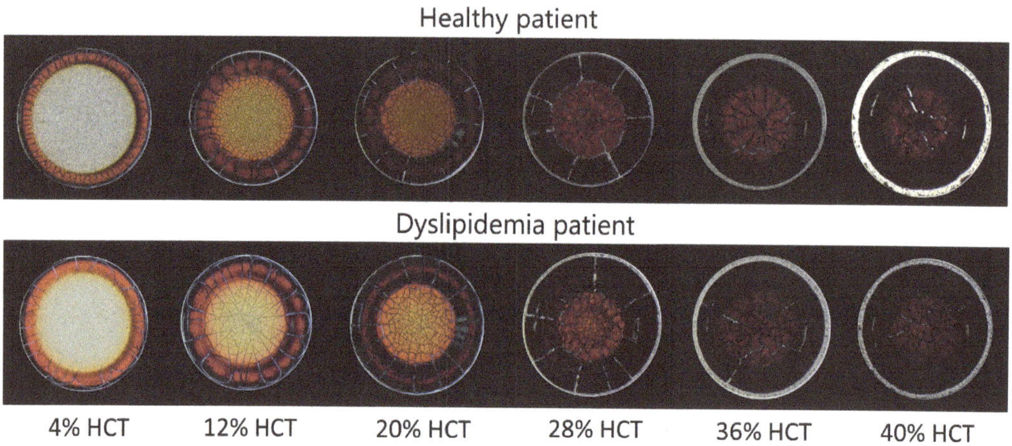


**Figure 1.** Different hematocrit concentrations. Dried blood drops with different HCT concentrations of a healthy patient (**upper**) and a patient with high levels of lipids (**bottom**). The diameters of the deposits are as follows: 2.86 ± 0.13 mm (4% HCT), 2.79 ± 0.11 mm (12% HCT), 2.84 ± 0.14 mm (20% HCT), 2.72 ± 0.12 mm (28% HCT), 2.68 ± 0.15 mm (36% HCT), 2.66 ± 0.13 mm (40% HCT) for the healthy patient and 2.93 ± 0.10 mm (4% HCT), 2.98 ± 0.15 mm (12% HCT), 2.92 ± 0.12 mm (20% HCT), 2.80 ± 0.13 mm (28% HCT), 2.62 ± 0.16 mm (36% HCT), 2.80 ± 0.14 mm (40% HCT) for a patient with high levels of lipids.

Deposits with low erythrocyte concentration (4% HCT) and normal lipid levels display a crown with clearly defined borders (yellow line in Figure 2a), accompanied by a high occurrence of radial cracks (blue arrow) and ortho-radial cracks (green arrow). Additionally, the central region (pink line in Figure 2a) contains numerous small aggregates, resulting in a granular surface texture. Conversely, a high lipid concentration produces a thicker crown with less defined borders and a reduced number of cracks.

Dried droplets with 12% HCT exhibit a crown lacking well-defined borders and display radial and orthoradial cracks with similar characteristics, as shown in Figure 2a. Droplets with normal lipid concentration produce a higher occurrence of random cracks (orange arrow) and mobile plaques (black arrow) in the central region compared to those from patients with dyslipidemia. Furthermore, the size of the plaques is significantly larger in the latter. Patterns in deposits from both the healthy patient and the patient with dyslipidemia, with a 20% HCT, are very similar to each other. They feature a crown with minimal radial and orthoradial cracks, along with a central region characterized by random cracks and plaques.

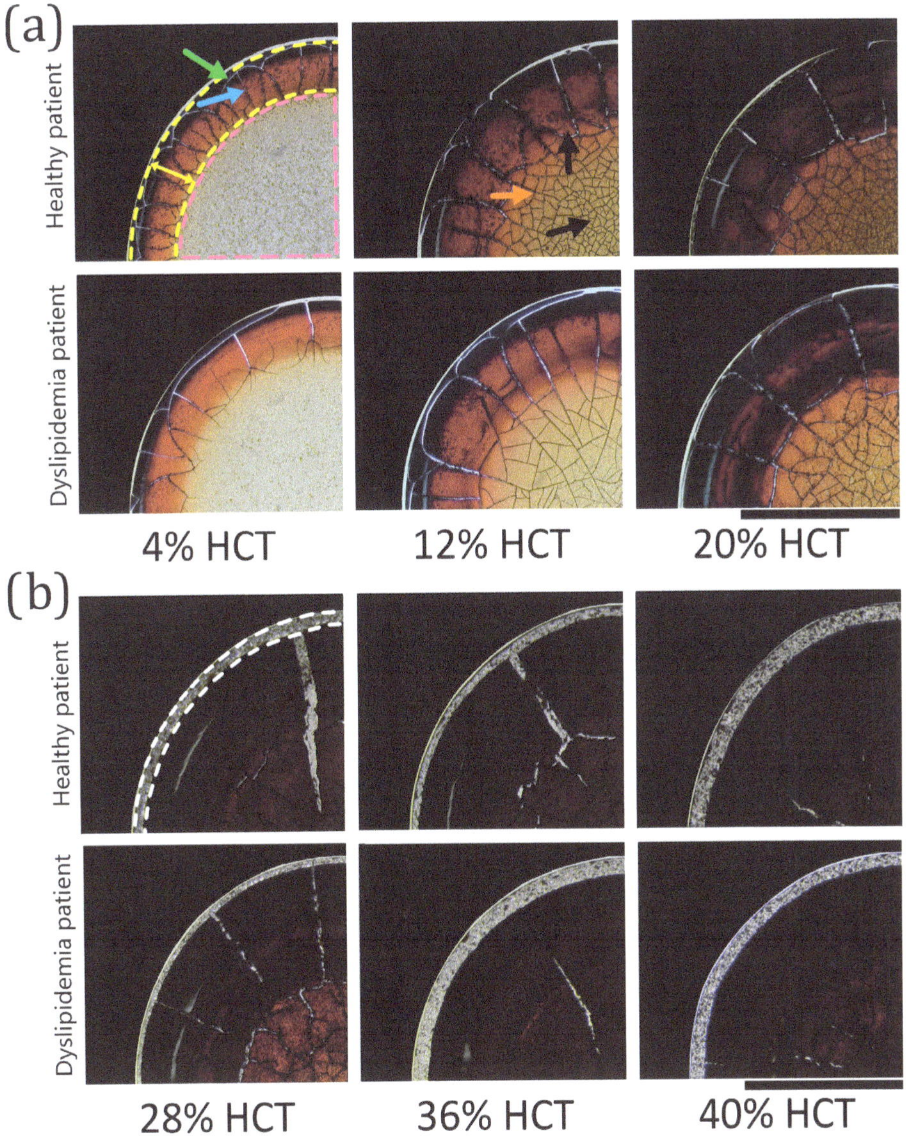


**Figure 2.** Patterns of dried blood droplets. (**a**) Deposits with low HCT concentration (4%, 12%, and 20%) from a healthy patient (**top**) and a patient with high lipid concentration (**bottom**). The yellow and pink lines determine the crown and central region, respectively. Radial cracks are denoted by the blue arrow, while ortho-radial cracks are indicated by the green arrow. Random cracks and mobile plaques are shown by the orange arrow and black arrows, respectively. (**b**) Dried droplets with elevated HCT concentrations (28%, 36%, and 40%) from a healthy patient (**top**) and a patient with high lipid concentration (**bottom**). The white line indicates the erythrocyte exclusion zone. The black bar indicates 1 mm.

Samples with a hematocrit of 28% produce random cracks and large plaques in the central region (Figure 2b). Distinct radial cracks are observed in both the central region and the crown of deposits obtained from the healthy patient (Figure 2b—top). In contrast, samples from the patient with Dyslipidemia generated a greater concentration of cracks confined to the crown region (Figure 2b—bottom). Notably, droplets with hematocrit concentrations greater than or equal to 28% do not induce the formation of ortho-radial cracks in the corona region.

Droplets with hematocrit levels of 36% and 40% generate a crown characterized by the formation of a minimal number of radial cracks (Figure 2b). Within the central region, a mixture of radial and random cracks extends beyond its boundaries. Moreover, noticeable variations in morphology are observed for the large, mobile plaques. Interestingly, the outer boundary of the corona exhibits a peripheral erythrocyte exclusion band, where the deposition of these cells is not possible (see white dotted line in Figure 2b). To demonstrate the high reproducibility of pattern formation in dried droplets of diluted blood, Figure 3 shows 24 deposits for each hematocrit (HCT) concentration, all generated under similar conditions of temperature and relative humidity.

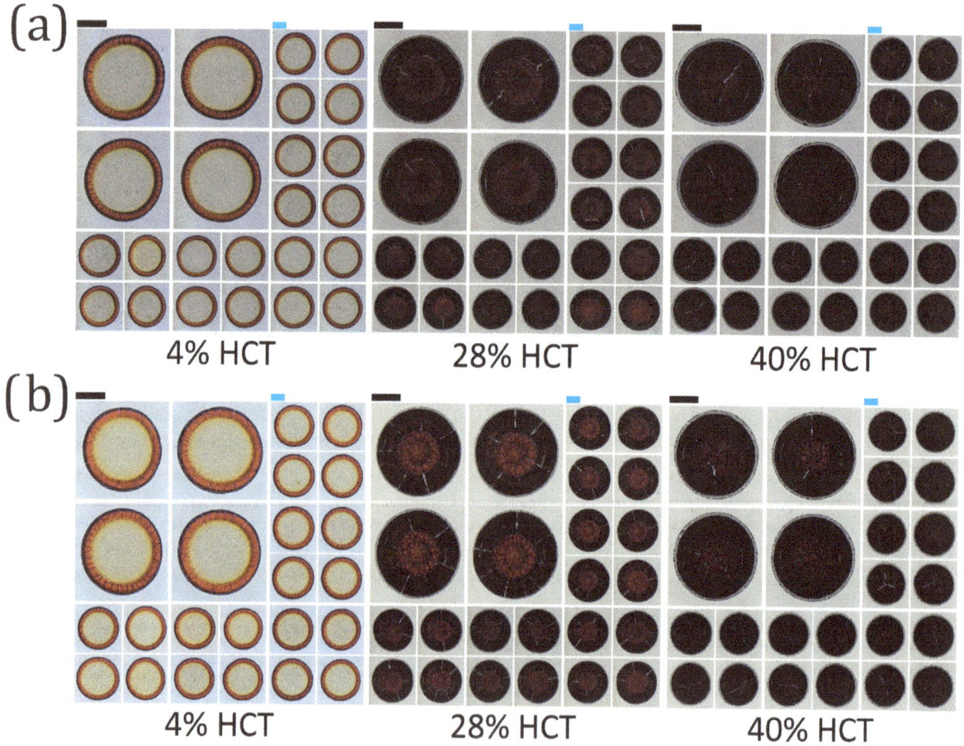


**Figure 3.** Reproducibility of patterns in dried blood drops. (**a**) Deposits from a healthy patient with different HCT concentrations (4%, 28%, and 40%). (**b**) Dried blood droplets from a patient with high lipid concentration with different HCT concentrations (4%, 28%, and 40%). The black and blue bar indicates 1 mm.

### 3.2. Texture Analysis of Dried Blood Droplets

We employed Gray Level Co-occurrence Matrix (GLCM) Entropy as a comprehensive approach to capture the intricate texture of dried droplets from both healthy and Dyslipidemia patients. Furthermore, in our pursuit of distinguishing between the groups of dried drops, we conducted the analysis on two distinct areas: one composed of the central region and the corona (entire deposit), and another considering only the central region.

Figure 4a shows the values of Entropy for entire deposits as a function of hematocrit concentration. The red and blue profiles correspond to dried drops of healthy and dyslipidemic patients, respectively. The image of the dried drop in the graph shows a deposit outlined by a green line indicating the analyzed region. Observe that concentrations of HCT < 20% give high Entropy values. In contrast, concentrations of HCT > 28% give low values of this parameter. This indicates that at high HCT concentrations, there are small heterogeneous regions in entire deposits. It should be noted that at 4% HCT the error bars indicate that it is possible to distinguish the dried blood drops of the healthy patient from those of the patient with Dyslipidemia. Entropy values calculated in the central region of the deposits are shown in Figure 4b. Deposits formed with HCT < 20% concentrations reach high Entropy values, while in those formed with HCT > 28% reach low values. This is because, at low HCT concentrations, cracks and plaques produce greater heterogeneity in the central region of the deposits compared to those generated with high concentrations. Comparing the error bars estimated from the standard deviation of the Entropy values of dried blood droplets with normal and high lipid concentrations, we observe the highest difference at 12% HCT, while the lowest at 20% HCT.

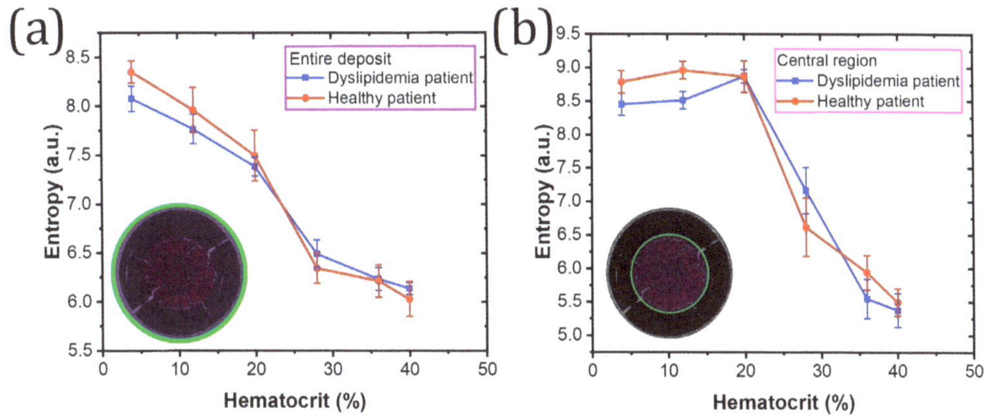


**Figure 4.** Texture analysis of dried blood drops. Entropy as a function of different HCT concentrations analyzed on two distinct areas: (**a**) entire deposits and (**b**) central region. The green circles enclosed area on the dried drops show the region analyzed. The error bars correspond to standard deviations from $n = 50$.

We used the Entropy values to perform a discriminant analysis based on Receiver Operating Characteristic (ROC curves). The ROC curves for 4% and 12% HCT, based on entropy values calculated in both regions (entire deposit and central region), exhibit a typical behavior of an excellent sensitivity to 1-specificity ratio, see Figure 5a. In contrast, the shape of the ROC curves for 36% and 20% HCT show a deficient analysis for the differentiation of elements of two groups, see Figure 5b.

The area under the ROC curve (AUC) is a quantity that shows the accuracy, of a classification method in differentiating between two groups. The area under the ROC curve (AUC) gives the probability that a deposit produced from a patient sample can be correctly classified as a dried drop of blood with elevated lipid levels. Figure 5c shows the accuracy of the texture analysis method in dried blood drops to differentiate between two groups. Entropy estimation on the entire deposit exhibits the highest accuracy at lower hematocrit concentrations (4%), whereas analyzing the central region yields superior accuracy for those formed with 12% HCT. In contrast, this quantity falls sharply for concentrations of 36% and 20% HCT, respectively.

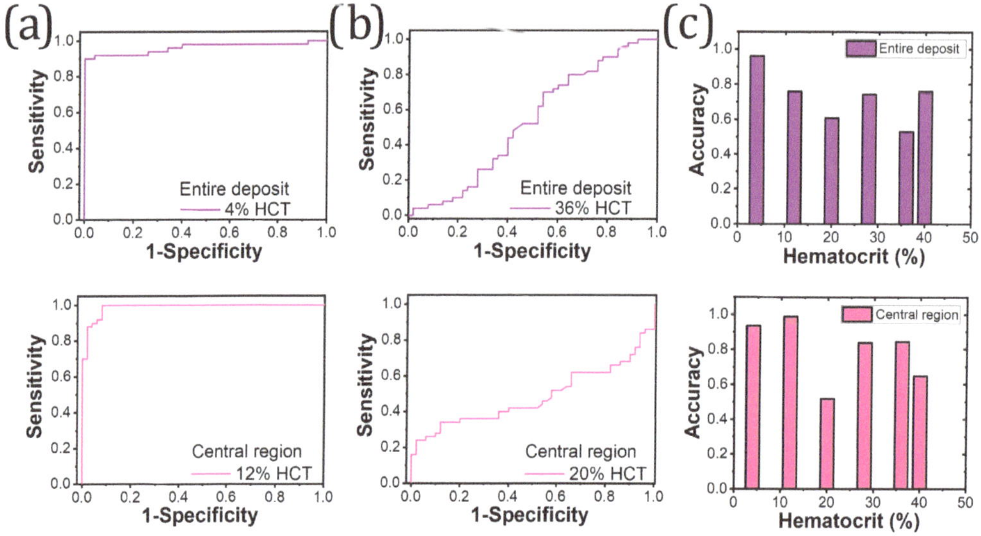


**Figure 5.** Receiver Operating Characteristic (ROC) and Accuracy. (**a**) ROC curve with excellent ratio sensitivity and 1-specificity at 4% HCT (**upper**) and 12% HCT (**bottom**). (**b**) ROC curve with deficient differentiation of elements of two groups at 36% HCT (**upper**) and 20% HCT (**bottom**). (**c**) Accuracy for identifying deposits with a high lipid concentration by Entropy analysis on the entire deposit (**upper**) and the central region (**bottom**).

## 4. Discussion

In this study, we examined patterns in dried droplets of diluted blood with ultrapure water to identify morphological features and distinguish between healthy blood and a blood sample with moderate hyperlipidemia. Through the analysis of GLCM-estimated entropy and ROC curves, we have identified that a high lipid concentration can be efficiently detected in deposits formed with low hematocrit concentrations, specifically in samples with a high weight percentage of ultrapure water. Ultrapure water reduces biomolecule concentration, promoting the formation of distinct aggregates, plaques, and cracks [43,44]. Previous studies have established a correlation between decreased blood analytes and increased structural diversity [45–47]. Higher dilutions have been found to enhance deposit adhesion to the substrate [45,47], and the presence of a prominent annular phenomenon has been observed as an indicator of water dilution in whole blood solutions [46]. Our findings align with these reports, indicating that reducing the saturation of blood elements in a patient with a high lipid concentration correlates with greater structural diversity.

The ability to detect and analyze patterns in dried blood droplets opens up possibilities for real-time monitoring of lipid concentrations. By leveraging this technique, it may be feasible to develop portable devices or smartphone apps that can capture images of dried blood droplets and perform pattern analysis on-site. This could enable individuals with hiperlipidemia or healthcare professionals to monitor lipid levels conveniently and promptly, allowing for timely interventions or adjustments in treatment plans. Moreover, the contactless nature of analyzing dried blood droplets for lipid concentration determination is a significant advantage. Traditional blood tests often require invasive procedures like venipuncture to collect samples, which can be uncomfortable. However, if the analysis of dried blood droplet patterns proves to be a reliable diagnostic method, it could offer a non-invasive alternative for diagnosing and monitoring hiperlipidemia. The study's results can serve as a starting point for the design and development of specialized devices or algorithms that can automate the analysis of dried blood droplet patterns, enabling faster and more precise diagnosis. For example, artificial intelligence (AI) and machine learning

(ML) can play a significant role in improving the technique described in the provided information. These algorithms could identify specific characteristics or combinations of characteristics that are most indicative of high lipid concentrations. By focusing on these features, the diagnostic method can be optimized for accuracy and efficiency. AI and ML models can be deployed on portable devices or integrated into smartphone apps, allowing for real-time analysis of dried blood droplets. This means that users can capture an image of the dried blood droplet and receive immediate feedback on their lipid levels. AI and ML models can continuously learn from new data, enabling ongoing improvement of the diagnostic method. As more samples and patterns are analyzed, the algorithms can refine their understanding and adapt to new patterns or variations.

Other techniques involve diluting whole blood samples [47,48], as well as plasma and blood serum [49–51], to evaluate test sensitivity and minimize interference from other blood components. Studies within this context have demonstrated that diluting blood serum with water does not affect the chemical composition and molecular structure of proteins in dried droplets [50,51]. Consequently, there are no changes in the components of the diluted serum that affect the formation of dried droplet patterns.

Different methods are available for measuring Lipid/cholesterol levels in blood and serum samples, including advanced chemistry analyzers, quantitative meters or novel biosensors. Routine chemistry analyzers utilize spectrophotometry, enzymatic reactions, fluorometric, or immunoassays to detect and quantify fatty substances accurately, making them suitable for precise clinical assessment [52]. However, Point-of-Care (POC) tests have made significant advancements, yet challenges remain in terms of improving accuracy, reliability, and limits of detection [53–55]. Our study aims to highlight the potential of analyzing dried droplets from diluted blood samples as a diagnostic alternative for moderate dyslipidemia. Table 2 provides an overview of measurement methods for cholesterol, triglycerides, low-density lipoprotein, and high-density lipoprotein.

**Table 2.** Briefly list of measurement methods of cholesterol, triglycerides, low-density lipoprotein, and high-density lipoprotein.

| Materials and Structures | Detection Methods | Sample Type | Target | Detection Limit or Accuracy | Refs. |
|---|---|---|---|---|---|
| Strip-based meter | N/A | finger-stick blood | Cholesterol, HDL, TG, LDL | 96% | [53] |
| Strip-based meter | N/A | finger-stick blood | Cholesterol, HDL, TG, LDL | 40% | [53] |
| Strip-based meter | N/A | finger-stick blood | Cholesterol, TG | 92% | [53] |
| Single-use strip | N/A | finger-stick blood | Cholesterol | 85% | [53] |
| Single-use strip | N/A | finger-stick blood | Cholesterol | 80% | [53] |
| Chromatogr. paper, 2-D | Colorimetric + Reagent | Serum Centrifuge | Cholesterol | 0.1 mM | [56] |
| PDMS + NC membrane, 3-D | Colorimetric + biomarker | Whole blood | Cholesterol | 11 mg $dL^{-1}$ | [57] |
| NC paper, 3-D | Electrochem + modified ED | Saliva | Cholesterol | 0.5 µg $dL^{-1}$ | [58] |
| Filter paper, 3-D | Colorimetric + biomarker | Whole blood | Cholesterol | N/A | [59] |
| flower-shaped lab-on-paper | Colorimetric + biomarker | Whole blood | Cholesterol, TG, LDL, HDL | 50 mg $dL^{-1}$, 70 mg $dL^{-1}$, 70 mg $dL^{-1}$, 60 mg $dL^{-1}$ | [59] |
| PMMA | Image Analysis | Diluted blood | Lipid | 95% | ** |

Triglycerides(TG), Low-density lipoprotein (LDL), and high-density lipoprotein (HDL), Polydimethylsiloxane (PDMS), Nitrocellulose (NC), Polymethyl methacrylate (PMMA). ** This report.

Prior to employing the analysis of dried droplets of diluted blood to detect a high lipid concentration, it is imperative to address critical considerations. Firstly, prompt capture of deposit images after drop drying is crucial to prevent oxidation-induced darkening. Secondly, the rapid degradation of dried drop samples with high hematocrit concentrations needs to be acknowledged. Lastly, this methodology still does not allow for the determination of triglyceride and cholesterol concentrations in the blood. However, the identification of certain advantages of the dried droplet texture analysis is evident.

The presented results highlight the potential of analyzing dried droplets of diluted blood samples in creating a diagnostic method for moderate dyslipidemia, suggesting its potential utility in diagnosing severe stages of the disease. The method requires low blood volumes, and further sample dilution may reduce the required blood sample volume even more. Furthermore, the calculated configurational entropy, both in the central region and the entire deposits, demonstrates high efficiency in differentiating between groups, eliminating the need to empirically observe specific morphological features in the deposits.

Patterns in dried drops depend on many parameters, such as moisture, viscosity, initial drop volume, and substrate type [60,61]. Since the initial droplet volume and relative humidity determine the drying time, the idea of knowing the effectiveness of this methodology using blood droplets of volumes less than 3 microliters and relative humidity less than 30% is very attractive. Moreover, creating a relationship between lipid concentration and texture parameters could provide a tool for patient monitoring and follow-up. We plan to perform such measurements in the future and report the results.

## 5. Conclusions

We present an experimental study analyzing the texture of dried droplet patterns from two diluted blood samples: one healthy and one with moderate hyperlipidemia. Our objective was to determine the optimal water-to-blood ratios for detecting a high lipid concentration. The observed patterns in dried droplets demonstrate a diverse range of complex structures, including plaques, random and radial cracks, granular patterns, and a peripheral erythrocyte exclusion band. These structural features vary across different regions and are influenced by the hematocrit concentration in the solution. Notably, the reduction of blood elements resulted in the generation of well-defined and reproducible patterns. The complexity of the patterns is quantified using Gray Level Co-occurrence Matrix Entropy (GLCM). Texture analysis of dried droplets achieves over 95% accuracy in detecting a high lipid concentration in solutions with 4% and 12% HCT.

Overall, we have probed that decreasing blood component concentrations can serve as an alternative approach for generating intricate patterns that function as indicators of pathology.

**Author Contributions:** M.A.-P.: Investigation, Methodology, Formal analysis. I.G.V.-T.: Investigation, validation. Y.J.P.C.: Supervision, Visualization, Writing—Original Draft. J.G.-G.: Conceptualización, Supervision, Project administration, Writing—Original Draft, Writing—Review and Editing. All authors have read and agreed to the published version of the manuscript.

**Funding:** This research received no external funding.

**Institutional Review Board Statement:** All subjects gave their informed consent for inclusion before they participated in the study. The study was conducted in accordance with the Declaration of Helsinki, and the protocol was approved by the Internal Ethics Committee of Collegiate Research Group Dynamical Systems and Complexity of the Autonomous University of Chiapas (01/FYM/RPR/016/23).

**Data Availability Statement:** Not applicable.

**Acknowledgments:** Y.J.P.C. wishes to acknowledge financial support by the CONACyt Postdoctoral fellowship.

**Conflicts of Interest:** The authors declare no competing interest.

## References


1. Yakhno, T.; Pakhomov, A.; Sanin, A.; Kazakov, V.; Ginoyan, R.; Yakhno, V. Drop drying on the sensor: One more way for comparative analysis of liquid media. *Sensors* **2020**, *20*, 5266. [CrossRef] [PubMed]
2. Gulka, C.P.; Swartz, J.D.; Trantum, J.R.; Davis, K.M.; Peak, C.M.; Denton, A.J.; Haselton, F.R.; Wright, D.W. Coffee rings as low-resource diagnostics: Detection of the malaria biomarker plasmodium falciparum histidine-rich protein-ii using a surface-coupled ring of ni (ii) nta gold-plated polystyrene particles. *ACS Appl. Mater. Interfaces* **2014**, *6*, 6257–6263. [CrossRef] [PubMed]
3. Wen, J.T.; Ho, C.M.; Lillehoj, P.B. Coffee ring aptasensor for rapid protein detection. *Langmuir* **2013**, *29*, 8440–8446. [CrossRef]
4. Trantum, J.R.; Wright, D.W.; Haselton, F.R. Biomarker-mediated disruption of coffee-ring formation as a low resource diagnostic indicator. *Langmuir* **2012**, *28*, 2187–2193. [CrossRef]
5. Kim, J.; Dowling, V.; Datta, T.; Pershin, Y.V. Whisky-born memristor. *Phys. Status Solidi (a)* 2022, *220*, 2200643. [CrossRef]
6. Carreón, Y.J.; Díaz-Hernández, O.; Escalera Santos, G.J.; Cipriano-Urbano, I.; Solorio-Ordaz, F.J.; González-Gutiérrez, J.; Zenit, R. Texture Analysis of Dried Droplets for the Quality Control of Medicines. *Sensors* **2021**, *21*, 4048. [CrossRef] [PubMed]
7. González-Gutiérrez, J.; Pérez-Isidoro, R.; Ruiz-Suárez, J. A technique based on droplet evaporation to recognize alcoholic drinks. *Rev. Sci. Instruments* **2017**, *88*, 074101. [CrossRef] [PubMed]
8. Carrithers, A.D.; Brown, M.J.; Rashed, M.Z.; Islam, S.; Velev, O.D.; Williams, S.J. Multiscale self-assembly of distinctive weblike structures from evaporated drops of dilute american whiskeys. *ACS Nano* **2020**, *14*, 5417–5425. [CrossRef] [PubMed]
9. Peschel, O.; Kunz, S.; Rothschild, M.; Mützel, E. Blood stain pattern analysis. *Forensic Sci. Med. Pathol.* **2011**, *7*, 257–270. [CrossRef]
10. Wang, F.; Gallardo, V.; Michielsen, S.; Fang, T. Fundamental study of porcine drip bloodstains on fabrics: Blood droplet impact and wicking dynamics. *Forensic Sci. Int.* **2021**, *318*, 110614. [CrossRef]
11. Sett, A.; Ayushman, M.; Dasgupta, S.; DasGupta, S. Analysis of the distinct pattern formation of globular proteins in the presence of micro-and nanoparticles. *J. Phys. Chem. B* **2018**, *122*, 8972–8984. [CrossRef]
12. Carreón, Y.J.; González-Gutiérrez, J.; Pérez-Camacho, M.; Mercado-Uribe, H. Patterns produced by dried droplets of protein binary mixtures suspended in water. *Colloids Surf. B Biointerfaces* **2018**, *161*, 103–110. [CrossRef]
13. Pal, A.; Gope, A.; Iannacchione, G.S. A comparative study of the phase separation of a nematic liquid crystal in the self-assembling drying protein drops. *MRS Adv.* **2019**, *4*, 1309–1314. [CrossRef]
14. Sefiane, K.; Duursma, G.; Arif, A. Patterns from dried drops as a characterisation and healthcare diagnosis technique, potential and challenges: A review. *Adv. Colloid Interface Sci.* **2021**, *298*, 102546. [CrossRef] [PubMed]
15. Yakhno, T.A.; Yakhno, V.G.; Sanin, A.G.; Sanina, O.A.; Pelyushenko, A.S.; Egorova, N.A.; Terentiev, I.G.; Smetanina, S.V.; Korochkina, O.V.; Yashukova, E.V. The informative-capacity phenomenon of drying drops. *IEEE Eng. Med. Biol. Mag.* **2005**, *24*, 96–104. [CrossRef] [PubMed]
16. Misyura, S. Different modes of heat transfer and crystallization in a drop of NaCl solution: The influence of key factors on the crystallization rate and the heat transfer coefficient. *Int. J. Therm. Sci.* **2021**, *159*, 106602. [CrossRef]
17. Pal, A.; Gope, A.; Iannacchione, G. Temperature and concentration dependence of human whole blood and protein drying droplets. *Biomolecules* **2021**, *11*, 231. [CrossRef]
18. Misyura, S. The dependence of drop evaporation rate and wettability on corrosion kinetics. *Colloids Surf. A Physicochem. Eng. Asp.* **2021**, *610*, 125735. [CrossRef]
19. Deegan, R.D.; Bakajin, O.; Dupont, T.F.; Huber, G.; Nagel, S.R.; Witten, T.A. Capillary flow as the cause of ring stains from dried liquid drops. *Nature* **1997**, *389*, 827. [CrossRef]
20. Pal, A.; Gope, A.; Iannacchione, G.S. Hierarchical Exploration of Drying Patterns Formed in Drops Containing Lysozyme, PBS, and Liquid Crystals. *Processes* **2022**, *10*, 955. [CrossRef]
21. Kim, H.; Boulogne, F.; Um, E.; Jacobi, I.; Button, E.; Stone, H.A. Controlled uniform coating from the interplay of Marangoni flows and surface-adsorbed macromolecules. *Phys. Rev. Lett.* **2016**, *116*, 124501. [CrossRef]
22. Misyura, S. The influence of key factors on the movement of a crystal and a non-crystalline particle on a free droplet surface. *Exp. Therm. Fluid Sci.* **2019**, *109*, 109883. [CrossRef]
23. Perrin, L.; Akanno, A.; Guzman, E.; Ortega, F.; Rubio, R.G. Pattern Formation upon Evaporation of Sessile Droplets of Polyelectrolyte/Surfactant Mixtures on Silicon Wafers. *Int. J. Mol. Sci.* **2021**, *22*, 7953. [CrossRef] [PubMed]
24. Chao, T.C.; Trybala, A.; Starov, V.; Das, D.B. Influence of haematocrit level on the kinetics of blood spreading on thin porous medium during dried blood spot sampling. *Colloids Surf. A Physicochem. Eng. Asp.* **2014**, *451*, 38–47. [CrossRef]
25. Brutin, D.; Sobac, B.; Loquet, B.; Sampol, J. Pattern formation in drying drops of blood. *J. Fluid Mech.* **2011**, *667*, 85–95. [CrossRef]
26. Mukhopadhyay, M.; Ray, R.; Ayushman, M.; Sood, P.; Bhattacharyya, M.; Sarkar, D.; DasGupta, S. Interfacial energy driven distinctive pattern formation during the drying of blood droplets. *J. Colloid Interface Sci.* **2020**, *573*, 307–316. [CrossRef]
27. Brutin, D. *Droplet Wetting and Evaporation: From Pure to Complex Fluids*; Academic Press: Cambridge, MA, USA, 2015.
28. Zeid, W.B.; Brutin, D. Influence of relative humidity on spreading, pattern formation and adhesion of a drying drop of whole blood. *Colloids Surf. A Physicochem. Eng. Asp.* **2013**, *430*, 1–7. [CrossRef]
29. Jain, K.S.; Kathiravan, M.; Somani, R.S.; Shishoo, C.J. The biology and chemistry of hyperlipidemia. *Bioorg. Med. Chem.* **2007**, *15*, 4674–4699. [CrossRef]
30. Bevk, M.; Kononenko, I. A statistical approach to texture description of medical images: A preliminary study. In Proceedings of the Proceedings of 15th IEEE Symposium on Computer-Based Medical Systems (CBMS 2002), Maribor, Slovenia, 4–7 June 2002; pp. 239–244.

31. Jain, S. Brain cancer classification using GLCM based feature extraction in artificial neural network. *Int. J. Comput. Sci. Eng. Technol.* **2013**, *4*, 966–970.
32. Tahir, M.A.; Bouridane, A.; Kurugollu, F. An FPGA based coprocessor for GLCM and Haralick texture features and their application in prostate cancer classification. *Analog Integr. Circuits Signal Process.* **2005**, *43*, 205–215. [CrossRef]
33. Mostaço-Guidolin, L.B.; Ko, A.C.T.; Wang, F.; Xiang, B.; Hewko, M.; Tian, G.; Major, A.; Shiomi, M.; Sowa, M.G. Collagen morphology and texture analysis: From statistics to classification. *Sci. Rep.* **2013**, *3*, 2190. [CrossRef]
34. Lombardi, J.; Pellegrino, J.M.; Soazo, M.; Corrêa, A.P.F.; Brandelli, A.; Risso, P.; Boeris, V. Mineral fortification modifies physical and microstructural characteristics of milk gels coagulated by a bacterial enzymatic pool. *Colloids Surf. B Biointerfaces* **2018**, *161*, 296–301. [CrossRef] [PubMed]
35. Xie, C.; Shao, Y.; Li, X.; He, Y. Detection of early blight and late blight diseases on tomato leaves using hyperspectral imaging. *Sci. Rep.* **2015**, *5*, 16564. [CrossRef] [PubMed]
36. Fawcett, T. An introduction to ROC analysis. *Pattern Recognit. Lett.* **2006**, *27*, 861–874. [CrossRef]
37. Munir, K.; Elahi, H.; Ayub, A.; Frezza, F.; Rizzi, A. Cancer diagnosis using deep learning: A bibliographic review. *Cancers* **2019**, *11*, 1235. [CrossRef] [PubMed]
38. Carreón, Y.J.; Gómez-López, M.L.; Díaz-Hernández, O.; Vazquez-Vergara, P.; Moctezuma, R.E.; Saniger, J.M.; González-Gutiérrez, J. Patterns in dried droplets to detect unfolded BSA. *Sensors* **2022**, *22*, 1156. [CrossRef]
39. Carreón, Y.J.; Ríos-Ramírez, M.; Moctezuma, R.; González-Gutiérrez, J. Texture analysis of protein deposits produced by droplet evaporation. *Sci. Rep.* **2018**, *8*, 9580. [CrossRef]
40. Pal, A.; Gope, A.; Kafle, R.; Iannacchione, G.S. Phase separation of a nematic liquid crystal in the self-assembly of lysozyme in a drying aqueous solution drop. *MRS Commun.* **2019**, *9*, 150–158. [CrossRef]
41. Shultz, E.K. Multivariate receiver-operating characteristic curve analysis: Prostate cancer screening as an example. *Clin. Chem.* **1995**, *41*, 1248–1255. [CrossRef]
42. Tharwat, A. Classification assessment methods. *Appl. Comput. Inform.* **2020**, *17*, 168–192. [CrossRef]
43. Gorr, H.M.; Zueger, J.M.; Barnard, J.A. Lysozyme pattern formation in evaporating drops. *Langmuir* **2012**, *28*, 4039–4042. [CrossRef] [PubMed]
44. Chen, G.; J Mohamed, G. Complex protein patterns formation via salt-induced self-assembly and droplet evaporation. *Eur. Phys. J. E* **2010**, *33*, 19–26. [CrossRef] [PubMed]
45. Pal, A.; Gope, A.; Obayemi, J.D.; Iannacchione, G.S. Concentration-driven phase transition and self-assembly in drying droplets of diluting whole blood. *Sci. Rep.* **2020**, *10*, 18908. [CrossRef] [PubMed]
46. Ramsthaler, F.; Schlote, J.; Wagner, C.; Fiscina, J.; Kettner, M. The ring phenomenon of diluted blood droplets. *Int. J. Leg. Med.* **2016**, *130*, 731–736. [CrossRef]
47. Iqbal, R.; Shen, A.Q.; Sen, A. Understanding of the role of dilution on evaporative deposition patterns of blood droplets over hydrophilic and hydrophobic substrates. *J. Colloid Interface Sci.* **2020**, *579*, 541–550. [CrossRef]
48. Bialkower, M.; Manderson, C.A.; McLiesh, H.; Tabor, R.F.; Garnier, G. Paper diagnostic for direct measurement of fibrinogen concentration in whole blood. *ACS Sensors* **2020**, *5*, 3627–3638. [CrossRef]
49. Huang, J.; Ali, N.; Quansah, E.; Guo, S.; Noutsias, M.; Meyer-Zedler, T.; Bocklitz, T.; Popp, J.; Neugebauer, U.; Ramoji, A. Vibrational spectroscopic investigation of blood plasma and serum by drop coating deposition for clinical application. *Int. J. Mol. Sci.* **2021**, *22*, 2191. [CrossRef]
50. Esmonde-White, K.A.; Esmonde-White, F.W.; Morris, M.D.; Roessler, B.J. Characterization of biofluids prepared by sessile drop formation. *Analyst* **2014**, *139*, 2734–2741. [CrossRef]
51. Lovergne, L.; Clemens, G.; Untereiner, V.; Lukaszweski, R.A.; Sockalingum, G.D.; Baker, M.J. Investigating optimum sample preparation for infrared spectroscopic serum diagnostics. *Anal. Methods* **2015**, *7*, 7140–7149. [CrossRef]
52. Schaefer, E.J.; Tsunoda, F.; Diffenderfer, M.; Polisecki, E.; Thai, N.; Asztalos, B. *The Measurement of Lipids, Lipoproteins, Apolipoproteins, Fatty Acids, and Sterols, and Next Generation Sequencing for the Diagnosis and Treatment of Lipid Disorders*; MDText.com, Inc.: South Dartmouth, MA, USA, 2000. Available online: http://europepmc.org/books/NBK355892 (accessed on 5 July 2023).
53. Kurstjens, S.; Gemen, E.; Walk, S.; Njo, T.; Krabbe, J.; Gijzen, K.; Elisen, M.G.; Kusters, R. Performance of commercially-available cholesterol self-tests. *Ann. Clin. Biochem.* **2021**, *58*, 289–296. [CrossRef]
54. Bastianelli, K.; Ledin, S.; Chen, J. Comparing the accuracy of 2 point-of-care lipid testing devices. *J. Pharm. Pract.* **2017**, *30*, 490–497. [CrossRef]
55. Lee, W.C.; Ng, H.Y.; Hou, C.Y.; Lee, C.T.; Fu, L.M. Recent advances in lab-on-paper diagnostic devices using blood samples. *Lab Chip* **2021**, *21*, 1433–1453. [CrossRef]
56. Zong, L.; Han, Y.; Gao, L.; Du, C.; Zhang, X.; Li, L.; Huang, X.; Liu, J.; Yu, H.D.; Huang, W. A transparent paper-based platform for multiplexed bioassays by wavelength-dependent absorbance/transmittance. *Analyst* **2019**, *144*, 7157–7161. [CrossRef] [PubMed]
57. Li, C.G.; Joung, H.A.; Noh, H.; Song, M.B.; Kim, M.G.; Jung, H. One-touch-activated blood multidiagnostic system using a minimally invasive hollow microneedle integrated with a paper-based sensor. *Lab A Chip* **2015**, *15*, 3286–3292. [CrossRef] [PubMed]
58. Lee, Y.J.; Eom, K.S.; Shin, K.S.; Kang, J.Y.; Lee, S.H. Enzyme-loaded paper combined impedimetric sensor for the determination of the low-level of cholesterol in saliva. *Sens. Actuators B Chem.* **2018**, *271*, 73–81. [CrossRef]

59. Park, C.; Kim, H.R.; Kim, S.K.; Jeong, I.K.; Pyun, J.C.; Park, S. Three-dimensional paper-based microfluidic analytical devices integrated with a plasma separation membrane for the detection of biomarkers in whole blood. *ACS Appl. Mater. Interfaces* **2019**, *11*, 36428–36434. [CrossRef]
60. Wang, M.; Zhu, J.; Zi, Y.; Huang, W. 3D MXene sponge: Facile synthesis, excellent hydrophobicity, and high photothermal efficiency for waste oil collection and purification. *ACS Appl. Mater. Interfaces* **2021**, *13*, 47302–47312. [CrossRef]
61. Wang, M.; Zi, Y.; Zhu, J.; Huang, W.; Zhang, Z.; Zhang, H. Construction of super-hydrophobic PDMS@ MOF@ Cu mesh for reduced drag, anti-fouling and self-cleaning towards marine vehicle applications. *Chem. Eng. J.* **2021**, *417*, 129265. [CrossRef]

processes

MDPI

Article

# Impact of Pulsed Electric Field Treatment on the Process Kinetics and Selected Properties of Air and Dehumidified Air-Dried Mushrooms

**Magdalena Dadan, Alicja Barańska, Aleksandra Matys, Katarzyna Rybak, Dorota Witrowa-Rajchert, Artur Wiktor and Małgorzata Nowacka ***

Department of Food Engineering and Process Management, Institute of Food Sciences, Warsaw University of Life Sciences (WULS-SGGW), 02-787 Warsaw, Poland; magdalena_dadan@sggw.edu.pl (M.D.); alicja_baranska@sggw.edu.pl (A.B.); aleksandra_matys@sggw.edu.pl (A.M.); katarzyna_rybak@sggw.edu.pl (K.R.); dorota_witrowa_rajchert@sggw.edu.pl (D.W.-R.); artur_wiktor@sggw.edu.pl (A.W.)

* Correspondence: malgorzata_nowacka@sggw.edu.pl

**Citation:** Dadan, M.; Barańska, A.; Matys, A.; Rybak, K.; Witrowa-Rajchert, D.; Wiktor, A.; Nowacka, M. Impact of Pulsed Electric Field Treatment on the Process Kinetics and Selected Properties of Air and Dehumidified Air-Dried Mushrooms. *Processes* **2023**, *11*, 2101. https://doi.org/10.3390/pr11072101

Academic Editor: Jan Havlík

Received: 13 June 2023
Revised: 7 July 2023
Accepted: 12 July 2023
Published: 14 July 2023



**Abstract:** The study examined the effects of pulsed electric field treatment on the kinetics and properties of convective-dried mushrooms using different drying agents. Increasing the drying air temperature reduced drying time, while the use of dehumidified air resulted in faster water removal. PEF treatment, depending on the parameters, shortened the drying time maximum by 12% or extended the drying time. The physical (dry matter content, rehydration properties, hygroscopic properties, and color) and chemical (polyphenols content and anti-oxidant activity) properties were analyzed. The dry matter contents of the mushrooms were influenced by the drying temperature, while PEF pre-treatment did not influence the rehydration and hygroscopic properties in both cases of drying using air humidity. However, the color parameters were affected by the drying method and energy input, with higher energy input leading to decreased lightness, increased redness, and color saturation. The chemical analyses revealed that the anti-oxidant compounds in the dried mushrooms were influenced by various factors, with PEF treatment and drying non-dehumidified air polyphenol content increasing, whereas dehumidified air caused more phenolic degradation if it was combined with PEF treatment. Anti-oxidant activity varied depending on the drying agent, with non-dehumidified air generally exhibiting better properties. The highest total polyphenol content and best anti-oxidant properties were obtained for the PEF pre-treated with 3 kJ/kg of energy and dried with non-dehumidified air at a temperature of 70 °C.

**Keywords:** mushrooms; pulsed electric field; convective drying; dehumidified air; bioactive compounds; rehydration rate

## 1. Introduction

Non-thermal technologies, such as a pulsed electric field (PEF), are increasingly becoming an alternative to conventional food processing methods. The use of PEF technology is economical, environmentally friendly, and allows for the inactivation of micro-organisms and enzymes. Furthermore, the use of a pulsed electric field in food processing improves mass transfer, minimizes the loss of nutrients during processing, and can also extend shelf life [1]. Moreover, when using PEF, it is possible to modify the texture and, thus, design the properties of the material [2].

The mechanism of action of PEFs is based on subjecting products to short-term electric impulses with a high-strength electric field, usually from 10 kV/cm [3]. The mutual interaction of electric charges causes local structural changes in the material, breaking or damaging the continuity of the cell membrane of the raw material. The main mechanism of PEF action is called electroporation or permeabilization and is based on the electrically induced formation of pores in a membrane, causing its increased permeability [4,5]. The impact of a PEF is dependent on the parameters, such as electric field strength and the shape and number of pulses, as well as their width and frequency [5,6]. Electroporation can be reversible or irreversible, depending, e.g., on the size of cells [6]. The irreversible electroporation causes permanent and irreversible disruption to the cell membrane and, therefore, has a positive effect on mass exchange or the inactivation of micro-organisms and can improve the drying process by accelerating the removal of water from the raw material [7].

Dellarosa et al. [8] analyzed the impact of a PEF on water distribution and loss in mushroom stalks. They reported that the PEF significantly disintegrated the tissue causing the redistribution of water from the intracellular to the extracellular compartments, which was comparable to the case of thermal treatment. Furthermore, the morphology and molecular weight of the polysaccharides in the cell wall of the mushrooms were modified due to PEF treatment. Many scientific papers have proven the acceleration of the drying process via pulsed electric field treatment. Depending on the matrix, a shortened drying time in the range of 8 to 30% was generally achieved [9]. In the work of Mirzaei-Baktash et al. [10], the drying time of mushroom slices was reduced by 7–25% for hot air drying and by 16–28% for the electrohydrodynamic method when a PEF was applied before the drying. However, the effect of a PEF is dependent not only on the PEF parameters but also on the drying method and drying conditions, and thus, both factors should be considered to guarantee a shorter drying time while also retaining the bioactive component contents.

Food drying is one of the oldest methods of food preservation, which plays a very important role in food processing. Due to the reduced water content and, thus, reduced water activity in a product, the growth of micro-organisms and the course of biochemical reactions are inhibited [11–13]. In the food industry, convective drying (air drying) is the most commonly used method on an industrial scale, which, according to numerous studies, is considered one of the most unfavorable methods of heat treatment. The material dried by convection is characterized by large physicochemical changes that affect the nature of chemical compounds and the structures of cells. Moreover, the energy consumption of this process is relatively high due to, e.g., the low thermal efficiency of dryers [13,14]. Nevertheless, this technique has many advantages, such as simplicity, low cost (of the dryers), and ease of process control [14], and is also frequently used in mushroom drying [15]. As Marçal et al. [15] summarized, mushrooms subjected to high-temperature drying are characterized by a change in their phenolic and organic acid profile and a decrease in polysaccharide content due to their conversion into oligosaccharides and Maillard reactions.

Conventionally, during convective drying, the material is dried by means of air of ambient humidity, which is heated to the set temperature. However, the problem is the lack of repeatability. Moreover, the course of the drying process is affected by the water content in the drying air. As a result of the reduced water content (dehumidified air), the potential of the heating medium to take moisture from the dried material increases, and due to the benefits that the use of dehumidified air can bring, it is, nowadays, more and more frequently used in experimental works [12].

Until now, the use of dehumidified air to support convective drying has not been studied in detail. The work of Matys et al. [12] showed some of the benefits of using dehumidified air, such as a reduction in drying time, significantly higher total phenolic content, and significantly better anti-oxidant activity. However, the authors noted this effect at a lower air temperature (55 °C and, in some cases, also 70 °C), and at 85 °C, the opposite effect was observed. Some properties and the shortening of the drying time were more pronounced when ultrasound treatment was carried out before drying with dehumidified air. Therefore, more studies need to be conducted concerning the combination of drying with dehumidified air and other pre-treatments, e.g., pulsed electric fields.

To the best of our knowledge, the combined influence of PEF treatment and drying with dehumidified air has not been investigated yet. Therefore, the aim of this study was to assess the possibility of using a PEF prior to convective drying with or without dehumidified air based on the drying kinetics and selected properties of mushrooms. The physicochemical properties (such as hygroscopic properties, rehydration properties, color, and dry matter content), as well as the chemical properties, were measured to analyze the influence of both the PEF and dehumidified air on the different properties of a material. Many publications have confirmed that mushrooms exhibit anti-oxidant potential due to the phenolic compound content, and this affects the nutritional value of dried mushrooms [16]. Furthermore, color is a crucial factor that has an impact on the acceptance of a product by consumers. A good rehydration ratio is important to obtain a soft, well-rewatered dried sample. Additionally, in order to obtain a stable product during storage, hygroscopic properties should not increase. What is important regarding PEF treatment—due to the electroporation phenomenon—is the leakage of the content of cells can increase the browning reactions, causing undesirable changes in color [17], which can significantly impact the acceptance of the product by consumers. Therefore, the maintenance or increase (in the case of bioactive components or rehydration properties) in these parameters is important.

## 2. Materials and Methods

### 2.1. Material

The white mushrooms (champignons) were purchased in a unit package of 500 g at a market in Warsaw. The purified raw material, without stems, was kept at room temperature and was used for further research.

### 2.2. Pulsed Electric Field (PEF) Pre-Treatment

Before drying, the raw material was pre-treated with a pulsed electric field using the PEF Pilot reactor (ELEA GmBH, Quakenbrück, Germany). The pre-treatment parameters were selected on the basis of the preliminary studies, during which the changes in the conductivity of the material CDI (cell disintegration index) were analyzed after PEF treatment [18], as is presented in Figure 1.

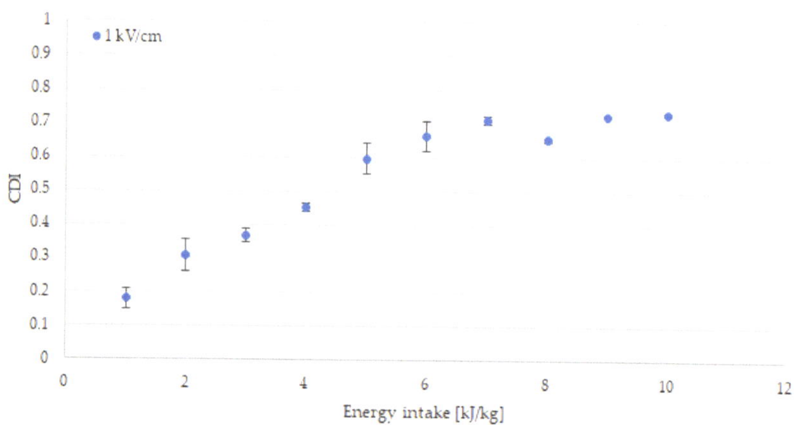


**Figure 1.** Cell disintegration index (CDI) after PEF treatment of mushrooms.

A range of specific energy input between 1 and 5 kJ/kg was selected for optimization, as it causes an increase in the CDI value but is below the "saturation" level; after exceeding this, the increase in energy intake does not significantly increase the CDI. The parameters of the pulsed electric field were set at the following values: the electrode voltage equaled 24 kV; the electric field strength stood at 1 kV/cm; the pulse frequency was at 20 Hz; the pulse width took the value of 7 μs. The number of rectangular pulses varied concerning the amount of supplied energy. According to the preliminary studies, the following energy values were set: 1, 3, and 5 kJ/kg due to the electroporation phenomenon. In order to set the appropriate PEF value, the mushroom caps were placed in an electrical treatment cell. The structure of the electrical treatment cell included a set of two parallel stainless-steel electrodes, which were distanced from each other by 24 cm. Then, the mushrooms were flooded with tap water (21 ± 1 °C, which served the function of a conductive medium) to a weight of 1 kg. The system prepared in this way was placed inside the reactor. After the treatment, the excess water from the mushrooms was removed using tissue paper, and the mushroom cups were cut into 5 mm-thick slices.

### 2.3. *Convective Drying*

The samples were dried in a laboratory convective dryer, which was integrated into an air dehumidification system, as was presented in the work by Matys et al. [12]. The air dehumidifier was composed of a cooling unit (MTA, TAEevo TECH020, Tribano, Italy) and a condensation-adsorption unit (ML270, Munters, Kista, Sweden). Air humidity after dehumidification was 1.5 $g/m^3$. The drying air temperature and velocity were set to 55, 70, and 85 °C and 2 m/s, respectively. The air flowed parallel to the layer of the mushrooms, which were laid on the sieve. The load on the sieve was 0.96 $kg/m^2$. The change in the mass (±0.1 g) of the sample was measured and recorded every minute. The process lasted until the samples reached a constant mass. The drying was performed in duplicate.

The drying curves were plotted as a relationship between the relative (dimensionless) water content and the time (MR = f(τ)), based on the recorded changes in the mass of the samples during drying:

$$MR = \frac{M_\tau}{M_0}, \quad (1)$$

where $M_\tau$ and $M_0$ correspond to the water content of the sample during drying [kg $H_2O$/kg d.m.] and the initial water content [kg $H_2O$/kg d.m.], respectively.

The drying time was defined by obtaining MR = 0.02 [19]. The change in drying time was calculated as a percentage change in the time of convective drying (CD) and convective drying with dehumidified air (DA).

### 2.4. *Dry Matter Content*

The dry matter content was determined via the gravimetric method at 70 °C for 24 h, according to the method of AOAC [20]. The measurement was performed for two repetitions.

### 2.5. *Hygroscopic Properties*

In order to determine the hygroscopic properties of the dried mushroom slices, a water vapor adsorption test was carried out. For this purpose, the dried mushrooms were weighed on an analytical scale and were then placed in a desiccator with a sodium chloride solution, giving a water activity of 0.75 [21]. The mass of the samples was remeasured after 1, 24, 48, and 72 h. The hygroscopic properties were determined as the mass of the sample over adsorption time in relation to the initial mass of the dried material. The measurement was performed for three replicates.

### 2.6. *Rehydration Properties*

Half a slice of the dried mushroom was placed in a beaker filled with 100 mL of distilled water. The beaker and all of its content were kept at room temperature (approx. 20 °C) for 1 h. After this time, the water was filtered through a sieve, and the rehydrated mushroom was slightly blotted and weighed. Then, the dry matter content was determined according to the methodology described in Section 2.4. The analysis was performed for three replicates.

The rehydration properties were determined on the basis of relative mass gain (Δm) and relative dry matter content (SSL) according to the following formulas:

$$\Delta m = \frac{m_\tau}{m_0}, \quad (2)$$

$$SSL = \frac{m_T dm_T}{m_0 dm_0}, \quad (3)$$

where $m_T$ is the mass of the rehydrated mushroom [g], $m_0$ stands for the mass of the sample before rehydration [g], $d_{mT}$ is the dry matter content in the rehydrated mushroom [%], and $d_{m0}$ is the dry matter content in the sample before rehydration [%].

### 2.7. Color

The fresh and dried samples were analyzed with a colorimeter (CR-5, Konica-Minolta, Tokyo, Japan) in order to determine their color in CIE L*a*b*system (light source: D65; standard observer: 2°; diameter: 8 mm). Each sample was analyzed over 10 replications. The total color difference (ΔE) and chroma (C*) were calculated based on the L*a*b* color parameters [13]:

$$\Delta E = \sqrt{(\Delta L^*)^2 + (\Delta a^*)^2 + (\Delta b^*)^2}, \quad (4)$$

$$C^* = \sqrt{(a^*) + (b^*)}, \quad (5)$$

where L* is the lightness, a* is redness/greenness, b* is yellowness/blueness of the samples, and ΔL*, Δa*, Δb* are the differences in the color parameters between the fresh and dried mushrooms.

### 2.8. Total Phenolic Content (TPC)

The determination of total phenolic content was carried out using the Folin-Ciocalteu method [18]—gallic acid was served as the standard. Grounded dried mushrooms were extracted with 80% ethanol solution in order to obtain polyphenolic compounds. In order to determine the TPC, 4.92 mL of distilled water, 0.18 mL of extract, and 0.3 mL of Folin-Ciocalteu's reagent were added to the test tubes. After 3 min, 0.6 mL of a supersaturated sodium carbonate solution was dispensed. The samples were mixed and then stored in the dark for 1 h. Finally, the absorbance at 750 nm was measured on a UV-VIS spectrophotometer (Heλios γ, Thermo Scientific) against a blank, i.e., extract-free. The results are expressed as milligrams of gallic acid per 1 g of dry matter. The analyses were performed in duplicate.

### 2.9. Anti-Oxidant Activity (DPPH and ABTS Assays)

Anti-oxidant activity was assessed on the basis of the degree of scavenging of the synthetic $DPPH^\bullet$ radicals and the $ABTS^{\bullet+}$ radical cations by the anti-oxidants extracted from the samples. For the procedure, the same ethanol extracts were used for the determination of total phenolic content. In order to perform the chemical analyses, free radical solutions and solutions for the measurements were prepared according to the methodology described in [11].

Ethanol extracts were placed in glass test tubes in the amount of 0.1, 0.2, 0.3, and 0.5 mL. The tubes were then filled up to a total volume of 2 mL with 80% ethanol solution. Then, 2 mL of DPPH solution was added and mixed thoroughly. After 30 min (room temperature, dark place), the absorbance was measured at a wavelength of 515 nm (Heλios γ spectrophotometer, Thermo Scientific) against an 80% ethanol solution. The analysis was performed twice.

The test tubes were dosed with ethanol extracts in the following volumes: 0.025, 0.05, 0.075, and 0.1 mL, and then 3 mL of ABTS solution was added thereafter. Everything was mixed and left for 6 min in a dark place. Subsequently, the absorbance was measured at a wavelength of 734 nm (Heλios γ spectrophotometer, Thermo Scientific) against an 80% ethanol solution. The analysis was conducted over two replications.

Anti-oxidant activity was expressed as the $EC_{50}$ for the DPPH and the ABTS radical methods. This indicator determines the dry matter content in the extract (mg d.m./mL) necessary to scavenge 50% of the initial amount of free radicals.

### 2.10. Statistical Analysis

The statistical analysis was performed using Statistica 13.3 software (TIBCO Inc., Palo Alto, CA, USA). The one-way analysis of variance (ANOVA) with Tukey's tests at a significance level of $\alpha = 0.05$ were used for this purpose. For the selection of the optimal temperature and PEF energy for non-dehumidified and humidified air, the DOE method was used (Statistica 13.3, TIBCO Inc., USA).

## 3. Results and Discussion

### 3.1. Drying Kinetics

Figures 2 and 3 show the kinetics of drying for the untreated and PEF-pre-treated mushrooms, from which water was removed with non-dehumidified and dehumidified air, respectively. As can be seen, the drying efficiency decreased over time. This phenomenon was related to the progressing difficulties in the water removal procedure [19]. By increasing the temperature of the drying air and therefore intensifying the water evaporation process [20,21], the drying time of the mushrooms was reduced (Table 1). For example, the drying time of the 85CD sample was 26% shorter than that of the 70CD sample. On the other hand, drying the 70CD sample to MR = 0.02 took 43% less time than the drying process of the 55CD sample (to the same relative water content). When comparing the two types of drying agents to each other—non-dehumidified and dehumidified air—it can be concluded that the use of air with reduced humidity ensured faster water removal than the non-dehumidified air. This phenomenon can be explained by the greater potential of the dehumidified air to absorb water from the given material [11]. So far, the effect of dehumidified air on convective drying has been investigated on fruit such as apples [11,22], kiwi [23], and quince [24]. The results appear to be consistent: reducing the humidity of the drying air reduces the drying time of a given product, which results from increasing the gradient of water vapor pressure between the drying air and the surface of the dried material. Moreover, a higher percentage reduction in drying time is achieved at lower air temperatures. For example, the drying time of the 55DA sample was 38% shorter than that of the 55CD sample. However, after increasing the temperature to 70 and 85 °C, the reduction was 26 and 10%, respectively. A pulsed electric field (PEF) is often used as a pre-treatment before drying. It is applied to, for example, maximize the efficiency of removing water from a given material. The application of a PEF did not reveal any distinct tendencies in the duration of the subsequent drying of the mushrooms. The highest reduction in drying time (14% in relation to the material untreated with a PEF and dried in the same conditions) was noted after providing energy at the amount of 3 kJ/kg to the mushrooms and then drying them with non-dehumidified air at 55 °C. It is consistent with the values obtained in the literature, in which the drying time reduction was in the range of 8–30% [9]. On the other hand, the drying time of the PEF3_55DA sample was 37% longer than that of the 55DA sample. By increasing the permeability of the cell membrane due to electroporation, the pulsed electric field led to a reduction in the convective drying time of parsnips, carrots [25], potatoes [26], onions [27,28], and red peppers [29]. The effectiveness of a PEF depends not only on the physical properties of a given material [30] but also on the applied parameters and method of drying. Therefore, both the PEF and the drying parameters should be adjusted to obtain a positive effect. Further research is necessary in this area.

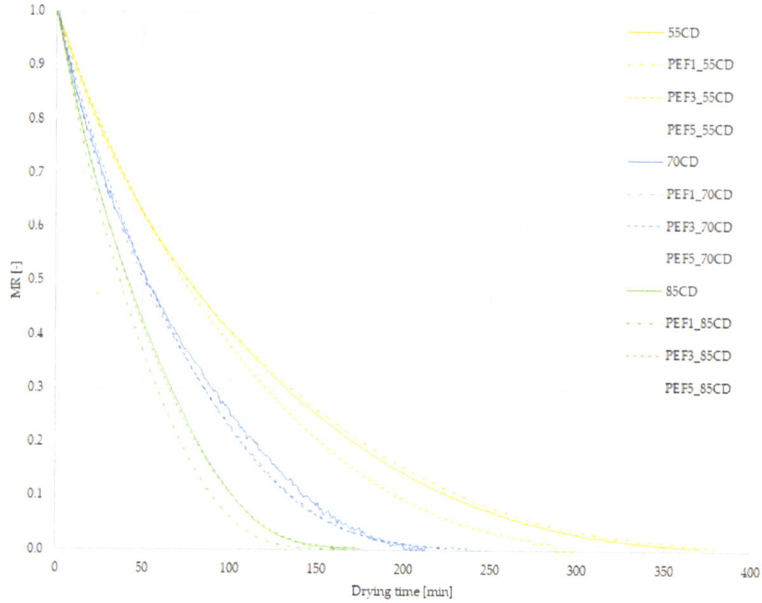


**Figure 2.** Drying kinetics of mushrooms obtained using the convective method with non-dehumidified air subjected to (or not) PEF.

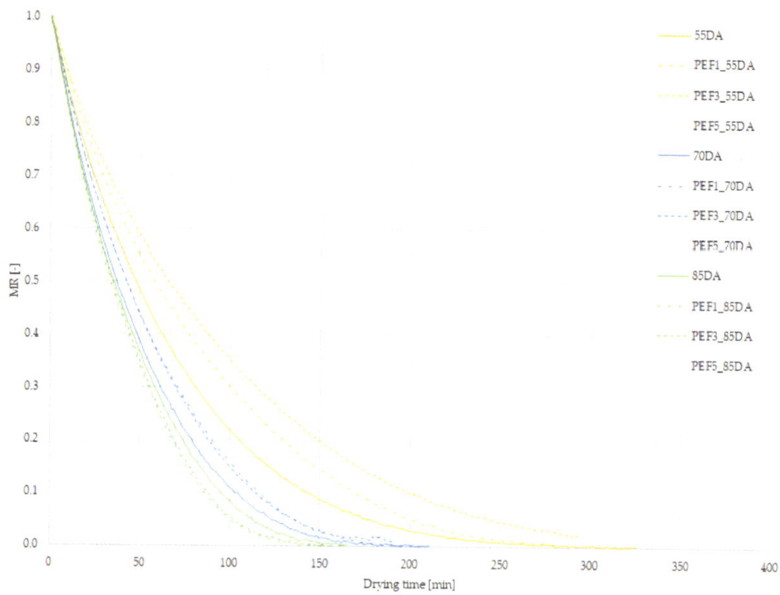


**Figure 3.** Drying kinetics of mushrooms obtained using the convective method with dehumidified air subjected to (or not) PEF.

**Table 1.** Drying time to MR = 0.02 of the convective-dried (CD) and dehumidified convective-dried (DA) mushrooms, including PEF pre-treatment and drying temperature. The drying time change was calculated as a percentage time change between CD and DA.

| Temperature [°C] | PEF Input [kJ/kg] | Drying Time [min] | | Drying Time Change [%] |
|---|---|---|---|---|
| | | CD | DA | |
| 55 | 0 | 336 | 209 | 38 |
| | 1 | 351 | 227 | 35 |
| | 3 | 288 | 287 | 3 |
| | 5 | 301 | 190 | 37 |
| 70 | 0 | 191 | 142 | 26 |
| | 1 | 194 | 152 | 22 |
| | 3 | 207 | 152 | 27 |
| | 5 | 212 | 173 | 18 |
| 85 | 0 | 140 | 126 | 10 |
| | 1 | 126 | 112 | 11 |
| | 3 | 139 | 115 | 17 |
| | 5 | 123 | 108 | 12 |

### 3.2. *Dry Matter Content*

The dry matter content of dried mushrooms with non-dehumidified air that were subjected to (or not) PEF treatment varied from 89.1 to 96.8% (Figure 4). It can be observed that both the temperature of drying and PEF pre-treatment significantly influenced the parameter. When comparing the non-treated samples to the PEF-treated samples at the same temperature of drying, it was observed that PEF affected the dry matter content, but the significance of this effect was dependent on the air temperature. Increasing the drying temperature improved the water evaporation rate, which resulted in higher dry matter content. However, regarding the lowest drying temperature (55 °C), PEF significantly decreased the dry matter content, irrespective of the energy input setting. In the case of the middle set of temperatures, the pre-treated samples were characterized by significantly lower dry matter content when compared to the non-treated samples, but only when 3 and 5 kJ/kg was applied. For the highest drying temperature, the application of the PEF to the sample did not significantly influence the dry matter content. Interestingly, there were no significant differences in dry matter content between the different energy input levels of the PEF-treated samples (1–5 kJ/kg), irrespective of the temperature. When taken together, these results suggest that drying temperature had a greater effect on this parameter than the use of the PEF.

The results of the dry matter content of the mushrooms dried with the application of the dehumidified air with or without PEF pre-treatment are presented in Figure 5, and they ranged from 83.9 to 97.6%. Similar to the variants dried with non-dehumidified air, the drying temperature had a significant effect on the dry matter content of the obtained mushrooms. However, it should be underlined that the temperature had the most significant influence on this parameter when 3 kJ/kg of energy was applied. When comparing the PEF-treated and the intact dried samples, it can be seen that the significant difference was noted only with the use of 3 kJ/kg when drying at 55 and 70 °C. With regard to the different energy inputs of the PEF pre-treatments, the only significant differences were observed for the mushrooms dried at the lowest drying temperature of 55 °C—the middle value of energy caused a significant decrease in dry matter content. For the other energy values—the results were statistically the same as for the non-treated sample.

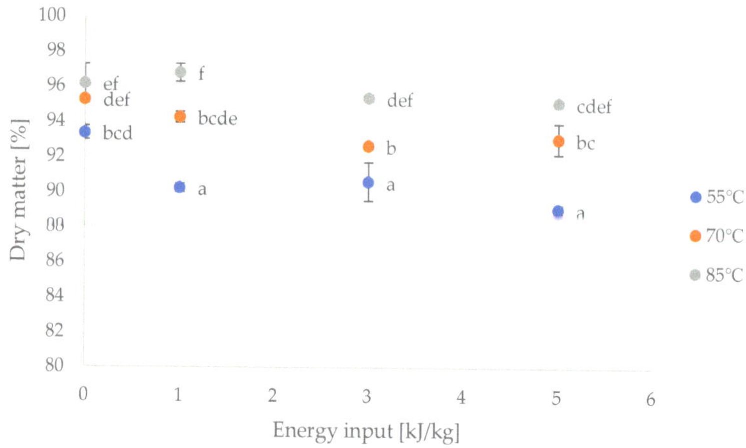


**Figure 4.** Dry matter [%] of mushrooms obtained using the convective method with non-dehumidified air subjected to (or not) PEF. The same letters (a–f) represent the homogeneous groups (α = 0.05).

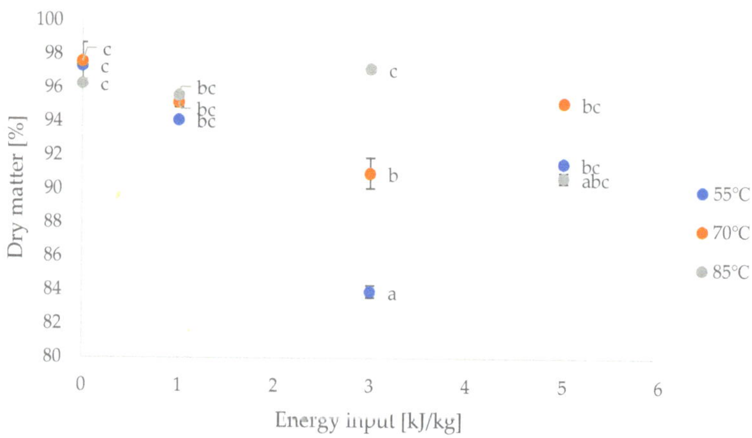


**Figure 5.** Dry matter [%] of mushrooms obtained using the convective method with dehumidified air subjected to (or not) PEF. The same letters (a–c) represent the homogeneous groups (α = 0.05).

The results of the dry matter content of the dried white button mushrooms presented in this study follow the data presented in the literature [31].

### 3.3. Hygroscopic Properties

The gain in mass of the dried white button mushrooms with non-dehumidified air after 72 h of moisture adsorption was significantly affected by drying temperature, and it ranged from 1.05 to 1.19 (Figure 6). Generally, with increasing drying temperature, the gain in mass of the samples increased. Moreover, the PEF pre-treatment did not have a significant effect on the mushrooms' hygroscopic properties, with the exception of the sample dried at the highest drying temperature and pre-treated with the highest energy input (85 °C, 5 kJ/kg), as well as with 3 and 5 kJ/kg and drying at 55 °C. In that case, the PEF-treated material was characterized by lower water vapor adsorption. The obtained results concur well with the findings of Wiktor et al. [32], who reported no difference in the hygroscopic properties of dried apples subjected to PEF pre-treatment. However, the results presented in this research are in contrast to the research of Rybak et al. [33], who freeze-dried PEF-treated red bell peppers and observed increased hygroscopic properties regarding the material after pre-treatment. Lammerskitten et al. [34] reported, on the other hand, that PEF pre-treatment decreased water vapor absorption in freeze-dried apples, as the sugar distribution profile of the samples changed as a result of electroporation.

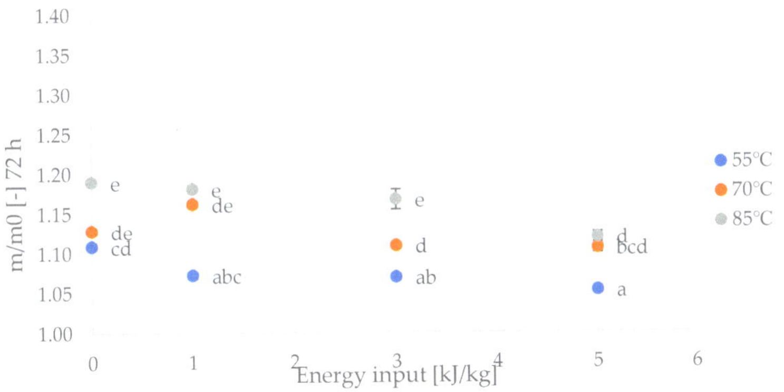


**Figure 6.** Hygroscopic properties ($m/m_0$) of mushrooms obtained using the convective method with non-dehumidified air subjected to (or not) PEF. The same letters (a–e) represent the homogeneous groups ($\alpha = 0.05$).

The gain in mass of the mushrooms after 72 h of treatment, which was obtained under low drying air humidity, significantly differed when dried at different drying temperatures, which is similar to the samples dried with non-dehumidified air (Figure 7). However, it should be underlined that the energy input of 3 kJ/kg had the greatest effect on the hygroscopic properties of the mushrooms dried at the lowest temperature, as the gain in mass of the sample was the lowest. Only in the case of this sample was the impact of the PEF significant. It can be concluded that, in general, the PEF energy value (1–5 kJ/kg) did not influence the hygroscopic properties in both cases of drying air humidity, and in most cases, the use of the PEF (when compared to the untreated mushrooms) did not change the hygroscopic properties. In fact, the unchanged or decrease in mass gain after water vapor adsorption is desirable, as the product can be stable during storage (does not adsorb as much moisture from the environment).

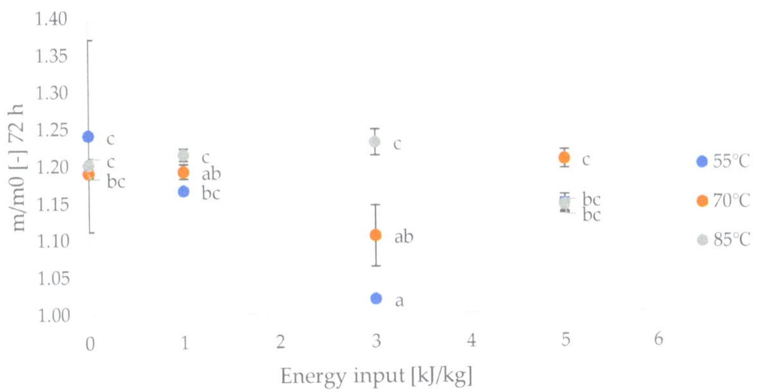


**Figure 7.** Hygroscopic properties ($m/m_0$) of mushrooms obtained using the convective method with dehumidified air subjected to (or not) PEF. The same letters (a–c) represent the homogeneous groups ($\alpha = 0.05$).

### 3.4. Rehydration Properties

The determination of the rehydration properties was performed in order to establish the ability of the drying sample to be re-imbibed with water [35,36]. On this basis, it can be determined, e.g., via the magnitude of the chemical and structural changes caused by the processing of the material [37,38]. Figure 8 shows the relative gain in mass (Δm) of the untreated and PEF-treated mushrooms dried with non-dehumidified air. As one can observe, all the analyzed samples did not differ statistically. On the other hand, Figure 9 shows the values of Δm of the untreated and PEF-pre-treated mushrooms dried with dehumidified air. A significant difference was noted only between the PEF3_55DA and PEF5_55DA samples. The PEF5_55DA sample exhibited a higher Δm. The different temperatures of the drying air and the variable amount of energy supplied to the mushrooms during PEF pre-treatment (except for the above-mentioned exception) did not lead to a clear, straightforward tendency in the relative mass-gain values.

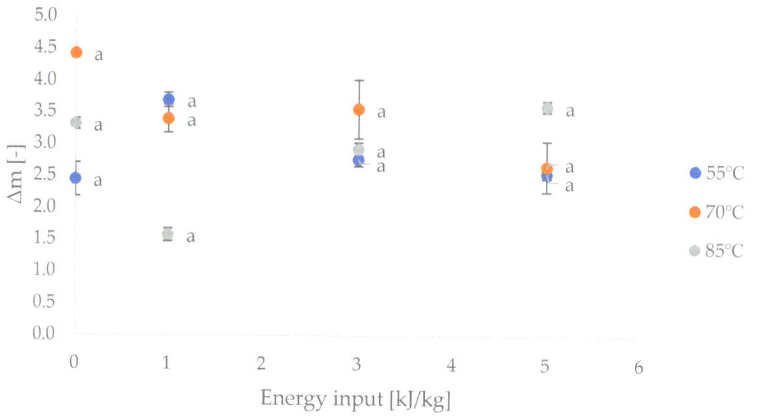


**Figure 8.** Relative gain in mass (Δm) of mushrooms obtained using the convective method with non-dehumidified air subjected to (or not) PEF. The same letter (a) represents the homogeneous groups (α = 0.05).

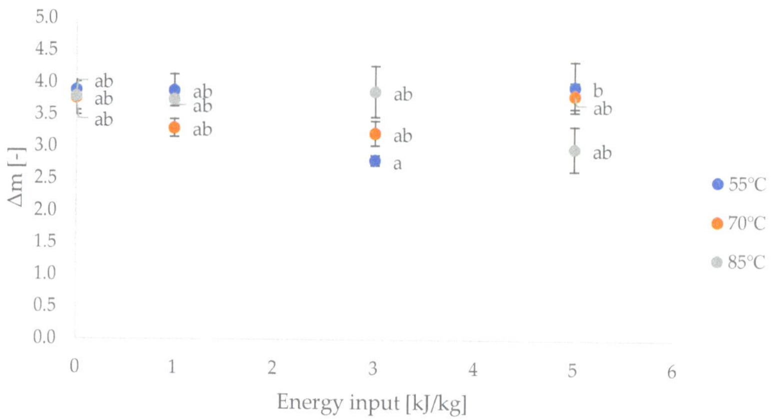


**Figure 9.** Relative gain in mass (Δm) of mushrooms obtained using the convective method with dehumidified air subjected to (or not) PEF. The same letters (a,b) represent the homogeneous groups (α = 0.05).

During rehydration, in addition to the water absorption process, the material tends to swell. The leaching of soluble solids into surrounding water takes place as well [35,39]. Figures 10 and 11 show the relative dry matter content after rehydration (to a dried mushroom before rehydration (SSL)) of the untreated and PEF-treated mushrooms dried by non-dehumidified and dehumidified air, respectively. The different temperatures of the drying air and the variable amount of energy supplied to the mushrooms during the PEF pre-treatment did not lead to a clear, straightforward tendency in the relative dry matter content. In most of the analyzed cases, the usage of dehumidified air resulted in obtaining dried mushrooms with higher SSL values. It means that the materials dried with dehumidified air lost less dry matter than those obtained with non-dehumidified air. Samples 55DA and PEF1_70DA were an exception to this; their non-dehumidified air-dried counterparts (55CD and PEF1_70CD) exhibited higher relative dry matter content.

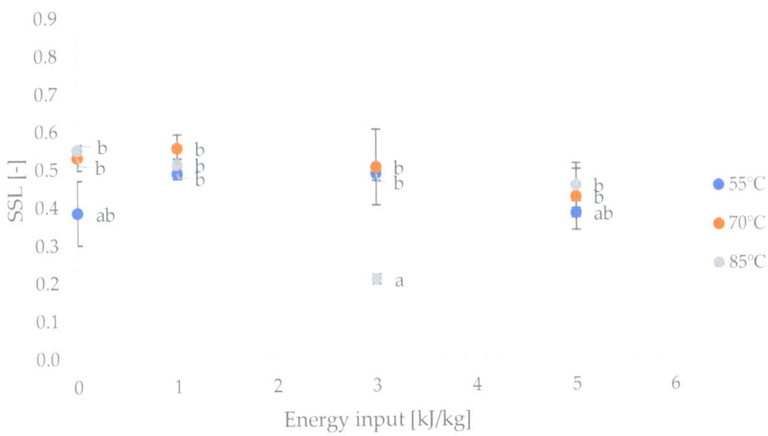


**Figure 10.** Relative dry matter content (SSL) of mushrooms obtained using the convective method with non-dehumidified air subjected to (or not) PEF. The same letters (a,b) represent the homogeneous groups (α = 0.05).

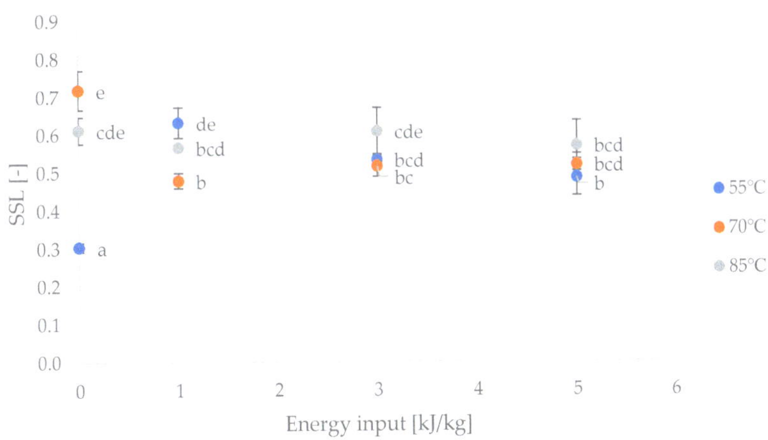


**Figure 11.** Relative dry matter content (SSL) of mushrooms obtained using the convective method with dehumidified air subjected to (or not) PEF. The same letters (a–e) represent the homogeneous groups (α = 0.05).

### 3.5. Color

Table 2 presents the color parameters of the analyzed mushrooms. The L* color parameter describes the lightness of the samples and ranged from 46.2 ± 3.1 to 79.9 ± 3.2 for those mushrooms dried using the convective method with non-humidified air. The significant effect of energy input was noted. The input of 1 kJ/kg did not affect any of the samples in comparison to the control. These results were the highest, indicating the lightest of all the samples. However, the increasing value of energy decreased the lightness of the samples dried at each of the drying temperatures. This is in agreement with Wiktor et al. [40], who observed the same relationship for dried carrots pre-treated with a PEF. Alam et al. [25] noted, as well, the darkening of the PEF pre-treated dried carrots and parsnips [25]. Electroporation, which led to leakages of the cellular content, such as enzymes, could well be responsible for this phenomenon. In contradiction to these observations, Won et al. [29], who applied PEF before drying red peppers, reported that the energy input did improve the color parameters of the samples. In general, the temperature of drying with non-dehumidified air did not affect the lightness; the only exception was the PEF_3_55C sample, which was characterized by significantly lower lightness than that of analogous samples dried at different temperatures. With regard to the samples dried with dehumidified air, the L* parameter varied from 58.7 ± 8.3 to 87.6 ± 3.2. In general, for this drying method, the effect of energy input was insignificant, as the variants pre-treated with the PEF did not differ (except for PEF1_55DA sample). However, the mushrooms subjected to the PEF were darker in comparison to the control samples, which, as aforementioned, was a result of the electroporation.

**Table 2.** The color parameters of the mushrooms obtained using the convective method with non-dehumidified air or dehumidified air subjected to (or not) PEF. The same letters ($^{a–d}$ and $^{A–D}$) in the columns represent the homogeneous groups (α = 0.05).

| Material | L* | a* | b* | C* | ΔE |
|---|---|---|---|---|---|
| 55CD | 79.9 ± 3.2 $^{d}$ | −0.6 ± 0.5 $^{a}$ | 17.1 ± 0.9 $^{a}$ | 17.1 ± 0.9 $^{a}$ | 10.3 ± 2.8 $^{a}$ |
| PEF1_55CD | 77.8 ± 2.4 $^{d}$ | +2.0 ± 0.6 $^{b}$ | 24.9 ± 1.4 $^{c}$ | 25.0 ± 1.4 $^{bc}$ | 16.4 ± 2.4 $^{a}$ |
| PEF3_55CD | 50.5 ± 3.5 $^{a}$ | +6.4 ± 1.5 $^{de}$ | 21.3 ± 1.6 $^{bc}$ | 22.3 ± 1.8 $^{bc}$ | 40.5 ± 3.5 $^{d}$ |
| PEF5_55CD | 44.9 ± 8.1 $^{a}$ | +7.0 ± 0.6 $^{de}$ | 21.6 ± 2.2 $^{bc}$ | 22.7 ± 2.2 $^{bc}$ | 46.2 ± 7.4 $^{d}$ |
| 70CD | 75.1 ± 8.2 $^{cd}$ | −0.8 ± 0.5 $^{a}$ | 12.3 ± 1.5 $^{a}$ | 12.3 ± 1.5 $^{a}$ | 14.5 ± 8.3 $^{ab}$ |
| PEF1_70CD | 78.7 ± 6.5 $^{d}$ | +0.1 ± 0.4 $^{ab}$ | 15.5 ± 0.7 $^{a}$ | 15.5 ± 0.7 $^{a}$ | 11.1 ± 6.4 $^{a}$ |
| PEF3_70CD | 61.9 ± 6.5 $^{bc}$ | +4.2 ± 1.8 $^{b}$ | 21.0 ± 2.9 $^{b}$ | 21.5 ± 3.0 $^{b}$ | 29.1 ± 6.9 $^{c}$ |
| PEF5_70CD | 46.2 ± 3.1 $^{a}$ | +8.0 ± 1.0 $^{f}$ | 22.1 ± 2.2 $^{bc}$ | 23.6 ± 2.2 $^{bc}$ | 45.1 ± 3.0 $^{d}$ |
| 85CD | 77.6 ± 7.1 $^{d}$ | −0.7 ± 0.3 $^{a}$ | 13.9 ± 0.7 $^{a}$ | 13.9 ± 0.7 $^{a}$ | 11.9 ± 7.1 $^{a}$ |
| PEF1_85CD | 70.8 ± 6.6 $^{cd}$ | +1.3 ± 0.5 $^{ab}$ | 16.8 ± 1.9 $^{a}$ | 16.9 ± 2.0 $^{a}$ | 19.2 ± 6.7 $^{ab}$ |
| PEF3_85CD | 64.3 ± 6.1 $^{c}$ | +4.9 ± 1.7 $^{cd}$ | 23.0 ± 2.5 $^{bc}$ | 23.5 ± 2.5 $^{bc}$ | 27.6 ± 5.9 $^{bc}$ |
| PEF5_85CD | 52.8 ± 3.0 $^{ab}$ | +7.7 ± 1.0 $^{f}$ | 24.1 ± 2.9 $^{bc}$ | 25.3 ± 3.0 $^{c}$ | 39.2 ± 3.1 $^{d}$ |
| 55DA | 87.1 ± 1.5 $^{B}$ | +0.2 ± 0.4 $^{A}$ | 14.3 ± 0.5 $^{A}$ | 14.3 ± 0.5 $^{A}$ | 2.9 ± 1.4 $^{A}$ |
| PEF1_55DA | 81.9 ± 3.4 $^{B}$ | +0.6 ± 0.4 $^{A}$ | 17.0 ± 1.1 $^{AB}$ | 17.0 ± 1.1 $^{ABC}$ | 8.5 ± 3.5 $^{AB}$ |
| PEF3_55DA | 61.6 ± 4.5 $^{A}$ | +4.5 ± 1.0 $^{BC}$ | 21.3 ± 3.2 $^{B}$ | 22.2 ± 3.4 $^{C}$ | 29.1 ± 4.8 $^{CD}$ |
| PEF5_55DA | 60.6 ± 6.9 $^{A}$ | +3.7 ± 1.4 $^{BC}$ | 18.4 ± 1.6 $^{AB}$ | 18.6 ± 1.9 $^{ABC}$ | 29.7 ± 7.9 $^{CD}$ |
| 70DA | 87.6 ± 3.2 $^{B}$ | 0.0 ± 0.6 $^{A}$ | 14.3 ± 2.8 $^{A}$ | 14.3 ± 2.8 $^{A}$ | 3.9 ± 2.4 $^{A}$ |
| PEF1_70DA | 68.8 ± 7.4 $^{A}$ | +3.4 ± 1.3 $^{B}$ | 18.0 ± 1.5 $^{AB}$ | 18.3 ± 1.6 $^{ABC}$ | 21.7 ± 7.4 $^{C}$ |
| PEF3_70DA | 59.4 ± 9.8 $^{A}$ | +5.4 ± 1.6 $^{BC}$ | 20.9 ± 6.0 $^{B}$ | 20.7 ± 5.5 $^{BC}$ | 32.8 ± 9.5 $^{D}$ |
| PEF5_70DA | 68.5 ± 7.2 $^{A}$ | +3.0 ± 1.7 $^{B}$ | 16.4 ± 2.6 $^{AB}$ | 15.8 ± 1.9 $^{AB}$ | 18.8 ± 3.1 $^{BC}$ |
| 85DA | 84.3 ± 1.9 $^{B}$ | +0.4 ± 0.3 $^{A}$ | 15.9 ± 0.8 $^{AB}$ | 15.9 ± 0.8 $^{AB}$ | 5.9 ± 1.8 $^{A}$ |
| PEF1_85DA | 58.7 ± 8.3 $^{A}$ | +4.3 ± 1.5 $^{BC}$ | 16.3 ± 1.6 $^{AB}$ | 16.9 ± 1.8 $^{ABC}$ | 31.4 ± 8.5 $^{CD}$ |
| PEF3_85DA | 64.5 ± 4.2 $^{A}$ | +3.2 ± 0.4 $^{B}$ | 16.2 ± 1.6 $^{AB}$ | 16.5 ± 1.5 $^{ABC}$ | 25.5 ± 4.1 $^{CD}$ |
| PEF5_85DA | 67.1 ± 3.1 $^{A}$ | +3.1 ± 0.6 $^{B}$ | 18.5 ± 0.9 $^{AB}$ | 18.8 ± 0.8 $^{ABC}$ | 23.3 ± 2.8 $^{CD}$ |

The redness of the samples (a*) varied from −0.8 ± 0.5 to +8.0 ± 1.0 and was significantly affected by the energy input of PEF pre-treatment with regards to the mushrooms dried conventionally. It can be observed that increasing the energy input increased the a* parameter, which indicated a higher saturation of red in the samples. This implies that enzymatic browning occurred as a result of the leak of the cellular content, which was the effect of electroporation. These observations correlate favorably with Wiktor et al. [40] and Alam et al. [25], who noted the same effect of PEF pre-treatments on carrot tissue and parsnips, respectively [25,41]. However, in another piece of research by Wiktor et al. [41] on carrots, the authors reported the opposite relationship between the energy input of the PEF and the a* color parameter [41]. As for the mushrooms dried with the application of dehumidified air, the values of the a* parameter ranged from $0.0 \pm 0.6$ to $+5.4 \pm 1.6$, and the effect of the PEF on this parameter was similar to the effect on lightness: PEF had less impact on the redness of the mushrooms. The effect of temperature and PEF energy on yellowness (b*) in the case of both drying media was similar to the case of the a* value.

The chroma (C*) of the samples dried using convective drying was significantly affected by PEF pre-treatment. It was observed that, with increasing energy input, C* increased, which underlined the importance of PEF pre-treatment, as C* defines the saturation of the color. However, with regard to the samples dried using dehumidified air, the PEF pre-treatment did not significantly influence C*. According to Tiwari et al. [42], samples that are characterized by a $\Delta E$ of higher than 2 have differences that are visible to an untrained observer. By taking this relationship into consideration, for all of the dried mushrooms, the changes in the colors were considered visible to the observer. Moreover, it can be noted that, with an increasing PEF pre-treatment energy input, $\Delta E$ increased as well, which was due to more prominent electroporation and, thus, enzymatic browning.

When taken together, these results suggest that the color parameters were significantly affected by PEF pre-treatment for both drying methods. It can be concluded that, in comparison to the previous data reported on PEF application to dried materials, the effect of this technique depends on the type of dried material and the parameters of both the drying and PEF.

### 3.6. Total Phenolic Content and Anti-Oxidant Activity

Table 3 presents the results of the chemical analyses concerning the content of anti-oxidant bioactive compounds in the obtained dried mushrooms. Polyphenols are unstable compounds that are prone to reacting with some factors and/or degradation. Their stability depends on enzymes, light, metal ions, oxygen, pH, and proteins, as well as temperature. In addition, they may interact with some food constituents [43]. Nevertheless, no straightforward tendency was observed concerning the TPC values and the temperature of the drying air. Interestingly, the introduction of a preliminary treatment in the form of a PEF led to, in some cases, a significant increase in TPC in the mushrooms dried with non-dehumidified air. The effect of a PEF on enzymes is not clear-cut. Depending on the treated matrix and the parameters applied, the PEF may both decrease and increase their activity [44]. Moreover, the extraction capacity of PEF-treated samples as a result of the structural damage of the cells may also increase [17]. However, in the case of drying with the second type of medium—air with reduced humidity—a slightly lower content of polyphenols was observed in the PEF-treated samples. The utilization of dehumidified air caused higher phenolic degradation during the drying process. It may be related to the higher partial pressure of oxygen present in the drying medium, which served as better conditions for enzymatic degradation. Among all the obtained dried materials, the PEF3_70CD sample showed the highest TPC (17.6 mg GAE/g d.m.).

**Table 3.** Total phenolic content (TPC) and anti-oxidant activity ($EC_{50}$ DPPH and $EC_{50}$ ABTS) of mushrooms obtained using the convective method with non-dehumidified air or dehumidified air subjected to (or not) PEF. The same letters ($^{a–d}$ and $^{A–E}$) in the columns represent the homogeneous groups ($\alpha$ = 0.05).

| Material | TPC [mg GAE/g d.m.] | $EC_{50}$ DPPH [mg d.m./mL] | $EC_{50}$ ABTS [mg d.m./mL] |
|---|---|---|---|
| 55CD | $11.5 \pm 0.5^{abc}$ | $0.84 \pm 0.02^{bcd}$ | $0.19 \pm 0.00^{ab}$ |
| PEF1_55CD | $12.4 \pm 0.6^{abcd}$ | $0.71 \pm 0.03^{ab}$ | $0.26 \pm 0.00^{cd}$ |
| PEF3_55CD | $12.0 \pm 0.2^{abc}$ | $0.76 \pm 0.02^{ab}$ | $0.27 \pm 0.01^{d}$ |
| PEF5_55CD | $12.7 \pm 0.1^{bcd}$ | $0.71 \pm 0.01^{a}$ | $0.25 \pm 0.00^{cd}$ |
| 70CD | $10.9 \pm 0.7^{ab}$ | $0.76 \pm 0.02^{ab}$ | $0.21 \pm 0.01^{abc}$ |
| PEF1_70CD | $12.7 \pm 0.5^{bcd}$ | $0.78 \pm 0.01^{ab}$ | $0.24 \pm 0.01^{bcd}$ |
| PEF3_70CD | $17.6 \pm 0.7^{e}$ | $0.70 \pm 0.01^{a}$ | $0.18 \pm 0.02^{a}$ |
| PEF5_70CD | $13.9 \pm 0.7^{d}$ | $0.79 \pm 0.01^{abc}$ | $0.23 \pm 0.00^{bcd}$ |
| 85CD | $10.6 \pm 0.2^{a}$ | $0.74 \pm 0.03^{ab}$ | $0.21 \pm 0.02^{abc}$ |
| PEF1_85CD | $13.1 \pm 0.3^{cd}$ | $0.75 \pm 0.00^{ab}$ | $0.23 \pm 0.00^{bcd}$ |
| PEF3_85CD | $13.2 \pm 0.1^{cd}$ | $0.97 \pm 0.03^{d}$ | $0.24 \pm 0.01^{cd}$ |
| PEF5_85CD | $13.2 \pm 0.2^{cd}$ | $0.91 \pm 0.05^{cd}$ | $0.26 \pm 0.01^{cd}$ |
| 55DA | $12.1 \pm 0.1^{C}$ | $0.60 \pm 0.02^{A}$ | $0.26 \pm 0.05^{A}$ |
| PEF1_55DA | $9.0 \pm 0.1^{ABC}$ | $0.98 \pm 0.03^{CDE}$ | $0.73 \pm 0.04^{BCD}$ |
| PEF3_55DA | $5.6 \pm 0.1^{A}$ | $0.97 \pm 0.01^{CDE}$ | $1.11 \pm 0.00^{D}$ |
| PEF5_55DA | $8.8 \pm 0.0^{ABC}$ | $0.98 \pm 0.00^{CDE}$ | $0.76 \pm 0.04^{BCD}$ |
| 70DA | $12.3 \pm 0.0^{C}$ | $0.60 \pm 0.02^{AB}$ | $0.61 \pm 0.04^{ABC}$ |
| PEF1_70DA | $5.9 \pm 0.1^{AB}$ | $1.26 \pm 0.05^{E}$ | $0.81 \pm 0.07^{BCD}$ |
| PEF3_70DA | $10.1 \pm 1.9^{C}$ | $0.65 \pm 0.17^{AB}$ | $0.52 \pm 0.21^{AB}$ |
| PEF5_70DA | $6.0 \pm 0.3^{AB}$ | $1.18 \pm 0.05^{DE}$ | $0.96 \pm 0.07^{CD}$ |
| 85DA | $11.1 \pm 0.2^{C}$ | $0.69 \pm 0.01^{ABC}$ | $0.69 \pm 0.06^{ABCD}$ |
| PEF1_85DA | $9.7 \pm 0.2^{BC}$ | $0.90 \pm 0.00^{ABCD}$ | $0.71 \pm 0.02^{ABCD}$ |
| PEF3_85DA | $10.1 \pm 0.2^{C}$ | $0.96 \pm 0.01^{BCDE}$ | $0.76 \pm 0.02^{BCD}$ |
| PEF5_85DA | $5.7 \pm 0.2^{AB}$ | $1.13 \pm 0.01^{DE}$ | $0.96 \pm 0.03^{CD}$ |

Table 3 also shows the calculated $EC_{50}$ values (DPPH and ABTS), which are used to interpret the anti-oxidant activity of the obtained dried mushrooms. The lower the values of these indicators, the higher the ability to scavenge free radicals [45]. As in the case of TPC, there was no straightforward tendency between the values of the $EC_{50}$ DPPH coefficient and the temperature of the drying air. When taking into account the type of the drying agent, in the vast majority of cases, the mushrooms dried with non-dehumidified air showed better anti-oxidant properties (lower $EC_{50}$ DPPH values) than the samples dried with dehumidified air. The exceptions were the mushrooms untreated with PEF before drying and the sample to which 3 kJ/kg of energy was applied before drying at 70 °C. In these cases, air with reduced humidity turned out to be more efficient. Most of the dried mushrooms obtained after the application of a pulsed electric field and drying with dehumidified air (55, 70, and 85 °C) exhibited lower anti-oxidant activity than the untreated samples dried under the same conditions. When the non-dehumidified air was used, a significant decrease was noted only for the PEF3_85CD and PEF5_85CD samples. The use of PEF treatment is connected to the risk of generating free radicals and reactive oxygen species [44], which can explain the observed tendency. Lower $EC_{50}$ DPPH values (better anti-oxidant properties) were reported for the PEF-pre-treated samples dried with non-dehumidified air at 55 °C (relative to the untreated sample: 55CD).

Additionally, in the case of the ABTS assay, no clear trend was observed between the values of the $EC_{50}$ ABTS coefficient and the temperature of the drying air. Nevertheless, the type of drying agent had a clear influence on the anti-oxidant activity. After the implementation of the air dehumidification system, the dried mushrooms with reduced anti-oxidant activity (with even four times higher $EC_{50}$ ABTS values) were obtained. As mentioned above, such drying conditions may have enforced the enzymatic degradation of the material. The application of the pulsed electric field did not affect or even slightly worsen the anti-oxidant activity of the samples after drying them with both types of drying agents. However, it seems that the simultaneous use of PEF and dehumidified air favored the loss of anti-oxidant potential. The best scavenging activity against ABTS radicals was demonstrated by the PEF3_70CD sample (0.18 mg d.m./mL), for which the highest TPC was also noted. The $EC_{50}$ ABTS values of samples 55CD, 70CD, and 85CD did not differ statistically from the value obtained by the PEF3_70CD sample (Table 3).

## 4. Cluster Analysis

A cluster analysis was conducted to compare the various technological variants based on process kinetics and product quality parameters (Figure 12). The analysis revealed three major groups: the first contains the mushrooms obtained using the convective method with dehumidified air, subjected to (or not) a PEF; the second group contains the samples obtained using the convective method with non-dehumidified air, subjected to (or not) PEF, as well as one with 55DA; the third group contains mixed samples. The results suggest that using the convective method with different humidity can affect the quality of the samples.

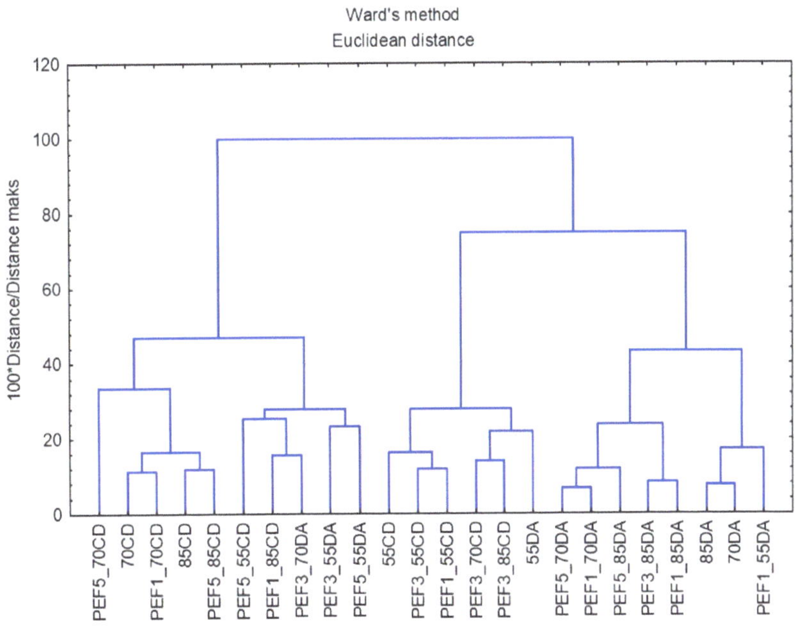


**Figure 12.** Cluster analysis of mushrooms obtained using pulsed electric field treatment and air or dehumidified air-dried mushrooms.

On the basis of the obtained results, statistical approximation profiles of the dried mushrooms obtained using the convective method with non-dehumidified and de-humidified air subjected to (or not) PEF, as well as its usability for obtaining the product over a short time and low $\Delta E$ values and high polyphenol content were made (Figures S1 and S2). The optimal properties for obtaining high-quality dried material over a short time were analyzed. For drying with non-dehumidified air, the best parameters were a PEF pre-treatment energy input of 3.5 kJ/kg and an air temperature of 77.5 °C, while for drying with dehumidified air, a PEF pre-treatment energy input of 1 kJ/kg and drying at 73 °C.

## 5. Conclusions

The study presents the impact of pulsed electric field treatment on process kinetics and the selected physical and chemical properties of convective-dried mushrooms. Furthermore, two types of drying media using non-dehumidified and dehumidified air were applied. Increasing the temperature of the drying air reduced drying time. Additionally, the use of dehumidified air resulted in faster water removal, from 10–37%. However, the use of PEF treatment, depending on the parameters, shortened the drying time maximum by 12% or extended the drying time. This means that, in the case of the use of 1–5 kJ/kg of specific energy input for PEF treatment, the parameters must be selected accordingly. However, in future work, the mushroom tissue should be subjected to treatment with higher PEF energies, and for this research, the effect of the treatment on the drying kinetics and quality of the final product must also be considered.

The physical parameters were dependent on different parameters. The dry matter content of the mushrooms was significantly influenced by drying temperature. The rehydration properties generally showed no significant differences between the untreated and PEF-pre-treated mushrooms dried with non-dehumidified or dehumidified air. The hygroscopic properties were also unaffected by the PEF, and a decrease in relative mass was noted, which means that this product can be stable during storage. However, the color parameters of the analyzed mushrooms were influenced by the drying method and energy input, with higher energy input resulting in decreased lightness (L*) and increased redness (a*), yellowness (b*), and chroma (C*) values, indicating enzymatic browning and greater color saturation. The effect of pulsed electric field (PEF) pre-treatment on the color parameters varied depending on the drying method and the specific characteristics of the mushrooms, with the PEF leading to a darker color. The results highlight the complex relationship between PEFs, drying, and color changes in different materials.

The results of the chemical analyses of the anti-oxidant compounds in the dried mushrooms depended on various factors. The use of pulsed electric field (PEF) treatment and drying with non-dehumidified air led to a slight increase in total polyphenol content, while dehumidified air caused higher phenolic degradation, especially when it was combined with PEF treatment. The PEF-treated samples under 3 kJ/kg of energy and dried with non-dehumidified air at a temperature of 70 °C had the highest total polyphenol content and the best anti-oxidant properties. The anti-oxidant activity of the dried mushrooms varied depending on the drying agent, with non-dehumidified air generally exhibiting better anti-oxidant properties.

**Supplementary Materials:** The following supporting information can be downloaded at: https://www.mdpi.com/article/10.3390/pr11072101/s1, Figure S1. Approximation profiles and usability for dried mushrooms obtained using convective method with non-dehumidified air (CD) subjected or not to PEF. Figure S2. Approximation profiles and usability for dried mushrooms obtained using convective method with dehumidified air (DA) subjected or not to PEF.

**Author Contributions:** Conceptualization, A.W., M.D. and M.N.; methodology, K.R., A.W., M.D., M.N. and D.W.-R.; software, A.M.; validation, K.R. and M.D.; formal analysis, A.B. and M.D.; investigation, K.R., M.D., A.M., A.B., A.W. and M.N.; resources, D.W.-R.; data curation, A.M., A.B. and K.R.; writing—original draft preparation, M.D., A.B., A.M., K.R. and M.N.; writing—review and editing, M.D., M.N., A.W. and D.W.-R.; visualization, A.B.; supervision, A.W. and D.W.-R.; project administration, M.N.; funding acquisition, A.W. All authors have read and agreed to the published version of the manuscript.

**Funding:** This project has received funding from the European Union's Horizon 2020 research and innovation program under grant agreement No 817683 (acronym FOX).

**Institutional Review Board Statement:** Not applicable.

**Informed Consent Statement:** Not applicable.

**Data Availability Statement:** The data presented in this study are available upon request from the corresponding author.

**Conflicts of Interest:** The authors declare no conflict of interest.

## References


1. Arshad, R.N.; Abdul-Malek, Z.; Roobab, U.; Munir, M.A.; Naderipour, A.; Qureshi, M.I.; Bekhit, A.E.-D.; Liu, Z.W.; Aadil, R.M. Pulsed electric field: A potential alternative towards a sustainable food processing. *Trends Food Sci. Technol.* **2021**, *111*, 43–54. [CrossRef]
2. Löffler, M.J. Generation and Application of High Intensity Pulsed Electric Fields. In *Pulsed Electric Fields Technology for the Food Industry*; Raso, J., Heinz, V., Eds.; Springer: Berlin/Heidelberg, Germany, 2022; pp. 55–106.
3. Mahnič-Kalamiza, S.; Miklavčič, D. The Phenomenon of Electroporation. In *Pulsed Electric Fields Technology for the Food Industry*; Raso, J., Heinz, V., Eds.; Springer: Berlin/Heidelberg, Germany, 2022; pp. 107–141.
4. Dellarosa, N.; Tappi, S.; Ragni, L.; Laghi, L.; Rocculi, P.; Rosa, M.D. Metabolic response of fresh-cut apples induced by pulsed electric fields. *Innov. Food Sci. Emerg. Technol.* **2016**, *38*, 356–364. [CrossRef]
5. Nowacka, M.; Dadan, M.; Janowicz, M.; Wiktor, A.; Witrowa-Rajchert, D.; Mandal, R.; Pratap-Singh, A.; Janiszewska-Turak, E. Effect of nonthermal treatments on selected natural food pigments and color changes in plant material. *Compr. Rev. Food Sci. Food Saf.* **2021**, *20*, 5097–5144. [CrossRef] [PubMed]
6. Mahnič-Kalamiza, S.; Vorobiev, E.; Miklavčič, D. Electroporation in food processing and biorefinery. *J. Membr. Biol.* **2014**, *247*, 1279–1304. [CrossRef]
7. Dellarosa, N.; Frontuto, D.; Laghi, L.; Rosa, M.D.; Lyng, J.G. The impact of pulsed electric fields and ultrasound on water distribution and loss in mushrooms stalks. *Food Chem.* **2017**, *236*, 94–100. [CrossRef]
8. Wiktor, A.; Parniakov, O.; Witrowa-Rajchert, D. Drying Improving by Pulsed Electric Fields. In *Pulsed Electric Fields Technology for the Food Industry*; Raso, J., Heinz, V., Eds.; Springer: Berlin/Heidelberg, Germany, 2022; pp. 385–397.
9. Mirzaei-Baktash, H.; Hamdami, N.; Torabi, P.; Fallah-Joshaqani, S.; Dalvi-Isfahan, M. Impact of different pretreatments on drying kinetics and quality of button mushroom slices dried by hot-air or electrohydrodynamic drying. *LWT* **2022**, *155*, 112894. [CrossRef]
10. Chen, X.D.; Mujumdar, A.S. Food drying fundamentals. In *Drying Technologies in Food Processing*; Chen, X.D., Mujumdar, A.S., Eds.; Blackwell Publishing Ltd.: Singapore, 2008; pp. 1–7.
11. Matys, A.; Wiktor, A.; Dadan, M.; Witrowa-Rajchert, D. Influence of Ultrasound and the Conditions of Convective Drying with Dehumidified Air on the Course of the Process and Selected Properties of Apple Tissue. *Foods* **2021**, *10*, 1840. [CrossRef]
12. Dadan, M.; Nowacka, M.; Wiktor, A.; Sobczynska, A.; Witrowa-Rajchert, D. Ultrasound to improve drying processes and prevent thermolabile nutrients degradation. In *Design and Optimization of Innovative Food Processing Techniques Assisted by Ultrasound*; Barba, F.J., Cravotto, G., Chemat, F., Lorenzo Rodriguez, J.M., Munekata, P.E.S., Eds.; Academic Press: London, UK, 2021; pp. 55–110. [CrossRef]
13. Dadan, M.; Nowacka, M. The assessment of the possibility of using ethanol and ultrasound to design the properties of dried carrot tissue. *Appl. Sci.* **2021**, *11*, 689. [CrossRef]
14. Marçal, S.; Sousa, A.S.; Taofiq, O.; Antunes, F.; Morais, A.M.M.B.; Freitas, A.C.; Barros, L.; Ferreira, I.C.F.R.; Pintado, M. Impact of postharvest preservation methods on nutritional value and bioactive properties of mushrooms. *Trends Food Sci. Technol.* **2021**, *110*, 418–431. [CrossRef]
15. Sledz, M.; Wiktor, A.; Rybak, K.; Nowacka, M.; Witrowa-Rajchert, D. The impact of ultrasound and steam blanching pre-treatments on the drying kinetics, energy consumption and selected properties of parsley leaves. *Appl. Acoust.* **2016**, *103*, 148–156. [CrossRef]
16. Association of Official Analytical Collaboration International. *Official Methods of Analysis of AOAC International*, 17th ed.; The Association of Official Analytical Chemists: Rockville, MD, USA, 2002.
17. Śledź, M.; Nowacka, M.; Wiktor, A.; Witrowa-Rajchert, D. Selected chemical and physico-chemical properties of microwave-convective dried herbs. *Food Bioprod. Process.* **2013**, *91*, 421–428. [CrossRef]
18. Rybak, K.; Wiktor, A.; Witrowa-Rajchert, D.; Parniakov, O.; Nowacka, M. The Quality of Red Bell Pepper Subjected to Freeze-Drying Preceded by Traditional and Novel Pretreatment. *Foods* **2021**, *10*, 226. [CrossRef] [PubMed]
19. Delgado, J.M.P.Q.; da Silva, M.V. Food Dehydration: Fundamentals, Modelling and Applications. In *Transport Phenomena and Drying of Solids and Particulate Materials*; Delgado, J.M.P.Q., Barbosa de Lima, A.G., Eds.; Springer: Berlin/Heidelberg, Germany, 2014; pp. 69–94. [CrossRef]
20. Lebovka, N.I.; Shynkaryk, N.V.; Vorobiev, E. Pulsed electric field enhanced drying of potato tissue. *J. Food Eng.* **2007**, *78*, 606–613. [CrossRef]
21. Liu, C.; Pirozzi, A.; Ferrari, G.; Vorobiev, E.; Grimi, N. Effects of Pulsed Electric Fields on Vacuum Drying and Quality Characteristics of Dried Carrot. *Food Bioprocess Technol.* **2020**, *13*, 45–52. [CrossRef]
22. Kaya, A.; Aydın, O.; Demirtaş, C. Drying Kinetics of Red Delicious Apple. *Biosyst. Eng.* **2007**, *96*, 517–524. [CrossRef]
23. Kaya, A.; Aydın, O.; Kolaylı, S. Effect of different drying conditions on the vitamin C (ascorbic acid) content of Hayward kiwifruits (*Actinidia deliciosa* Planch). *Food Bioprod. Process.* **2010**, *88*, 165–173. [CrossRef]
24. Kaya, A.; Aydin, O.; Demirtas, C.; Akgün, M. An experimental study on the drying kinetics of quince. *Desalination* **2007**, *212*, 328–343. [CrossRef]
25. Alam, M.R.; Lyng, J.G.; Frontuto, D.; Marra, F.; Cinquanta, L. Effect of Pulsed Electric Field Pretreatment on Drying Kinetics, Color, and Texture of Parsnip and Carrot. *J. Food Sci.* **2018**, *83*, 2159–2166. [CrossRef]

26. Liu, C.; Grimi, N.; Lebovka, N.; Vorobiev, E. Convective air, microwave, and combined drying of potato pre-treated by pulsed electric fields. *Dry. Technol.* **2019**, *37*, 1704–1713. [CrossRef]
27. Ostermeier, R.; Giersemehl, P.; Siemer, C.; Töpfl, S.; Jäger, H. Influence of pulsed electric field (PEF) pre-treatment on the convective drying kinetics of onions. *J. Food Eng.* **2018**, *237*, 110–117. [CrossRef]
28. Ostermeier, R.; Parniakov, O.; Töpfl, S.; Jäger, H. Applicability of Pulsed Electric Field (PEF) Pre-Treatment for a Convective Two-Step Drying Process. *Foods* **2020**, *9*, 512. [CrossRef]
29. Won, Y.-C.; Min, S.C.; Lee, D.-U. Accelerated Drying and Improved Color Properties of Red Pepper by Pretreatment of Pulsed Electric Fields. *Dry. Technol.* **2015**, *33*, 926–932. [CrossRef]
30. Yamada, T.; Yamakage, K.; Takahashi, K.; Takaki, K.; Orikasa, T.; Kamagata, J.; Aoki, H. Influence of Drying Rate on Hot Air Drying Processing of Fresh Foods Using Pulsed Electric Field. *IEEJ Trans. Electr. Electron. Eng.* **2020**, *15*, 1123–1125. [CrossRef]
31. Sharma, B.H.P.D. Drying Characteristics of Button Mushroom. *Int. J. Curr. Microbiol. Appl. Sci.* **2021**, *10*, 503–512. [CrossRef]
32. Wiktor, A.; Landfeld, A.; Matys, A.; Novotná, P.; Dadan, M.; Kovářiková, E.; Nowacka, M.; Mulenko, M.; Witrowa-Rajchert, D.; Strohalm, J.; et al. Selected Quality Parameters of Air-Dried Apples Pretreated by High Pressure, Ultrasounds and Pulsed Electric Field—A Comparison Study. *Foods* **2021**, *10*, 1943. [CrossRef] [PubMed]
33. Rybak, K.; Parniakov, O.; Samborska, K.; Wiktor, A.; Witrowa-Rajchert, D.; Nowacka, M. Energy and Quality Aspects of Freeze-Drying Preceded by Traditional and Novel Pre-Treatment Methods as Exemplified by Red Bell Pepper. *Sustainability* **2021**, *13*, 2035. [CrossRef]
34. Lammerskitten, A.; Wiktor, A.; Mykhailyk, V.; Samborska, K.; Gondek, E.; Witrowa-Rajchert, D.; Toepfl, S.; Parniakov, O. Pulsed electric field pre-treatment improves microstructure and crunchiness of freeze-dried plant materials: Case of strawberry. *LWT* **2020**, *134*, 110266. [CrossRef]
35. Witrowa-Rajchert, D.; Lewicki, P.P. Rehydration properties of dried plant tissues. *Int. J. Food Sci. Technol.* **2006**, *41*, 1040–1046. [CrossRef]
36. Sacilik, K.; Elicin, A.K. The thin layer drying characteristics of organic apple slices. *J. Food Eng.* **2006**, *73*, 281–289. [CrossRef]
37. Khraisheh, M.A.M.; McMinn, W.A.M.; Magee, T.R.A. Quality and structural changes in starchy foods during microwave and convective drying. *Food Res. Int.* **2004**, *37*, 497–503. [CrossRef]
38. Ashtiani, S.-H.M.; Sturm, B.; Nasirahmadi, A. Effects of hot-air and hybrid hot air-microwave drying on drying kinetics and textural quality of nectarine slices. *Heat Mass Transf.* **2018**, *54*, 915–927. [CrossRef]
39. Dehghannya, J.; Farshad, P.; Heshmati, M.K. Three-stage hybrid osmotic–intermittent microwave–convective drying of apple at low temperature and short time. *Dry. Technol.* **2018**, *36*, 1982–2005. [CrossRef]
40. Wiktor, A.; Nowacka, M.; Dadan, M.; Rybak, K.; Lojkowski, W.; Chudoba, T.; Witrowa-Rajchert, D. The effect of pulsed electric field on drying kinetics, color, and microstructure of carrot. *Dry. Technol.* **2016**, *34*, 1286–1296. [CrossRef]
41. Wiktor, A.; Sledz, M.; Nowacka, M.; Rybak, K.; Chudoba, T.; Lojkowski, W.; Witrowa-Rajchert, D. The impact of pulsed electric field treatment on selected bioactive compound content and color of plant tissue. *Innov. Food Sci. Emerg. Technol.* **2015**, *30*, 69–78. [CrossRef]
42. Tiwari, B.K.; Patras, A.; Brunton, N.; Cullen, P.J.; O'Donnell, C.P. Effect of ultrasound processing on anthocyanins and color of red grape juice. *Ultrason. Sonochem.* **2010**, *17*, 598–604. [CrossRef] [PubMed]
43. Deng, J.; Yang, H.; Capanoglu, E.; Cao, H.; Xiao, J. Technological aspects and stability of polyphenols. In *Polyphenols: Properties, Recovery, and Applications*; Woodhead Publishing: Swaston, UK, 2018; pp. 295–323. [CrossRef]
44. Wiktor, A.; Pratap-Singh, A.; Parniakov, O.; Mykhailyk, V.; Mandal, R.; Witrowa-Rajchert, D. PEF as an alternative tool to prevent thermolabile compound degradation during dehydration processes. In *Pulsed Electric Fields to Obtain Healthier and Sustainable Food for Tomorrow*; Barba, F.J., Parniakov, O., Wiktor, A., Eds.; Academic Press: Cambridge, MA, USA, 2020; pp. 155–202. [CrossRef]
45. Matys, A.; Witrowa-Rajchert, D.; Parniakov, O.; Wiktor, A. Application of pulsed electric field prior to vacuum drying: Effect on drying time and quality of apple tissue. *Res. Agric. Eng.* **2022**, *68*, 93–101. [CrossRef]